Wired TV

Wired TV

Laboring Over an Interactive Future

EDITED BY DENISE MANN

Rutgers University Press
New Brunswick, New Jersey, and London

Library of Congress Cataloging-in-Publication Data
Wired TV : laboring over an interactive future / edited by Denise Mann.
    pages   cm
Includes bibliographical references and index.
ISBN 978-0-8135-6454-8 (hardcover : alk. paper) — ISBN 978-0-8135-6453-1 (pbk. : alk. paper) — ISBN 978-0-8135-6455-5 (e-book)
 1. Television broadcasting—United States. 2. Interactive television—United States.
 3. Social media—United States. 4. Mass media—United States.   I. Mann, Denise, editor of compilation.
HE8700.8.W57 2013
384.3'1—dc23   2013013408

A British Cataloging-in-Publication record for this book is available from the British Library.

Vincent Brook, "Convergent Ethnicity and the Neo-Platoon Show: Recombining Difference in the Postnetwork Era" originally appeared in *Television & New Media* 10, 4 (July 2009): 331–353.

Derek Johnson, "Authorship Up for Grabs: Decentralized Labor, Licensing, and the Management of Collaborative Creativity" is adapted from material in Derek Johnson, *Media Franchising: Creative License and Collaboration in the Culture Industries* (New York: New York University Press, 2013). Thanks and acknowledgment go to NYU Press for the kindness in allowing this material to be repurposed here.

This collection copyright © 2014 by Rutgers, The State University
Individual chapters copyright © 2014 in the names of their authors
All rights reserved
No part of this book may be reproduced or utilized in any form or by any means, electronic or mechanical, or by any information storage and retrieval system, without written permission from the publisher. Please contact Rutgers University Press, 106 Somerset Street, New Brunswick, NJ 08901. The only exception to this prohibition is "fair use" as defined by U.S. copyright law.

Visit our website: http://rutgerspress.rutgers.edu

Manufactured in the United States of America

Contents

	Acknowledgments	vii
	Introduction: When Television and New Media Work Worlds Collide DENISE MANN	1
1	Authorship Up for Grabs: Decentralized Labor, Licensing, and the Management of Collaborative Creativity DEREK JOHNSON	32
2	In the Game: The Creative and Textual Constraints of Licensed Video Games JONATHAN GRAY	53
3	Going Pro: Gendered Responses to the Incorporation of Fan Labor as User-Generated Content WILL BROOKER	72
4	Labor of Love: Charting *The L Word* JULIE LEVIN RUSSO	98
5	The Labor Behind the *Lost* ARG: WGA's Tentative Foothold in the Digital Age DENISE MANN	118
6	Post-Network Reflexivity: Viral Marketing and Labor Management JOHN T. CALDWELL	140
7	Fan Creep: Why Brands Suddenly Need "Fans" ROBERT V. KOZINETS	161

8	Outsourcing *The Office* M. J. CLARKE	176
9	Convergent Ethnicity and the Neo-Platoon Show: Recombining Difference in the Post-Network Era VINCENT BROOK	197
10	Translating Telenovelas in a Neo-Network Era: Finding an Online Home for MyNetwork Soaps KATYNKA Z. MARTÍNEZ	223
11	The Reign of the "Mothership": Transmedia's Past, Present, and Possible Futures HENRY JENKINS	244

Notes on Contributors — 269
Index — 273

Acknowledgments

I undertook this collection thinking it would be a quick and easy endeavor—a cleansing of the palate, if you will—after the long and arduous task of completing a single-author book. After putting this project aside multiple times to attend to various pressing demands, I looked up and realized that nearly a dozen scholarly books had been published on the topic of television in the digital age. The presence of such strong and compelling works—all from unique and valuable perspectives—made the task of compiling this collection that much more daunting. Therefore, I'd like to thank the first-rate editorial team at Rutgers University Press and give a special acknowledgment to editor Leslie Mitchner for her early support of the project in its proposal stage and for her helpful comments, encouragement, and provocative suggestions on the manuscript when I finally delivered it.

Situating this project in such a cluttered research space was made easier because of the original and insightful work of my contributors. In particular, I'd like to single out Henry Jenkins, who added yet another responsibility to his already overflowing "to do" box by agreeing to co-chair our annual Transmedia Hollywood conference. This forum provided the perfect backdrop for researching this book's themes by bringing together cutting-edge media makers and thought leaders from diverse corners of the entertainment industry and academia to debate issues attending the future of entertainment. The first conference year was the most daunting. We had timed our launch in Los Angeles to coincide with the arrival of hundreds of media scholars in March 2010 to attend the annual Society of Cinema and Media Studies (SCMS) conference. Given our zealous desire to share the riches of our proximity to Hollywood, we also assembled a SCMS workshop at the Bonaventura Hotel and dubbed it "The Geek Elite Debates the Future of Entertainment." The workshop convened a truly elite group of television writer-producers—Carlton

Cuse and Damon Lindelof (*Lost*), Tim Kring (*Heroes*), Kim Moses (*The Ghost Whisperer*), Javier Grillo-Marxuach (*The Middleman*), and Mark Warshaw (*Smallville, Heroes*). All of them were laboring over the central questions that inform this collection: how to challenge themselves and their creative teams to embrace the transformative implications of the Internet and Web 2.0 in the context of an often rigid, opaque, and risk-adverse bureaucratic system like the network television industry. I remain indebted to each of these prominent creators for their ongoing support as I continued to ply them with questions and requests for follow-up interviews while assembling this collection.

Next, I'd like to thank my UCLA colleagues: Teri Schwartz, dean of the School of Theater, Film, and Television, and William McDonald, chair of the Department of Film, TV, and Digital Media; my immediate workplace family, UCLA Producers Program faculty members Barbara Boyle, Ben Harris, and Myrl Schreibman, as well as my extended workplace family of over thirty busy, top-tier industry professionals, who agree to teach classes on behalf of our next-generation producers. I'd like to thank my adopted workplace family, the Cinema and Media Studies faculty members Janet Bergstrom, John Caldwell, Allyson Fields, Stephen Mamber, and Chon Noriega. In addition, I'd like to thank all the MA, MFA, and PhD students at UCLA for their direct and indirect contributions to this project through their insightful works and original ideas delivered during my seminars. In particular, I'd like to thank my research assistant and doctoral student Jessica Fowler for her careful copyediting and helpful suggestions. In addition, I'd like to thank several current and recent UCLA doctoral students—M. J. Clarke, Lindsay Giggey, Felicia Henderson, Drew Morton, Jennifer Porst, and Ben Sampson—whose exacting research and writing on the contemporary entertainment industry make teaching such a continued source of pleasure.

Wired TV

Introduction

When Television and New Media Work Worlds Collide

DENISE MANN

In a 2005 trade article entitled "The End of Television (As You Know It)," a cable executive reduced the vast cultural-industrial transition then under way to a singular, technologically driven event—the incongruous conjoining of two black boxes—by stating, "The computer has crashed into the television set."[1] In fact, the situation is far more challenging and elusive to describe, given the glacial pace at which vast bureaucratic organizations like the networks embrace change and the epochal nature of the impact of the Internet and Web 2.0 companies on the traditional television industry. Complicating matters for media scholars engaged in the perilous task of analyzing the operation of network bureaucracies during a period of transformation is John T. Caldwell's warning: "'*The industry*' is not a monolith controlled by 5–6 giant conglomerates, but a series of dense rhizomatic networks of sub-companies held at a safe distance, loosely structured to flexibly adapt to new labor markets, new digital technologies, and consumer unruliness."[2]

Max Weber's turn-of-the-twentieth-century theories regarding bureaucratic structure and its formative influence on the genesis of capitalism inform a recurring strain of later critiques of the workplace as a site of cultural sameness, from Frankfurt School theorists Theodor Adorno and Max Horkheimer to American sociologists David Reisman (*The Lonely Crowd*) and William Whyte (*The Organization Man*).[3] The legacy of sameness can be seen in today's post-network entertainment industry insofar as the charismatic and rationalist authorities placed in charge of these complex organizations tend to avoid risk associated with change, despite mounting evidence that their aging, linear system is failing. Weber described how additional layers of managers typically analyze the performance of employees in their respective divisions while also drafting and dispersing reports across the entire organization. In the case of today's Disney-ABC Television group, that workforce numbers more than seven thousand.[4]

Unraveling these complex bureaucracies and entrenched hierarchies is not easy, prompting most scholars to rely too heavily for insight on the public utterances of network heads. While acknowledging the obvious benefits of "ethnographic fieldwork [which] can provide rich insights that speculative theorizing misses," Caldwell warns media scholars to be aware that what "high-level professionals say has almost always been scripted and rehearsed."[5] To triangulate their findings, the authors in this collection move beyond the public statements of network personnel in the trade press to include textual analyses of specific television franchises and, when possible, analyses of the self-statements of a variety of production personnel, all in an effort to understand the changing workplace rules and bureaucratic structures operating at the major broadcasting units. That said, even the well-rehearsed sound bites of high-level professionals can shed light on a particular situation, albeit in unintended ways, so long as the scholar considers the underlying meanings of these discursive statements. For example, even though network presidents routinely assure journalists and the public of their interest in digital new media, most have been conditioned to see their future largely as a virtual (pun intended) replica of their present reality, that is, simply a shift from broadcast to online viewing. For instance, when asked to describe the future of broadcasting at a public event in late 2011, CW Television Network president Mark Pedowitz stated that "TV viewing . . . will eventually migrate online with full commercial load."[6] This off-the-cuff comment is symptomatic of the larger problem facing the networks as they transition from a broadcast to a digital era: their tendency to return to old patterns and profit centers, which reinforces their reluctance to rethink their creative and business models.[7]

Even though digital video recorders (DVRs) are typically cited as the biggest threat to the traditional ad-supported television industry, given the consumer's ability to fast-forward past the expensive thirty-second spots that are the networks' primary source of income, it is more likely that the infiltration of broadband into the majority of homes and the popularity of Web 2.0 social media companies like YouTube, Facebook, and Twitter have jeopardized the smooth operation of the networks' existing hierarchies and interlocking bureaucracies since 2005. These two landmark events radically altered consumer behavior, prompting network gatekeepers and their advertising partners to demonstrate an uncharacteristic openness to engaging outsiders—writers, producers, digital makers, technology experts, entrepreneurs, and even unaffiliated, aspirational prosumers—anyone, it seems, who could offer even temporary solutions to contain this catastrophic shift. Notably, the big media companies even began to fund scholarly research into the economic, technological, and social implications of "connected" or "smart" television experiences, resulting in a number of lively and engaging cultural-industrial analyses of this new consumer behavior and media engagements. Of course, what the big media funding organizations were hoping to get from the scholars was a series of white papers offering a list of quick fixes to these commercial threats to their future dominance of the television industry.[8]

Wired TV is one of the growing number of studies concerned with the current television industry and its attempts to grapple with various threats (and enhancements) to the linear broadcast system. This collection departs from previous media industry studies in its central focus on the sociology of creative labor in post-network Hollywood and, in particular, its analyses of various subgroups of creative outsiders, many of whom are eager to bring change to the old-fashioned broadcasting system. The networks began to experiment in earnest with the traditional television experience from 2005 to 2010 by collaborating with three important groups: members of the Hollywood creative community who wanted to expand television's storytelling worlds and marketing capabilities by incorporating social media; members of the Silicon Valley tech community who helped the networks recast television distribution for the digital era; and super-fans who were eager and willing to use social media story extensions to proselytize on behalf of a favorite network series.

Among the labor strikes over the past seventy years in Hollywood, the 2007–2008 Writers Guild of America (WGA) strike was noteworthy for its focus on digital content (both original and derivative) tied to major television scripted series and lucrative reality series. Both formats benefit media corporations by providing low-cost content (which often doubles as branded marketing) and harm writers by ignoring the labor involved. Despite the mutually beneficial collaborations that took place between television writers and network marketers in the pre-strike moment when series showrunners (executive or supervising producers) oversaw a number of inventive, popular, derivative short videos, contests, and alternate reality games (ARGs) tied to valuable series like *Lost*, *Heroes*, *The Office*, *30 Rock*, and *The Ghost Whisperer*, tensions mounted when the writers watched the networks sell ad time on their original online creations without offering them additional salary or royalties. The clash between writers and management finally boiled over into a strike. At its conclusion, the guild's new Minimum Basic Agreement (MBA) outlined the digital rights of writers and granted showrunners ultimate responsibility for derivative digital content linked to their series. Since the strike, however, the

FIG. I-1 The *Lost* writers, led by Damon Lindelof and Carlton Cuse, labored over an array of digital experiments. (Photo courtesy of Disney-ABC)

networks have continued to exploit the thin line between content and promotions in ways large and small. Transmedia producer Mark Warshaw observed that the only type of digital content that secured any real protection in the post-strike environment was the webisode, because of the need to use Screen Actors Guild (SAG) members from the series proper, as seen in "The Lost Diaries" or *30 Rock*'s "The Donaghy Files" webisodes.[9]

The *Lost* franchise represents one of the more visible manifestations of the ABC network's revised policies toward collaboration with these three groups.[10] The *Lost* writers, led by Damon Lindelof and Carlton Cuse, collaborated with the ABC marketing team to create a wide array of digital experiments, including "The Lost Experience," "The Lost Diaries," and "Lost Untangled." Several of these online platforms used the self-reflexive, irreverent, tonally dissonant, and amateur aesthetic once associated with user-generated content and fan mash-ups. At the height of *Lost*'s success, Cuse explained his and Lindelof's philosophy toward fan engagement and interactivity: "What we've been able to do, which I think is different than most network shows, is leave certain things ambiguous and open to interpretation. And that allows people to get on the boards and theorize about what's meant by a given story or scene, or move in the show's direction. It allows people to feel participatory about the process."[11] In contrast, less well-funded and well-known digital creators find themselves shut out by Hollywood's institutionalized gatekeepers because their work lacks clear-cut paths to monetization or ignores copyright restrictions. For example, the two amateur creators known only as "the Fine brothers" used a voiceover narrator and positioned the licensed action figures of Jack, Kate, Sawyer, and others to parody the convoluted *Lost* series. The Fine brothers were forced to watch helplessly from the sidelines as their unofficial production, "The Lost Parodies," was displaced by "Lost Untangled," a virtually identical, albeit officially sanctioned social media campaign.[12]

What motivated the Fine brothers and other digital outsiders to work long hours without compensation? In *Nice Work if You Can Get It* (2009) and *No-Collar* (2003) Andrew Ross documented the fervor of young hires at digital agencies like Razorfish, which offer overworked creators small salaries, no benefits, and little to no job security. Many of the pre- and post-dot.com sweatshops that Ross surveyed have benefited greatly from this system designed to trap hapless laborers intent on securing creative work without having to navigate Hollywood's onerous battery of agents, managers, and attorneys. Still, there may be another way of interpreting this behavior. A young hire at one of the network's digital marketing groups was told to produce behind-the-scenes celebrity interviews, which would appear on the network website and accompany the broadcast of a major awards show. This recent film school graduate, while initially thrilled to have a creative opportunity inside a major Hollywood entertainment group, quickly began to see the job as a trap. To engineer his exit, he spent countless off-job hours to develop a Facebook app, first by partnering with a technology expert, then by taking night courses to learn how to pitch his idea to venture capital investors. In essence, the recent network hire was hoping to trade the predictability and security of a bureaucratic job (even one in the relatively new digital marketing division) for the excitement and risk associated with a start-up. Large numbers of energetic, multitasking millennials are abandoning the enervating sameness of traditional Hollywood creative jobs and hoping to adopt the get-rich lifestyle of a dot.com entrepreneur. How can media scholars make sense of this latest movement of young professionals seeking work in creative industries outside of Hollywood proper?

Whereas the networks and television studios are frequently portrayed in the popular press as rigid and inflexible bureaucracies, their Silicon Valley counterparts—companies as diverse as Apple, Google, Facebook, YouTube, Netflix, and Amazon—have benefited from a unified discourse that portrays them as nimble, competent, and risk-taking organizations. To further cement this image, first Google and then Facebook backed a large-scale public relations push to position themselves as viable alternatives to television's traditional thirty-second ad, even without clear-cut evidence of their effectiveness in reaching consumers. According to Brad Smallwood, Facebook's head of measurement and insights, "The Twitters, Yahoos, Microsofts—and Google obviously as well—we want them all to be a part of this process. If we can demonstrate confidence in online, we will see a good share of those dollars end up on Facebook."[13] One of the peripheral benefits of this decade-long onslaught of revolutionary discourse surrounding Silicon Valley companies has been the creation of a generation of digital entrepreneurs who are eager to launch their own risky start-ups or join established ones, often with little or no compensation.

In the past decade, in particular, "*the industry*" has expanded to include large numbers of younger, smaller, independent-minded sub-companies and units—transmedia production companies, digital marketing agencies, and even lone digital artists—that have intentionally aligned themselves with Silicon Valley's work ethos over Hollywood's. In some cases, the new start-ups insist on both a geographic separation from Hollywood and a symbolic one by setting up shop in the newly defined tech corridors of Los Angeles and New York and naming these locations "Silicon Beach" and "Silicon Valley," respectively; in other cases, they simply align themselves with the self-discipline and innovation associated with Silicon Valley versus the constraints of institutionalized Hollywood.[14]

One explanation for contemporary Hollywood's mishandling of valuable media franchises can be traced to the origins of the franchise system in business in general in the United States. In *Media Franchising*, Derek Johnson notes the strong similarities between network television's franchise system and that of fast-food restaurants like McDonald's in the 1950s. The McDonald brothers successfully franchised their "Speedee" drive-in restaurant by assigning individual vendors to run additional shops in Downey, California, and Phoenix, Arizona. The brothers established a profitable assembly-line restaurant and self-serve model by paring down their offerings to a simple menu of hamburgers, French fries, and shakes. Following a similar pattern, Johnson observes, the television networks have historically distributed their broadcast series using a nationwide system of affiliates, which are owned and operated by independent vendors.[15] Furthermore, Johnson sees strong links between early fast-food and gas station franchises and the studios' use of media franchises as another form of collaboration and partnership in the realm of production and content.[16] This joint effort occurs when studios assign their own in-house, corporate licensing executives to oversee the work of outside vendors who have been contracted to create computer games, comics, novelizations, and other licensed goods derived from the studios' original intellectual property, such as a television series or theatrical film.

In *Transmedia Television* M. J. Clarke offers another perspective on why production personnel may be seeking alternatives to creative jobs in traditional mass media by exposing the routinized sameness and unplanned inefficiencies inherent in the Hollywood system.[17] Interviews with licensed vendors, a group of creators who have traditionally held positions of weakened affiliation with the Hollywood system, gave Clarke valuable insights about their counterparts, the network licensors. The vendors inadvertently revealed that the licensors were ill equipped or unwilling to manage the creativity of outside agencies. Unlike writers, directors, and actors who are hired by the studios on "work-for-hire" contracts and guaranteed a salary, benefits, insurance, and other niceties based on studio agreements with the labor guilds, the vendors must pay their own production costs, worker benefits, and insurance. Furthermore, most see their creative work devalued by licensing executives, who focus narrowly on budget and schedule, treating these derivative works as publicity for the digital video (DVD) launch of a movie or television series.

Both Johnson's and Clarke's scholarly works shed light on the insufficiencies of studio licensing divisions and, by extension, other unexplored bureaucratic practices in the traditional television industry—a major concern of this volume. Media scholars can confirm these insights by interviewing other marginalized production personnel, including low-paid transmedia producers and digital laborers, many of whom are eager to enhance their status as creative pioneers by talking to scholars. Expanding on Caldwell's analogy between Hollywood and the U.S. military—the presence of subcontractors that constitute a "para-industry"—one can suggest that these self-appointed creative trailblazers function like military scouts who undertake risky reconnaissance missions into unknown territories, gather intelligence, and report their findings to higher-ups.[18] Similarly, media scholars, who are often squeamish about engaging directly with the enemy—in this instance, high-powered heads of media corporations—can benefit greatly from encounters with members of this creative advance guard during "debriefing" sessions, otherwise known as interviews.

This type of scholarly work has been facilitated by the fact that most transmedia producers and do-it-yourself (DIY) digital media makers use the Internet and social media as their bully pulpit to show off their wares and to advance the principles of mass collaboration, collectivity, and interactivity to reinvent entertainment. Among these potent new thought leaders are high-profile transmedia producers like Jeff Gomez, who advocated for the Producers Guild of America's (PGA) newly sanctioned transmedia producer credit. Gomez's company, Starlight Runner Entertainment, is characterized on its website as a producer of transmedia franchises. Its clients include entertainment companies and their intellectual properties, such as Showtime's *Dexter*, Disney's *Tron*, and 20th Century Fox's *Avatar*; toy manufacturers, such as Hasbro's Transformers and Mattel's Hot Wheels; and consumer brands, such as Coca-Cola's "The Happiness Factory." The company offers guidance to studio executives (in licensing, consumer products, and marketing) by providing a story bible (of biblical proportions) to help them expand the story worlds associated with their valuable media franchises through the development of

FIG. I-2 Jeff Gomez, CEO of Starlight Runner Entertainment, educates media professionals about the benefits of transmedia storytelling. (Photo courtesy of Jeff Gomez, Partner, Starlight Runner Entertainment)

books, comics, graphic novels, video games, and alternate reality experiences.[19] In other words, Starlight Runner provides the type of creative training and story oversight that is missing from perhaps all of a studio's divisions except the creative development divisions.

Another cutting-edge, New York–based company is Campfire, an independent digital marketing agency launched by the creators of the *Blair Witch Project* (1999) and devoted to creating social media–inspired platforms that double as marketing campaigns. The executives describe their use of transmedia storytelling across multiple platforms and media formats as a new means of managing brand launches.[20] In other words, companies like Campfire must educate prospective clients about the benefits of a digital marketing campaign that incorporates transmedia storytelling in order to distinguish themselves from the traditional Madison Avenue advertising firms that focus on expensive thirty-second television spots and conventional print campaigns. Both Starlight Runner and Campfire have taken great pains to educate the public and prospective clients about the artistic and commercial benefits of transmedia storytelling in order to assert their own role as creative storytellers and as a way to win clients. These proclamations are part of what Caldwell describes as "industrial self-reflexivity," that is, the effort by Hollywood's production personnel to describe their activities to the public in a way that justifies their careers and the work created by their companies.[21] This type of critical industrial discourse appears virtually on the company websites and executive blogs of Starlight Runner and Campfire, but also in the real, physical spaces of university lecture halls and conference meeting rooms as more and more of these industrial trailblazers are invited to debate the future of entertainment in industry and academic forums.[22]

Notably, most self-descriptions by transmedia producers reveal a split focus between commercial and ideological goals. On the one hand, Starlight Runner's official website celebrates the company's successful campaigns on behalf of high-profile, paying clients (Coca-Cola, Hasbro, Disney, Showtime); on the other, the lectures and unofficial blogs of Starlight Runner executives Jeff Gomez, Mark S. Pensavalle, and Caitlin Burns seek to differentiate the goals of transmedia producers from those of Hollywood proper. Most media scholars will empathize with Burns's celebrations of independent creative voices that encourage marginalized groups like young women or gays to read media against the grain.[23] However, Jeff Gomez's revolutionary proclamations on the company website and trade industry blogs tend to celebrate the opposite, namely, studio blockbuster digital marketing campaigns that successfully incorporate transmedia storytelling. For instance, Gomez commented on Disney's inventive "use of the entire Marvel Universe in the marketing of the upcoming *Avengers* film."[24]

The sometimes uneasy coexistence of commercial and ideological goals is true of most individuals and companies that have assumed the moniker of transmedia producer. High-profile transmedia production companies like Starlight Runner are paid large sums of money to expand valuable studio-owned franchises like *Tron* or *Pirates of the Caribbean*, whereas DIY start-ups are forced to use Kickstarter, social media marketing, and other crowdsourcing activities to raise money to create and sell their social experiments. The latter group—often unpaid or semiprofessional creators engaged in speculative productions of original or experimental web content—pride themselves on their ability to operate outside Hollywood's heavily conglomerated industry. They tend to follow a different group of mentors or guides, such as DIY filmmaking advocates like Peter Broderick, Jon Reiss, and Ted Hope, or transmedia gurus like creator and spokesperson Lance Weiler, all of whom have their own websites, books, and speaking tours.[25] However, the dividing line between those aligned with Hollywood and those positioned in opposition to it is starting to blur as high-profile Hollywood insiders like David Fincher use crowdsourcing to fund and sell their own experimental works. Here, again, the media scholar must read against the grain, recognizing how Fincher sought to differentiate himself from traditional Hollywood by using a promotional video to accompany his Kickstarter campaign, in which he distanced himself from his celebrity as an Oscar-winning feature film director of commercially successful franchises (*Seven*, *The Social Network*, *The Girl with the Dragon Tattoo*) by mocking his knowing complicity with Hollywood marketing practices.[26]

Despite all this attention to transmedia storytelling and other future entertainment models taking place just outside Hollywood's storied gates, the networks have shunned many of these game-changing affiliations, unless the transmedia producer's interests and abilities happen to align with the networks' traditional programming, marketing, and licensing practices. This stubborn streak could become the networks' undoing going forward. In *Wikinomics*, an analysis of traditional corporate bureaucracies that have adapted to mass collaborative workplace cultures, Don Tabscott and Anthony D. Williams marvel at the huge strides that decades-old organizations like IBM, Boeing, BMW, and Proctor & Gamble have made in the structure and operation of their divisions as they incorporate the collectivity associated with the Internet. At the same time, they question the ability of closed shops—corporations that remain proprietary about trade secrets and promote from within the company—to survive the new millennium.

With the exception of the five-year window of experimentation and collaboration between 2005 and 2010, the networks have maintained a largely proprietary stance concerning their internal operations along three registers: by controlling the release of information about their creative and business practices via corporate public relations departments; by micro-managing their series and licensed content, such as computer games, novelizations, and the like; and finally, by limiting fan engagement with television content via restrictive software and legal boilerplates attached to online promotions. Each of these corporate roadblocks has further distanced the networks from future forms of creative collaboration, confining them to an aging system of interlocking bureaucracies, entrenched hierarchies, and strategic partnerships.

Prosumer Television Experiments Go Pro

The networks' experiments with transmedia storytelling and social media engagements between 2005 and 2010 were tolerated in large part because they were introduced by Hollywood insiders who were also proven hitmakers, including head television writers like J. J. Abrams, Carleton Cuse, Damon Lindelof, and Tim Kring. Notably, these skilled creative personnel were eager to break free from the rigidity of the Hollywood system by embracing the traits of innovation and dexterity on display among Silicon Valley's social media advance guard, namely, the creators of YouTube, Facebook, and Twitter.[27] However, shortly after the WGA strike, the networks wrested control of much of this creative activity in the digital space by defining it as marketing and by ramping up their own in-house marketing units to oversee its production. The network managements' knee-jerk response to the conclusion of the power struggle appears short-sighted in retrospect. Because network digital marketers are now in charge of creative content with very little input from series' writer-producers and their teams, the results have been a jumble of uninspired brand integrations, behind-the-scenes interviews, and contests rather than compelling and immersive story expansions. The in-house digital marketing divisions are further constrained by the need to coordinate their activities with the traditional network bureaucracies (that is, current programming, development, marketing, and licensing), which still tend to treat digital marketing and interactive advertising as largely untested practices with unproven approaches to monetization. In short, rather than facilitate collaboration with creative partners, the networks have added an additional bureaucratic layer, which further impedes the type of mass collaboration and long-tail logic embraced by such innovative, entrepreneurial trailblazers as Google, Amazon, Apple iTunes, eBay, and Netflix, all of which are flourishing in the digital age.

The model of creative collaboration among fans, series writers, and network marketers precedes the *Lost* case by several decades. The relationship began unofficially in the mid-1960s when *Star Trek* series creator Gene Roddenberry encouraged a letter-writing campaign by fans to try to persuade the network to save the series. These mass collaborative activities increased exponentially in the Internet era once fans had the technical capability to expand their reach to an online community. Fans of cultish, hyper-serialized franchises of the late 1990s like *The X-Files* (1993–2002), *Babylon 5* (1994–1998), *Buffy the Vampire Slayer* (1997–2003), and *Angel* (1999–2004) were particularly active, using these new technologies to poach, mash-up, and share insights with series writers and with other members of the online community. Although fans were initially regarded by the networks (and even some writers) as harmless annoyances, once they started using blogs, wikis, video sharing, Facebook, and other digital resources, the networks were forced to take notice.

One of the key innovators in this space was J. J. Abrams, who chose the newly formed WB Television Network (The WB) over powerful NBC as the home of his first series, *Felicity* (1998–2002). Knowing that the weblet desperately needed programming targeted to the young, predominantly female niche of twelve- to thirty-four-year-olds, Abrams and his advisors expected that The WB would put hefty marketing muscle into selling his series.[28] Taking up the challenge, Lew Goldstein and Robert Bibb generated interest in *Felicity* and all of The WB teen dramas through a series of cross-promotional tie-ins with retail outlets, selling not just DVDs but also related music, clothing, and, not least, The WB network brand.[29] By the time Abrams sold his second series, *Alias* (2001–2006), to ABC, he and his writing and producing team were seasoned veterans of this type of cultish, multi-platform franchise designed to engage a participatory audience.

The list of *Alias* writers who went on to become creative pioneers on other fan-friendly, hyper-serialized television franchises includes Jesse Alexander (*Smallville, Lost, Heroes*), Jeff Pinkner (*Lost, Fringe*), Alex Kurzman (*Fringe*), Roberto Orci (*Fringe*), and Jeff Bell (*X-Files, Angel, Day Break*). Jesse Alexander, Mark Warshaw, and other creative personnel started writing character blogs such as "Dawson's Desktop" and creating webisodes and other story expansions such as "The Chloe Chronicles" and "Smallville Legends" for cultish, fan-friendly, hyper-serialized series like *Dawson's Creek* (1998–2003) and *Smallville* (2001–2010). Abrams went on to co-create *Lost* with Damon Lindelof, who consulted with two of his senior mentors, Carlton Cuse and Tim Kring, who became showrunners on their own transmedia series, *Lost* and *Heroes* (2006–2010), respectively.

Abrams and his Bad Robot production company also contributed indirectly to *Heroes*' use of social media to engage fans when Tim Kring, Jesse Alexander, and other members of the *Heroes* creative team launched an author-driven fan website called 9th Wonders, modeled on Abrams's Fuselage. Furthermore, Jeph Loeb, Jesse Alexander, and several other *Smallville* alumni recommended that Kring hire digital producer Warshaw to work on the *Heroes* web materials. NBC, which had recently hired Vivi Zigler to run its digital marketing unit, engaged Warshaw to serve as a liaison between the *Heroes* writers and the marketers. Because Warshaw had worked so closely with the writers (his office was situated adjacent to the writers' room), he joined them on the picket line during the WGA strike of 2007–2008—an act of worker loyalty that contributed to his dismissal and signaled the growing confusion over whether writers or network marketers should oversee these digital efforts. At CBS, the husband-and-wife showrunner team of Kim Moses and Ian Sanders hired interns and other low-level digital laborers to help them produce interactive content to shore up the flagging ratings for their series *The Ghost Whisperer* (2005–2010) and convince their stodgy bosses to keep the series on the air.[30]

As should be clear at this point, many of the television writers and producers who were innovators in the online space took their inspiration from inventive fan mash-ups, slash fiction, blogs, and other early prosumer activities. The networks did not become fully invested in these online storytelling activities until they recognized their auxiliary value as inexpensive grassroots promotions that could build positive word-of-mouth about their series and enhance DVD sales by supplementing their bonus features. For example, ABC marketing head Mike Benson seized the opportunity to partner with the *Lost* showrunners Cuse and Lindelof on a number of online experiments, including "The Lost Experience" alternative reality game and "The Dharma Initiative" and "The Lost Untangled" recaps. Needless to say, writers who engaged in these digital story expansions also recognized their secondary function as a means for networks to provide advertisers with detailed consumer information via online registrations. Even though several of Abrams's series (*Undercover, Alcatraz*) have experienced uneven ratings, his production company Bad Robot remains deeply committed to this multifaceted approach to creating and selling his new series (*Fringe, Revolution*). Furthermore, his team frequently frontloads enigmatic storytelling "rabbit holes" into each new feature product (*Cloverfield, Super Eight*, and so on) and works closely with studio and network marketing teams, as well as with Madison Avenue advertising firms and Silicon Valley tech partners, to create highly interactive, branded entertainment experiences. For example, a new iPhone app that allows users to create their own special effects was showcased in a tongue-in-cheek promotional video that featured an impervious announcer dodging an exploding helicopter, missiles, and a barrage of assault weapons.[31] Notably, the Bad Robot production office invokes a pop culture explosion of old and new gadgetry—old-fashioned typewriters, a green phone without a dial face or digits—to inspire the team to use new technologies in service of manufacturing enigmatic story worlds and marketing strategies and not as ends in and of themselves.

Transmedia producers recognize the hybrid status of social media experiments as both content and marketing. As a *New York Times* journalist explains, "In an era of bloggers, e-mail and smartphones that can record video, it takes an extra effort to stem the overflow of premature information that might render a project stale by the time it comes out."[32] Despite the pervasiveness of viral marketing ploys and the increasingly spreadable nature of media, transmedia producers at companies like Bad Robot choose to prioritize narrative mystification over marketing saturation, even though the latter has been the explicit goal of Hollywood media corporations since the mid-1970s. Paradoxically, the networks that allow the Bad Robot team to create social media experiences around major franchises (*Lost, Fringe, Revolution*) stake their commitment to Abrams on his reputation as a proven success story rather than as a transmedia provocateur. The networks allow his team to finance and produce social media experiences tied to his television series and films so long as those efforts align with the networks' traditional on-air marketing efforts.

This early experimentation and collaboration between network marketers and television creative personnel taught the networks how to craft effective digital marketing campaigns from within their organizations. In the process, they have shifted responsibility from high-paid television writing teams to low-cost digital labor in divisions like NBC Digital Entertainment and New Media. Although showrunners are invited to discuss digital expansions to their series at NBC, the digital marketing team's contact with the head writers is typically limited to two meetings, one shortly after the commencement of work on the series in late spring and again in early fall before the launch. In the first meeting, the NBC digital marketers work with the showrunners to generate a grab bag of ideas for the digital campaign; after they meet again in the fall, the digital marketers will hand off a short list to their team to execute. Because most showrunners are busy throughout the summer ramping up multiple episodes in advance of the fall season, the writer-producers welcome the help from the admittedly capable and inventive NBC digital marketing team.[33]

Although in-house digital marketing units have streamlined the process, freeing up time for busy showrunners *and* traditional network marketers, it is important to consider what may have been lost in the process. By reasserting bureaucratic oversight of online promotions and simultaneously stepping back from the type of complex, hyper-serialized, transmedia storytelling seen in highly visible global franchises like *Smallville*, *Lost*, and *Heroes*, the networks may be ignoring the powerful pull of interactive narratives, which can be crafted best by skilled storytellers.[34] In other words, after dipping a toe into the rabbit hole of stylistic and narrative structural experimentation and collaboration, which resulted in numerous awards and media attention, several networks have scrambled back to the familiar ground of their traditional analog business by using online promotions simply as an adjunct to their broadcast series, designed like *TV Guide* with public-relations-driven, celebrity interviews for ease of use by a mass audience.

Flexing Their Digital Marketing Muscle

In 2007 Quincy Smith, the former head of CBS Interactive, chastised tradition-bound network marketers: "So you've got to get smarter about what those cuts of your content are going to be online. Should it just be the TV promos? No, they see that on TV. So 'Seven-Minute *Sopranos*' and those equivalents come to mind in a major way. '500 Reasons to Love *Jericho*,' a thousand montages of *Ghost Whisperer* in the way teenagers love."[35] According to Smith, traditional television promotions must be replaced with more interactive, participatory forms of digital storytelling. Starting in the late 1990s, first Fox and, later, The WB and the United Paramount Network (UPN) experimented with new ways to reach the younger, millennial audience with previously mentioned examples like "Dawson's Desktop" and "The Chloe Chronicles." In contrast, the big three networks maintained a rigid focus on the mass audience, albeit with a special interest in the eighteen-to-forty-nine audience. That is, until 2004, when ABC aggressively promoted three promising, new series—*Lost*, *Desperate Housewives*, and *Grey's Anatomy*—to try to pull itself out of fourth place. Because ABC's ratings were so bad at the start of 2004, the network could not sell new fall shows with traditional on-air promotions strategically run during episodes of successful series. Hence, the network marketing team was forced to use a variety of inexpensive, inventive, grassroots promotional devices that extended the series' storylines. They bought magazine and bus-stop ads for the fictional Oceanic Airlines to promote *Lost* and printed acerbic aphorisms about suburban life on dry-cleaning bags to advance *Desperate Housewives*. These clever, nontraditional promotions gave then marketing chief Mike Benson's team an opening to engage in online narrative experimentation as well, especially in the case of the *Lost* franchise.

Most marketing executives are quick to point out that television creators should not be allowed to determine a marketing campaign because they lack objectivity about their own series. By 2006, however, network marketers began to use this argument to justify their decision to oversee all digital productions. The in-house digital marketing divisions hired large numbers of low-level digital laborers, including writers' assistants (sometimes through diversity programs), to write or in some cases direct these online campaigns.[36] The marketers could claim that they were meeting the guidelines of the WGA that required them to consult first with the series showrunners. However, by hiring low-level writers to serve as liaisons between the showrunners and marketers, the networks were putting young workers in an awkward situation, especially during a strike, by forcing them to choose allegiance to talent or management.

Prior to the May 2006 upfronts (introduction of the coming season), Jeff Zucker, then head of NBC-Universal Entertainment, announced to advertisers his commitment to digital platforms. "TV 360 is our entire approach to TV," he stated. "Programming can be developed for linear applications, but every program has to have a broadband component or a mobile application. We are going to the upfront in the next few weeks with that being our mantra."[37] *Heroes* became the flagship series for NBC's "360" plan. Creator Tim Kring found himself locked into network negotiations with key brands like Nissan, the primary sponsor of the franchise, in which he agreed to incorporate product integrations across multiple platforms of the *Heroes* story universe, including the series proper, the online graphic novels, the free iTunes download of the pilot, as well as "Heroes 360," the initial name given to a host of online contests and experiences available on the network website.[38] At the end of the day, Kring admitted that "none of that trickled down to the show."[39] Instead, the product integrations were seen as part of the marketing for the series and a source of additional ad revenue, not as a supplement to the production budget.

Zucker put former marketer and development executive Vivi Zigler in charge of series development from 2005 to 2006, replacing longtime development head Ted Frank in the process.[40] After cementing her relationships with the powerful showrunners, Zigler headed the new NBC-Universal Digital Entertainment and New Media division from 2006 to 2012. The good news is that during her long run Zigler streamlined the network's digital division and led her team to critical acclaim, winning several Webby awards for innovative online campaigns for *Heroes*, *30 Rock*, *The Office*, and *Kath & Kim*. The bad news is that by hiring large numbers of low-paid digital laborers, the division took over many of the creative tasks previously held by guild-represented writer-producers.

In the aftermath of the WGA strike, there has been a steady decline in the power and privilege of television writers who identify themselves as transmedia storytellers and who are committed to reinventing television for the Web 2.0 era. For example, Jeph Loeb, Alexander, and Warshaw lost their jobs on *Heroes* in the aftermath of the WGA strike despite (or perhaps because of) their pivotal role as innovators in the interactive space. Something similar happened at ABC when the network reversed its original position on transmedia storytelling franchises shortly after *Lost* ended its six-year run in 2010. Both ABC head Steve McPherson and marketing head Mike Benson were fired.

By 2012, ABC's mandate had returned to its more traditional focus on bringing viewers back to the linear broadcast. Albert Cheng, Disney-ABC's executive vice president of digital media, described the complex research metrics used by his team to prove that the type of fan-friendly, derivative content-promotion hybrids associated with the *Lost* era had failed to reach a mass audience. Rather than intuit that spreadable media can exponentially expand the network's access to digital natives where they live—online—ABC is using its marketing research to justify the decision to discontinue these early social experiments. The digital marketing team has also reverted to a more conventional approach to online promotions, using them primarily as digital billboards designed with a mass audience in mind. Something analogous has occurred at NBC, according to Jennifer Gillan: "After the strike was settled, NBC's ambivalence about online content and the transmedia storytelling approach to broadcast TV series continued to grow."[41]

Cable's Two-Tiered Approach

Unlike the broadcast networks, the cable networks have demonstrated a greater willingness to hire outside companies and digital consultants like Starlight Runner and Campfire to help differentiate themselves from the networks. Starting in the 1990s, the premium networks began to use catchy marketing slogans like "It's Not TV, It's HBO" and Showtime's "No Limit." HBO's earliest original series, such as *Sex and the City*, *The Sopranos*, and *The Wire*, were routinely celebrated in the popular press and scholarly publications as quality television, suggesting that both sets of critics internalized the cable networks' marketing speak. The intimation of quality was initially designed to entice the older, sophisticated, yuppie audience, a group more inclined to re-create the art-house theatrical experience at home by purchasing expensive flat-screen televisions rather than by engaging with the series online. However, once FX, AMC, and other basic cable networks began to create grittier, more violent, and risqué fare like *The Walking Dead*, *Sons of Anarchy*, and *American*

Horror Story, the premium networks jumped to pursue this audience as well by upping the taboo quotient of sex and violence in original series like *True Blood*, *Dexter*, and *Game of Thrones*. (Notably, FX's catchphrase seeks to outdo all other slogans with the claim that "There Is No Box.")

To help re-energize their aging brands and to engage this new, younger audience, HBO and Showtime have hired savvy transmedia companies Starlight Runner and Campfire to develop the type of complex, demanding, social media experiences favored by digital natives. Starlight Runner's Jeff Gomez alludes to infrastructural differences between the broadcasters and cable networks that may explain why the networks tend not to hire outside companies like his: "Cable networks, particularly the movie channels, don't have as many third party concerns, nor do they have access to as many in-house resources. This is why Starlight Runner Entertainment has had more luck at consulting on transmedia implementations with networks like Showtime. We are providing expertise that may not be on hand, either on the show end or the marketing/digital end, and our experience at balancing the delicate relationships between marketing and creative allows us to maximize the efficacy of the multi-platform content."[42] Gomez goes on to explain the unique needs of cable networks: "It's also true that cable networks, which often cater to extremely loyal niche fan bases, are more often interested in unique experiments that have resulted in vanguard techniques that will become standard practice for all of television in the years to come. HBO's second screen experience around *Game of Thrones*, for example, or SyFy's use of intrinsic webisodes for *Battlestar Galactica* were significant innovations that are more common today. In both cases, third party partners were recruited to produce the content."[43]

On the distribution front, the cable networks have been quicker to embrace the long-tail logic outlined by former *Wired* magazine editor Chris Anderson. In contrast, the studios and networks still rely on a blockbuster model, that is, a saturated marketing and distribution strategy focused on the opening weekend (or fall season, in the case of the networks). Small, quirky series like HBO's *Enlightened*—about a neurotic woman who suffers a nervous breakdown that she believes is a spiritual awakening—would not command enough advertising dollars to make sense for the traditional networks; but the critically acclaimed series helps to augment HBO's older subscriber base while its streaming site, HBO-Go, enables it to pursue a more mobile niche using highly interactive social experiences.[44]

The cable networks are able to embrace these massive infrastructural changes more effectively than the broadcasters because of two industrial buffers: a subscriber base that makes them less dependent on declining advertising dollars and programming that targets niche audience groups, thereby keeping production costs down. Brian Swarth, vice president of digital marketing at Showtime, has acknowledged the inherent difference between the

FIG. I-3 Campfire's website showcased its marketing campaign for *True Blood* (HBO), which successfully aligned with the storytelling and aesthetic goals of series creator Alan Ball. (Photo courtesy of Michael Monello, Partner, Campfire)

ad-supported network business and cable's subscription model when it comes to embracing social media: "To date, much of the industry's emphasis on social TV has been around driving viewers to linear premieres. And for a basic ad supported network, that's critical. But Showtime's business model (premium subscription based TV) allows us to be more flexible in terms of how and when we use social TV to engage with our viewers. For some of our shows, a large percentage of our viewership comes from non-linear platforms. So we are always thinking about new models of engagement in time shifted viewing experiences using social TV."[45] Whereas the networks have moved most of their digital marketing activities to in-house divisions, the premium cable networks (HBO, Showtime) and the basic cable networks (FX, A&E, AMC, USA, Discovery) have hired several outside transmedia production companies, digital consultants, and social media marketing experts to support their in-house efforts.[46]

The Networks Have Dibs on Digital Distribution

Controlling distribution has always been an essential element of the television business and is seen as the networks' primary means of bolstering their $60 billion-a-year income from advertising. That imperative explains why the notoriously self-reliant networks turned to a number of Silicon Valley outsiders for advice on how to incorporate digital distribution starting in 2005, when broadband was available in the majority of homes. The overnight success of

Napster, a music search and peer-to-peer filing service, served as a wake-up call for the traditional television industry. Starting in 2001, Napster, according to *Wired* magazine, "allowed people to use the Internet to do what they had done for years in neighborhoods, schoolyards and concert venues: They swapped music."[47] After watching what happened to the music industry when it resisted change at the start of the new millennium, the networks began to respond, albeit at a glacial pace. ABC was the first, engaging in a revolutionary pay-television pact with Apple iTunes in 2005 and offering free streaming of current primetime shows like *Lost*, *Grey's Anatomy*, and *Desperate Housewives* on ABC.com. Both deals were engineered by ABC's digital media head, Albert Cheng, with the support of his boss, Disney-ABC Entertainment Group head Anne Sweeney, and her boss, Disney Corporation head Robert Iger. In his statements to the press, Cheng revealed his underlying goal of maintaining his power base in the traditional network hierarchy while also aligning himself and his team with the start-up traits of nimbleness and gumption. He explained, "I see us as a Silicon Valley startup within a big company."[48] In fact, Cheng had risen up the ranks of Disney, previously serving as a senior vice president of business strategy and development. Cheng's background in technology and distribution may explain his division's focus on these two areas in contrast to Vivi Zigler's NBC-Universal Digital Entertainment group, which, given her background in marketing and development, is more focused on digital marketing content. The two ABC digital distribution deals created a ripple effect across the Hollywood entertainment landscape, disrupting Hollywood's business-as-usual relationships with advertisers, affiliates, and the labor guilds. Notably, Chuck Slocum, one of the lead negotiators at WGA, suggested in a private interview that the ABC-Apple deal represented the first volley shot by the big media companies that ultimately resulted in the WGA strike of 2007–2008.

The next network to rethink distribution for the digital age was CBS, which made the bold decision to hire the previously mentioned Silicon Valley rainmaker, Quincy Smith, who formerly worked for big media banking firm Allen & Co., "where he put together a slew of Silicon Valley tie-ups, including Google's (GOOG) $1.6 billion pickup of YouTube in 2006."[49] CBS hired the entrepreneurial boy wonder to run the network's digital shop and manage a series of acquisitions. CBS's position on digital distribution was somewhat different from that of its peers. Rather than channel content through the CBS.com website, the company created the CBS Audience Network in 2007 and forged deals with multiple streaming sites, including AOL, Microsoft, Comcast, Joost, Bebo, and Veoh. The second phase of CBS's Audience Network plans was a pre-Facebook effort to embrace social networking. CBS partnered with early social media experimenters, including Clearspring, Ning, and RockYou!, which meant that users could share CBS clips with each other on their blogs, wikis, and community pages.[50] Although CBS public relations releases on Smith are uniformly positive, his own statements suggest a workplace culture clash between the hotshot dealmaker (called "The Energizer Bunny" by one industry blog) and the conservative network suits.

During one 2007 interview, the outspoken Smith bluntly reminded the networks that their bloated industry was not impervious to new, independent, online competitors: "It's arrogant to assume that our content is actually what's going to be what works online. The Audience Network is on a level playing field with *Lonely Girl*. That's our competitor. Most media just assumes 'Well, if you're not going to have my content, you're going to die.'"[51] In other words, Smith argued, the networks' brand, high production values, expensive casts, and big-name stars were not going to affect how Web browsers perceived their streamed television series if new, inexpensive, but more exciting content like *Lonely Girl 15* (or *Angry Birds*, to use a more current example) captured the attention of a young, impatient, multitasking web audience. Smith's answer to the question of online distribution was not "TV on CBS.com" but rather "TV everywhere."[52] The network brand no longer matters, Smith argued, despite the lengthy focus on branding that began in the 1980s in reaction to the threat posed by the five hundred channels available on cable; rather, Smith urged, CBS must pursue each of the distinct demographics on the sites they frequent. Furthermore, Smith dismissed the decision by the other three networks to create their own streaming site, Hulu, as a waste of time, given that YouTube, Netflix, Amazon Unbox, and other sites had already done a better job of aggregating users. Notably, Smith was gone from CBS by 2009, a mere two years after his appointment. He explained in a press conference that he would return to Silicon Valley, set up his own shop, and retain CBS as one of his clients, suggesting that the company preferred to hear his provocative ideas from a safe, geographic remove.

NBC-Universal head Jeff Zucker announced Hulu in 2007, the same year that Smith launched the CBS Audience Network. Co-owned by three of the four major networks (CBS opted out), Hulu was designed to provide television audiences with a legal alternative to the popular YouTube. Whereas Smith's effort to shake up the status quo at CBS represented a proactive vision of the future, Zucker's decision to create a new, streaming portal appears reactive in comparison. Zucker ranted that YouTube had achieved its tipping point on the back of NBC-Universal when an anonymous user streamed NBC's *Saturday Night Live* rap-parody "Lazy Sunday" on YouTube shortly after its broadcast on December 5, 2005. The short video went viral instantly, turning YouTube into an overnight success story.[53] Although YouTube was best known in the beginning for its amateur videos of sneezing pandas and cats playing the piano, the popular site had played fast and loose with copyrighted material, allowing its low-end, user-generated content to coexist with illegal downloads of expensive, professionally generated Hollywood content. Shortly after Google acquired YouTube for $6.5 billion in a much-ballyhooed stock trade deal on October 9, 2006, Google insisted YouTube turn "pro" by abiding by the Digital Millennium Copyright Act and instantly pulling down content if it received complaints from copyright holders. Google was intent on policing YouTube so that it could continue to do business with Hollywood and Madison Avenue.

The networks' initial response to YouTube was reactive and shortsighted, preventing them from realizing that they could have used the popular site to sneak release their broadcast clips and promos. On the other hand, YouTube's unruly launch of an online distribution site without regard for Hollywood's preexisting rules of engagement gave the young company a major head start in the race toward the future of entertainment. By 2008, the Google-owned YouTube, once the reviled wild child of the new digital landscape, had amassed 40.9 percent of online users compared with Hulu's 2.9 percent.[54] In 2011, YouTube challenged television's throne by announcing an original channels experiment, commonly referred to as "YouTube's 100 Channels." Eager to position itself as distinct from network television but also capable of replacing it, YouTube used inflated rhetoric to describe the venture, calling it "a new digital video platform that will rival television programming."[55] YouTube is using this platform to transition from an earlier focus on user-generated content to semiprofessional content and to help it compete for television's advertising dollars. Paradoxically, YouTube is accomplishing this goal by appropriating network television's age-old curatorial category of "channels."[56] Further identifying itself with instigators of historical change in the entertainment industry, YouTube compared its "100 Channels" to the advent of cable television's five hundred channels in the late 1980s and early 1990s.[57] Most significantly, Google-YouTube, Microsoft, AOL, Yahoo!, and Hulu each demonstrated its status as a potential contender for the network throne by holding the first New York–based digital upfronts on April 24, 2012.[58]

Hulu was launched on March 12, 2008. Notably, the networks opted to hire another Silicon Valley outsider, Amazon executive Jason Kilar, to run the new online portal. To maximize his entrepreneurial style and can-do spirit, Kilar insisted on carte blanche to run the shop without interference from the network owners. Kilar's plan for Hulu appeared to be to convert television viewers into online viewers by replicating television's free, ad-supported model and by organizing content according to genre and popularity—essentially the same type of curatorial tools used by Amazon, Netflix, and other long-tail services. Kilar was making steady progress when, in 2010, the networks started to panic over such a long-range goal. Instead, they wanted to secure some short-term profits and so demanded that Kilar convert Hulu into a two-tiered system. The first tier would remain the same, providing free, ad-supported content.

The second tier would offer premium content for a monthly fee, in a fashion similar to premium cable. In essence, the network bosses had pulled the rug out from under Kilar's blue-sky vision of the future. Kilar used his blog to denounce his bosses publicly for their failure to recognize that consumers prefer to pick and choose when and where to watch media and to condemn the cable bundle as an outmoded concept. Some industry pundits called this outburst Kilar's Jerry Maguire moment—a public statement of principle that was sure to get him fired. Although Kilar hung onto his job until 2013, the public clash revealed the considerable chasm dividing the corporate-minded networks from their freewheeling, entrepreneurial, tech industry counterparts.

Hulu Plus was launched on November 17, 2010, against Kilar's better judgment. Even though Zucker had helped to inaugurate Hulu as NBC-Universal's vision of the future, the network executive was also the first to flinch a mere nine months later, when he warned his peers not to be seduced by "digital pennies" over broadcasting's "analog dollars." In a public declaration of weak support for Hulu a year into its operation, Zucker conceded to journalists that it was now earning "digital dimes." Shortly after the launch of Hulu Plus, when asked if he had left analog dollars on the table by recklessly pursuing digital pennies, Zucker was more circumspect than glib. His statement provides an apt summary of the state of the networks in 2010, with one foot on the sandy shoals of the digital future and one foot on the once terra firma of the analog past: "You can't bury your head in the sand in the advance from tech. You have to try different things, because technology is not going to go away. But are we making enough money from digital? No. But we have to keep trying and we're making progress."[59] Zucker was fired in late 2010, shortly after cable giant Comcast acquired NBC-Universal. Many of his critics predicted that "the savvy Comcast brass would recognize how badly the NBCU topper had 'Zucked-up' his job."[60] Comcast seems intent on shutting down Hulu's vision of television's online future by requiring all users to verify that they've paid their cable bill, signaling yet another effort to shore up television's traditional business as usual.

While Hulu's shift to a "free television" model online was being challenged by its network owners, new contenders for the throne emerged, including Netflix, which experienced a slow but steady increase in online subscribers, rising to 27 million U.S. subscribers and 6 million international subscribers at the end of 2012.[61] These statistics, however, represent only the number of consumers who have subscribed rather than the number of hours they commit to using these streaming resources. Although U.S. viewers spend an average of 7 hours per person per month consuming content online via Netflix, Amazon, Hulu, and YouTube, traditional television viewing still dominates with a total of 160 hours of viewing time per person per month.[62] These statistics give the networks yet another reason to stay committed to their aging business model despite mounting evidence that consumer behavior is undergoing a major revolution.

Policy Wars: Who Will Control the Pipes?

As the networks struggle to reassert the primacy of in-house oversight of their valuable media franchises on all fronts—production, marketing, and distribution—larger policy battles are being waged in Washington over the fate of free broadcast television. The networks versus cable and satellite carriers like Comcast and Dish Network represent opposing sides of an aging policy on retransmission that was put in place by the Federal Communications Commission (FCC) in 1992. The policy requires the networks to make their high-end content available to viewers who subscribe to cable and satellite services; at the same time, these pay-television carriers must compensate the networks for less desirable local programming such as the news, which is still protected under the outdated rubric of "for the public good." Proponents of the policy argue that these fees represent the last defense for the networks in an era of declining ad revenues. The cable and satellite carriers counter that the policy gives the networks undue leverage, allowing them to raise rates for key sporting events by as much as 300 percent.[63] The carriers are starting to retaliate by creating "blackouts" of high-profile boxing matches or football games or by pulling popular Nickelodeon, MTV, or Comedy Central shows from their schedule—unruly behavior that Washington policymakers would like to curb. Although the FCC currently has the back of the old-fashioned broadcasters and their long-time partners, the affiliates and the advertisers, it may be only a matter of time before policymakers shift their allegiance to the new power-brokers—the cable and satellite companies.

The networks' future looks even more uncertain and circumscribed if one considers the factions fighting for control of the digital landscape. Policymakers may be protecting the networks in their short-term battle with cable and satellite carriers over the fate of ad-supported, free television; however, in her recent research, Jennifer Holt warns that the FCC is favoring private tech companies like Verizon, Google, and other broadband service providers in the battle over the global reach of the Internet. The tech companies are eager to replace the current, democratic access to the Internet with a tiered approach to content delivery that will be under their control. With this considerable leverage on their side, the private tech companies could demand that consumers pay additional fees to have specific content delivered more quickly, thereby undermining the existence of net neutrality in favor of privileged access to content determined by those who control the pipes into the home.[64]

The Essays: Work Worlds in Transition

This collection would not have been possible without the proliferation of recent noteworthy scholarly accounts of the post-network television industry's relationship to new media. These include works by Henry Jenkins (*Convergence Culture* and *Spreadable Media*), Lynn Spigel and Ian Olsson (*Television after TV*), Jennifer Gillan (*Television and New Media*), Amanda Lotz (*The Television Will Be Revolutionized*), Jonathan Gray (*Show Sold Separately*), Sharon Marie Ross (*Beyond the Box*), and James Bennett and Niki Strange (*Television as Digital Media*), to name just a handful. Each of these works has contributed greatly to this collection's varied attempts to capture an industry in transition as the networks struggle to acclimate themselves to a steady flow of new digital technologies, social media companies, and changed consumer behavior. The essays in this collection use a multidisciplinary approach to explain the impact of today's complex entertainment industry practices on creative personnel, marketers, and fans. Among the issues considered are new consumer behaviors (multitasking, downloading and sharing of digital content), new digital technologies (DVRs, smart phones, iPads, and so on), new governmental policies (the retransmission policy of 1992, the Digital Millennium Copyright Act of 1998, and the FCC broadband policy statement of 2005), new social media companies (Facebook, YouTube, Twitter), new types of fan engagement (mash-ups, wikis, and the like), new storytelling strategies (ARGs, web series), the buying and selling of global franchises, a targeted approach to ethnically diverse audiences, and the failure of studio licensing departments to manage creativity and collaborative authorship effectively.

Although the unifying methodological focus of the essays in this collection is media industries scholarship, many of the essays incorporate aspects of production studies and audience studies when possible to examine this industry in transition from the ground up. For instance, several essays grapple with the production inefficiencies that continue to hobble network television, including outdated development, consumer products, and licensing departments. Jonathan Gray and Derek Johnson examine how old-fashioned television licensing departments are undermining the creative potential of licensed computer games tied to important network series like *24*, *The Simpsons*, and *Battlestar Gallactica*, among others, by treating them as promotional materials for DVD boxed sets rather than as valuable creative works. According to Gray, even if the networks view this issue from a purely industrial, profit-motive perspective, it makes little economic sense to maintain these outdated practices.[65] In his essay, Johnson demonstrates how licensed vendors have become a denigrated production category; as such, their creative autonomy is curtailed even as they receive insufficient creative oversight from busy television show-runners or from licensing executives, who lack story training. Furthermore, studios are expanding the category of licensee to include prosumers engaged in promotions. Essays by Will Brooker, Julio Russo, and myself consider the cultural-industrial implications of having networks assert corporate control over both writer-created and fan-created derivative content that doubles as digital promotions. Brooker describes gendered responses to Videomaker and FanLib—two corporate attempts to turn fans into promotional partners. Russo examines Showtime's use of OurChart.com and "You Write It" as two efforts by the media corporation to convert the free labor of a politicized lesbian, gay, bisexual, and transgender (LGBT) fan community into corporate work-product in order to market its series *The L Word*. Whereas the focus in Brooker's and Russo's chapters is on the fan community's response to network-created digital promotions, my chapter examines the competing demands placed on the low- to mid-level writers who were assigned to create "The Lost Experience" ARG—an interactive, story-driven experience that doubles as promotion for the *Lost* series.

The next two chapters explore the inevitable inversions of content and marketing that are taking place as network executives and brand managers seek alternatives to the thirty-second spot in the age of DVRs and shifting consumer behavior. John T. Caldwell's chapter speaks to the unholy collusion of content creation and marketing that results when network and cable series like *Entourage* and *Extras* shift the on-screen focus of their storytelling to the behind-the-scenes machinations of the media industry workplace. This reflexivity-as-marketing churn is undermining skilled labor as corporations harvest consumer-made content. Robert Kozinet's chapter reveals how the increasingly blurred line dividing content and marketing has been taken to its next logical step, now that brand marketers are redirecting fan fervor over a favorite television series or game to such banalities as potato chips and tennis shoes.

M. J. Clarke focuses on the networks' current efforts to expand their repertoire of program genres by buying popular, pre-tested, foreign format sales and turning them into highly successful American franchises, such as *The Office*, *American Idol*, and *America's Got Talent*.[66] Clarke observes the self-reflexive industrial irony implicit in *The Office*, whose inefficient workers invoke the stagnation and lack of innovation inherent in aging bureaucracies like the paper industry (and, by extension, the network television industry).

The chapters by Vincent Brook and Katynka Z. Martinez question whether the broadcast industry's historical obligation to serve the public good requires them to produce ethnically diverse representations of the population. Brook chronicles the legacy of "multiculti" casting from the 1970s and 1980s in series such as *M*A*S*H*, *Hill Street Blues*, and *St. Elsewhere*, and considers whether today's multiracial and multi-gendered casts, primetime soap narrative structures, and, most significantly, the interracial romance in *Lost*, *Heroes*, *Grey's*

Anatomy, and *Ugly Betty* are responding to global economic forces, sociocultural changes, or outside pressure from public interest groups.[67] Martinez documents a different corporate approach—Fox Entertainment Group's failed effort to target a commercially significant niche audience of bicultural Hispanics by cobbling together local affiliates left in play after the newly created CW weblet was formed and by offering a superficially updated telenovela format. Fox's goal was to capture one of the fastest-growing demographics in the United States, with "Nielsen recently forecast[ing] that the purchasing power of U.S. Hispanics will increase to $1.5 trillion in 2015 from $1 trillion in 2010."[68] However, the Fox network, like the rest of the other networks, had little experience in targeting niche audiences and failed to get the new initiative off the ground.

Finally, Henry Jenkins's "The Reign of the 'Mothership': Transmedia's Past, Present, and Possible Futures" invokes a blue-sky future filled with promise based on the creative contributions of valuable, independent, sub-companies like Starlight Runner Entertainment and Campfire, which represent a new breed of independent producer dedicated not just to rethinking storytelling in the context of existing media franchises, but also to taking these on-screen engagements into the world at large through experimental practices like ARGs. At the same time, Jenkins explores the conservative countermeasures adopted by the Hollywood system to manage these new creative forces by reasserting the primacy of the "mothership"—namely, the traditional film or television series—over and above the more radical re-envisioning of interactive, expansive storytelling and audience engagement that is being proposed by creative outsiders.

In conclusion, each of the essays in this collection is committed to addressing the myriad of new digital threats and the equal number of digital opportunities that have become part and parcel of today's post-network era. Each charts the various ways, large and small, that network television sought to embrace the multi-screen, connected experience but often fell short by reverting to bureaucratic operations tied to their former strength as broadcasters. That said, as several of the essays in this collection demonstrate, the networks gleaned invaluable lessons starting in the mid-2000s by engaging not just skilled production personnel but also a formerly untapped, grassroots constituency of nonprofessionals (an active and engaged community of fans), whose admiration for specific media franchises made them willing partners. Fans, therefore, should be seen not simply as a source of detailed consumer information for advertisers via online registration but also as a means to keep the complex ideas on display in these vast story worlds in circulation. The debate over whether transmedia storytelling and digital marketing experiences are distinct has become moot in an era when digital natives are hardwired to *share* content with their community of social media contacts on a mutual, virtual playground—a radical change in the role and function of today's television creative personnel that both media companies and labor guilds need to address.[69] The good news for both storytellers and marketers is that these new social experiences can intensify consumers' participatory engagement with media, helping to make sure these works remain part of the cultural zeitgeist far longer than any on-air promotion for a single linear broadcast can. To facilitate the creation of these new business and creative approaches to the television experience of the future, the broadcast networks would be well advised to recognize the value of the skilled collaborators who are adept at creating these types of expansive, participatory story universes and pay them accordingly. What will be the future of the television industry? Stay tuned (or at least online).

Notes

1   David Kiley and Tom Lowry, with Ronald Grover, "The End of TV (As You Know It)," *Business Week*, November 21, 2005, http://www.businessweek.com/magazine/content/05_47/b3960075.htm.
2   John T. Caldwell, "Para-Industry: Researching Hollywood's Blackwaters," *Cinema Journal* 52, 3 (spring 2013): 157–165.
3   M. J. Clarke, "Outsourcing *The Office*," this volume.
4   Meg James, "Disney's ABC Television Group to Cut 5% of Workforce," *L.A. Times*, January 30, 2009, http://articles.latimes.com/2009/jan/30/business/fi-abc30.
5   Caldwell, "Para-Industry."
6   Quoted in Nellie Andreeva, "HRTS: Broadcast Network Chiefs Reflect On 'Crazy' Pitch Season," *Deadline Hollywood*, October 11, 2011, http://www.deadline.com/2011/10/hrts-broadcast-network-chiefs-reflect-on-crazy-pitch-season/#more-181647.
7   Pedowitz likened CW's low ratings to the top scores on social media grids like "Get Glue" and concluded that the key "is to find a way to get them [the audience] to come back and view the shows live." Lacey Rose, "CW President Mark Pedowitz Opens Up about Ratings, Profits, and 'Vampire Diaries,'" *Hollywood Reporter*, May 2, 2012, http://www.hollywoodreporter.com/news/cw-mark-pedowitz-la-complex-netflix-vampire-diaries-318772.
8   See UC Santa Barbara, Carsey-Wolf Center, Media Industries Project, "Connected Viewing Initiative Research Team," http://www.carseywolf.ucsb.edu/mip/connected-viewing-initiative-research-team.
9   Author's interview with Mark Warshaw, former transmedia producer, *Heroes*, November 17, 2008.
10 See, for instance, Roberta Pearson, ed., *Reading Lost: Perspectives on a Hit Television Show* (London: I. B. Tauris, 2009).
11 Matt Hoey, "All Who Wander Are Not Lost: Showrunners Carlton Cuse and Damon Lindelof Lead a New Generation of Writers into Terra Incognita TV," *WGA Written By*, September 2006, http://www.wga.org/writtenby/writtenbysub.aspx?id=2195.
12 Jill Weinberger, "A Tangled Web: The Fine Brothers Comment on ABC's *Lost Untangled*," *Gigaom*, February 9, 2009, http://gigaom.com/video/a-tangled-web-the-fine-brothers-comment-on-abcs-lost-untangled/.

13. Cotton Delo, "Facebook Measurement Chief Advocates New Standard for Gauging Reach," *Ad Age*, August 27, 2012.
14. See, for instance, Gabriel Metcalf, "Hollywood vs. Silicon Valley," *The Urbanist*, July 2012, http://www.spur.org/publications/library/article/hollywood-vs-silicon-valley.
15. Derek Johnson, *Media Franchising: Creative License and Collaboration in the Culture Industries* (New York: New York University Press, 2013).
16. Ibid.
17. M. J. Clarke, *Transmedia Television: New Trends in Network Serial Production* (New York: Continuum, 2012).
18. Caldwell, "Para-Industry."
19. See the Starlight Runner Entertainment website: http://www.starlightrunner.com/about.
20. See the Campfire website: http://campfirenyc.com/about/.
21. John Thornton Caldwell, *Production Culture: Industrial Reflexivity and Critical Practice in Film and Television* (Durham, NC: Duke University Press, 2008).
22. For instance, Campfire's Steve Coulson and Starlight Runner's Jeff Gomez have appeared on panels at MIT's "Futures of Entertainment" conferences. Starlight Runner's Caitlin Burns appeared at UCLA/USC's 2011 "Transmedia, Hollywood" conference.
23. Caitlin Burns, "Mysteries of Girls' Media," http://girlsmediamystery.blogspot.com/; Caitlin Burns et al., "All Things Fangirl," http://www.allthingsfangirl.com/.
24. Christopher Palmeri, "'Avengers' Gamers on Facebook Get First Look at Movie Spot," *Bloomberg*, April 5, 2012, http://www.starlightrunner.com/news.
25. See Manohla Dargis, "Declaration of Indies: Just Sell Yourself!" *New York Times*, January 14, 2010, http://www.nytimes.com/2010/01/17/movies/17dargis.html?_r=0 (on Broderick and Reiss), and Tom Cheshire and Charlie Burton, "Transmedia: Entertainment Reimagined," *Wired*, July 10, 2008, http://www.wired.co.uk/magazine/archive/2010/08/features/what-is-transmedia.
26. Borys Kit, "David Fincher Leads a Kickstarter Campaign for 'Goon' Animated Movie (Video)," *Hollywood Reporter*, October 12, 2012, http://www.hollywoodreporter.com/heat-vision/david-fincher-leads-a-kickstarter-378574.
27. See Denise Mann, "The Labor Behind the *Lost* ARG: WGA's Tentative Foothold in the Digital Age," in this collection. Cuse emulates the "real paradigm shift" that YouTube founder Chad Hurley created.
28. David Kronke, "Are You 12 to 34? If So, The WB's Got You Covered," *Daily News*, 2002, http://www.thefreelibrary.com/ARE+YOU+12+TO+34%3F+IF+SO,+THE+WB%27S+GOT+YOU+COVERED.-a094026682.
29. "Kmart and the WB Television Network Announce Exclusive Marketing Agreement," *PR Newswire*, May 20, 2004, http://www.prnewswire.com/news-releases/kmart-and-the-wb-television-network-announce-exclusive-marketing-agreement-74139452.html.
30. For a more detailed discussion of Sander and Moses's work on *The Ghost Whisperer*, see Henry Jenkins, "The Reign of the 'Mothership': Transmedia's Past, Present, and Possible Futures," this collection. Also see Jennifer Gillan, *Television and New Media: Must-Click TV* (New York: Routledge, 2011).
31. Ben Fritz, "Action Movie FX: Bad Robot App Delivers Big Bangs for Your Buck," *Los Angeles Times, Hero Complex*, January 6, 2012, http://herocomplex.latimes.com/2012/01/06/action-movie-fx-bad-robot-app-delivers-big-bangs-for-your-buck/.

32 Frank Bruni, "Filmmaker J. J. Abrams Is a Crowd Teaser," *New York Times*, May 26, 2011, http://www.nytimes.com/2011/05/29/magazine/filmmaker-j-j-abrams-is-a-crowd-teaser.html.
33 Author's interview with Carole Panick Angelo, VP, NBC Digital Entertainment and New Media, August 13, 2008. In a 2013 follow-up, Angelo said NBC places greater focus on video content in 2013 (both story-driven and unscripted, behind the scenes) than in 2008.
34 Notably, the NBC digital marketing group initiates and produces all the webisodes associated with the "Revolution Revealed Series" (unscripted) and the "Grimm Webisodes" (scripted). The graphic novels tied to *Revolution* feel corporate and alienating when compared with the evocative, moody, artist-generated graphic novels tied to *Heroes*; the comedic, broadly appealing "Grimm Webisodes" ignore the dark, edgy feel of *Grimm*, the television series.
35 Louis Hau, "Q&A: CBS Interactive's Quincy Smith," *Forbes*, September 13, 2007, http://www.forbes.com/2007/09/12/cbs-internet-video-biz-media-cx_lh_0912smith.html.
36 Yule Caise won the NAACP/NBC Fellowship in Screenwriting and was later hired as a new media producer and staff writer and digital director on *Heroes*.
37 Claire Atkinson, "NBC's Upfront Goes Digital: Jeff Zucker Says Network Sell Will Be 'TV 360,'" *Ad Age*, April 26, 2006, http://adage.com/article/media/nbc-s-upfront-digital/108822/.
38 See the *Heroes* wiki for a more complete list of product integrations, http://heroeswiki.com/Product_placement.
39 Catharine P. Taylor, "What NBC's Heroes Got Out of Product Integration: Marketing, Not Money," CBSNews.com, June 15, 2010, http://www.cbsnews.com/8301–505123_162–43745459/what-nbcs-heroes-got-out-of-product-integration-marketing-not-money/.
40 Author's interview with Ted Frank, executive vice president, NBC Entertainment Strategy and Programs, April 30, 2009.
41 Gillan, *Must-Click TV*, 232.
42 Quoted in Ibid.
43 Author's e-mail interview with Jeff Gomez, Starlight Runner Entertainment, February 4, 2013. Also see Campfire's *Game of Thrones* campaign: http://storycode.org/case_study/steve-coulson-game-of-thrones.
44 See the HBO-Go website: http://www.hbogo.com/#home/.
45 Quoted in Natan Edelsburg, "An Inside Look at Showtime's Social TV Strategy," *Lost Remote*, July 3, 2012, http://www.lostremote.com/2012/07/03/an-inside-look-at-showtimes-social-tv-strategy/.
46 Author's interview with John DiMinico, a branding strategist who has worked with FX and AMC (*Dirt, Justified, Damages, Nip/Tuck, American Horror Story*), August 18, 2009. See http://johndiminico.com/John_Di_Minico/John_DiMinico_%26_Associates_home.html.
47 Brad King, "The Day the Napster Died," *Wired*, May 15, 2002, http://www.wired.com/gadgets/portablemusic/news/2002/05/52540?currentPage=all.
48 Quoted in Chuck Salter, "Brave New Mouse," *Fast Company*, June 1, 2007, http://www.fastcompany.com/magazine/116/features-brave-new-mouse.html.
49 Peter Kafka, "CBS Digital Boss Quincy Smith Plans His Next Deal: His Own M&A Shop," *All Things Digital*, July 11, 2009, http://allthingsd.com/20090511/cbs-digital-boss-quincy-smith-plans-his-next-deal-his-own-ma-shop/.

50 "CBS Daytime TV Goes Online," *CBS.com*, http://emol.org/tv/cbs/daytimetv
    streaming.html.
51 Louis Hau, "Q&A: CBS Interactive's Quincy Smith," *Forbes*, September 13, 2007,
    http://www.forbes.com/2007/09/12/cbs-internet-video-biz-media-cx_lh_
    0912smith.html.
52 Ibid.
53 Peter Kafka, "NBC CEO Jeff Zucker: Hulu Will Start Breaking Even 'Soon,'" *All
    Things Digital*, May 29, 2009, http://d7.allthingsd.com/20090528/d7-interview
    -nbc-universal-ceo-jeff-zucker/.
54 Heather Hopkins, "YouTube Tipping Point Question," *Hitwise*, November 24,
    2008, http://www.experian.com/blogs/hitwise/2008/11/24/youtube-tipping
    -point-question/ .
55 Ryan Nakashima, "YouTube Launching 100 New Channels," *USA Today*, Octo-
    ber 29, 2011, http://www.usatoday.com/tech/news/story/2011-10-29/youtube
    -original-programming/50997002/1.
56 Ibid.
57 Ibid.
58 Mel Carson, "Digitas NewFront 2012 Unites Digital Content Big Hitters in NYC,"
    Microsoft Advertising, April 24, 2012, http://community.microsoftadvertising.com/
    msa/en/global/b/blog/archive/2012/04/27/digitas-newfront-2012-digital-content
    -nyc.aspx.
59 David Kaplan, "NBCU's Zucker: In an On-Demand World, Content Matters
    More Than Schedules," *PaidContent*, November 18, 2010, http://paidcontent.org/
    article/419-nbcus-zucker-in-an-on-demand-world-content-matters-more-than
    -schedules/.
60 Nikki Finke, "Jeff Zucker Fired by Steve Burke," *Deadline*, September 24, 2010,
    http://www.deadline.com/2010/09/cnbc-zucker-wont-make-comcast-merger/.
61 Lucas Shaw, "Netflix Stock Surges as Company Posts Unexpected Q4 Profit," *The
    Wrap*, January 23, 2013, http://www.thewrap.com/media/article/netflix-stock
    -surges-company-posts-q4-profit-74416.
62 Goldman Sachs Global Investment Research, January 21, 2013.
63 Doug Halonan, "Senate Panel to Probe Broadcast, Cable Programming Feud," *The
    Wrap*, July 16, 2012, http://www.thewrap.com/tv/column-post/senate-panel-probe
    -broadcast-cable-programming-feud-47996.
64 Jennifer Holt, "Access of Evil: Google, Verizon and the Future of Net Neutrality,"
    In Media Res, August 25, 2010, http://mediacommons.futureofthebook.org/imr/
    2010/08/25/access-evil-google-verizon-and-future-net-neutrality.
65 Jonathan Gray, *Show Sold Separately: Promos, Spoilers, and Other Media Paratexts*
    (New York: New York University Press, 2010), 188.
66 See "TV Formats," *CNN*, March 28, 20120, http://www.cnn.com/2012/03/28/
    showbiz/tv/tv-formats-sale/index.html.
67 Also see "SAG Diversity Conference Assists the Underrepresented and Under-
    employed," October 9, 2010, http://www.sag.org/sag-diversity-conference-assists
    -underrepresented-and-underemployed.
68 Jeanine Poggi, "ABC and Univision Plan English-Language News Channel for
    Hispanics: Website Planned for This Summer, TV Network Next Year," *Ad Age*,
    May 7, 2012, http://adage.com/article/mediaworks/abc-univision-plan-english
    -language-news-channel/234601/?utm_source=mediaworks&utm_medium=
    newsletter&utm_campaign=adage.

69 One digital native who works at a major talent agency describes the institutional brick wall she faces whenever she sends her clients—directors who specialize in low-cost, digital, social media experiences—to discuss feature projects with studio executives in the story departments: the studios send the directors to their marketing department. Conversely, the studios don't hesitate to ask these same directors to produce inventive digital marketing that doubles as brand-supported content in a classic case of heads we win, tails you lose.

1
Authorship Up for Grabs
• • • • • • • • • • • • • • • • • • • •
Decentralized Labor, Licensing, and the Management of Collaborative Creativity

DEREK JOHNSON

In one of the most dramatically tense storylines offered by the "reimagined" television series *Battlestar Galactica* (2003–2009), the crew of the titular spacecraft encounters another military battlestar, the *Pegasus*, which had also escaped the Cylon attack that destroyed their homeland and the rest of the Colonial Fleet. This joyful reunion gives way to a power struggle, however, when the two commanders recognize their competing ideals and incompatible plans for the surviving civilization. *Galactica* and *Pegasus* briefly formed one big happy fleet, but they soon served as two poles in a line of tension between competing claims to authority. Ultimately, the claim made by series protagonist Commander Adama is upheld, and the newly introduced Admiral Cain is conveniently killed by a Cylon agent.

Within the franchise that is *Battlestar Galactica*—operating at an institutional and cultural level, rather than the diegetic—similar tensions over creative authority within shared, collaborative structures can be observed. Authorship and creative power, like the command of a battlestar fleet, come up for grabs when interested individuals from across a wide range of institutional lines share creative resources and collaborate in the process of sustaining an intellectual property. In media franchising, production is multiplied across a range of sites. Within the single industry of television, the original 1978 *Battlestar Galactica* supported a remake in 2003 and the *Caprica* spin-off in 2010. Yet multiplication can also happen across institutional and cultural lines, where the property is proliferated by authorized licensees as well as by consumers who make bottom-up contributions to the content network via fan fiction, online video, and grassroots video game production. The media franchise, therefore, is a cultural form defined and animated by a multiplicity of production cultures, wherein differing market motivations, competing visions, and unequal positions of institutional power ensure that creative collaboration is a negotiated process rather than the neat outcome of economic synergy and cross-promotion. The result of *Battlestar*'s decentralized network of creative labor is not a singular, monolithic brand, but a struggle over creative authorship by diverse interests.

Nevertheless, *Battlestar* has been accompanied by discursive claims about singular authorship and centralized creative power. Writers such as Michael Newman and James Longworth consider television to be an authored medium in which auteur "showrunners" control and unify the collaborative creative work of writing staffs.[1] As Derek Kompare argues, online podcasts and other sites of extra-textual television discourse have constructed executive producer and developer Ronald D. Moore as the authorial voice of the contemporary *Battlestar* franchise.[2] Moore explained in his podcasts that in the collaborative medium of television his role as showrunner was "to maintain the voice of the show, as it were. The show has a voice. And it's my voice."[3] Through such declarations, Moore positioned himself as the singular creative force driving *Battlestar*, suggesting that an authorial unity emerged from his supervision of what was otherwise a collaborative process.

As that collaboration multiplied beyond television, however, Moore's role in harmonizing multiple creative voices became less certain. Moore had minimal contact with licensed creators. His interface with writers of *Battlestar* comic books, for example, consisted largely of a passive approval that required very little active participation or intervention. Brandon Jerwa explained that his pitches for licensed comic books were "run directly through Ron Moore's office," but his goal as a licensed writer was actually to receive as little corrective input as possible. "My editor told me that I was getting the fewest notes of any Galactica [comic] writer," Jerwa boasted. "That was great."[4] Though the goal was apparently to minimize Moore's input and interference in these licensed products, Moore remained a central authority against which other producers had to negotiate their own creative positions. Despite feeling strongly, for example, that the Cylon Number Six character needed a real name, Jerwa regarded such questions as "something that's more Ron Moore's call."[5] Like Commander Adama, Moore's singular power persisted in the face of decentralized and potentially competing claims to creativity authority.

This essay explores the contradiction between singular authorship and the decentralized creativity of networked production cultures as demonstrated by the franchising of a television series like *Battlestar*, especially as the creative process extends to encompass grassroots production cultures animated by the socially collaborative architectures of the Internet and other digital technology. How did creative authority come up for grabs in these collaborative, networked structures at the same time that it was industrially and discursively ascribed to single sites of creative agency? As Don Tapscott and Anthony Williams argue, digital technology has enabled its users to produce cultural works outside of corporate control through the open, collaborative, self-organized, and peer-based creative processes of "wikinomics."[6] With the creative labor of fans increasingly serving corporate uses on the official *Battlestar* website—suggesting Tiziana Terranova's notion of "free labor"—industry franchise strategies have embraced decentralized creative infrastructure at the same time that singular creative authorities such as Moore have confounded it.[7] These diverse production cultures networked by franchising complicate Axel Bruns's notion of "produsage." Bruns defines produsage as the participatory use of common property within an ongoing, communal processes of creation, a merging of production and consumption that he places in opposition to traditional industrial production models.[8] In franchises like *Battlestar*, however, divergent modes of industrial production and grassroots produsage—from Moore to comic writers to amateur consumers—shared the intellectual property actually owned at an institutional level by NBC-Universal. Franchising here recombined industrial production and produsage so as to put traditional authorship and the collaboration of wikinomics into tension. Within the franchise structure, both industrial producer and amateur producer became, in effect, franchisees—enfranchised, so to speak, to act as agents of production. The centrally empowered Moore did not need to give producers any more thought than he did professional licensees, but all were collaborators in the larger franchise.

To explore the tensions in these creative networks composed of decentralized production cultures this essay will first provide historical context for considering the emergence of wikinomics- and produsage-based modes of creative collaboration within existing industrial patterns. Based on evidence drawn from producers' archives at the University of California, Los Angeles, and the University of Southern California, the essay will argue that *Battlestar* shares a licensing tradition with properties like *Star Trek* (the franchise in which Moore cut his own creative teeth in the 1990s) in that bureaucratic management structures contractually subordinated licensed collaborators to a central creative authority. However, this essay will also examine how, despite maintaining this tradition, *Battlestar* experimented with new licensing arrangements that incorporated amateur sites of creativity into the relations of collaborative production, using them to sustain the franchise's online components. The new possibilities of produsage and the collaborative social networks of wikinomics thus began to be merged into these existing industrial licensing models. Finally, this essay will examine a third site of emerging produser creativity and collaboration outside of the contractual licensing agreements of the industry. *Diaspora*, a freeware *Battlestar* game made by fans of the television series (http://diaspora.hard-light.net/), offers a unique opportunity to uncover not only the processes and practices by which open-source produsage can work but also the ways it functions in tension with parallel industrial modes of production. Through interviews with the makers of the game, it will be possible to see how the new collaborative creativity supported by wikinomics and produsage negotiates persistent claims to proprietary ownership and creative authority over intellectual property. Even as the decentralized creative labor of produsage challenges industrial licensing models, the notion of central creative authority persists as a contested battleground where ownership and the right to contribute to culture are negotiated.

The Creative Tradition of Television Licensing

The primary means of networking content across media has historically been through licensing arrangements in which an intellectual property owner leases to a third party the right to make use of that property in another market. NBC-Universal, for example, contracts with Dynamite Entertainment to make comic books based on *Battlestar* and with Diamond Select Toys to make action figures. Although media critics often consider licensing in relation to questions of media ownership, the practice can also be very productively considered from a creative standpoint. Licensing contracts codify the management of intellectual property across media markets, governing the creative process by formalizing the dominance of the intellectual property owner. That is not to say that creativity is necessarily stifled by the contractual relations of licensing; rather, the third party must negotiate a very specific set of circumstances in which the licensor has final right of *approval*. In the case of the comic books based on the television series *24*, for example, IDW Comics editor Andrew Harris explained that licensor 20th Century Fox "has been both careful in its scrutiny of the story but also extremely accommodating in allowing our creators to exercise their creative vision."[9] Similarly, video game licensees report that creative innovation is possible with franchise properties, but "you have to go through a whole process of running them through licensing to see if it fits ... continuity and all that stuff."[10] Licensing imposes a set of constraints upon creativity, where approval comes from outside the site of production. As game writer Danny Bilson explains, "You're always under the licensor's knuckles so you can't do anything they haven't already done. You can't push it too far."[11]

Although scholars like Avi Santo, Simone Murray, and M. J. Clarke have offered excellent analyses of media licensing,[12] the field of media studies has more generally over-emphasized issues of conglomeration, where cross-ownership and the promises of synergy make these kinds of external creative relationships unnecessary. Nevertheless, informal approval structures persist in conglomerate creative collaboration. Subsidiaries with vested interest in one property closely monitor and oversee its creative uses by other divisions. Paul Levitz, former publisher of the Time Warner subsidiary DC Comics, likened the process to "giving your children over to others," explaining, "Whenever we do a movie, a TV show, or even a radio program based on our heroes, the DC team watches over the production, fretting, fussing, and worrying ... about every step the creative team takes away from the 'canonical' path laid out by our history."[13] So, even though DC properties were not out-licensed, so to speak, when they left their original site of production, they circulated within Time Warner with similar creative constraints attached.

The issue of approval, however, raises the question of who has the power to approve. A brief analysis of the *Star Trek* franchise—the setting in which Ronald D. Moore worked from 1989 to 2001—reveals an institutional shift from oversight of television licenses at the creative level to management of them at an executive, bureaucratic level. In the 1960s, television producers like *Trek*'s Gene Roddenberry considered the activities of licensors as part of his day-to-day decision making as a television producer. Licensing was a creative realm, not just the domain of executives for the Desilu studio producing *Trek*. As early as 1964, Roddenberry expressed an interest in laying a foundation to support licensed extensions, instructing production designer Pato Guzman to "give some thought to a *distinctive emblem* for our ships and the uniforms of our crewman. You may have been the one who suggested a week or so ago that this would have the side advantage of giving us a merchandising *trademark*" (emphasis in original).[14] Set and costume designs thus took licensing into consideration, as did prop production, with the interlocking configuration of the hand phaser, phaser pistol, and phaser rifle designed for its attractiveness to the toy market.[15] Perhaps more significantly, Roddenberry conceptualized licensing as a means of sustaining the series, seeing partnerships with licensees as a means of "creative and design assistance."[16] In exchange for fabricating props and/or models not accounted for in the relatively meager budget for the series, Roddenberry would agree to sign away merchandising rights. For example, when the cost of building a shuttlecraft model and set proved prohibitively expensive (estimated at $12,000 internally in April 1966), the producers turned to toy model manufacturer AMT. By September, AMT had agreed to produce a $24,000 shuttle (50 percent recoverable from Desilu's royalty) in exchange for the rights to market a corresponding model kit to consumers.[17]

In addition to using licensees as a creative resource to bring *Trek* to fruition on television in the 1960s, the producers also worked with licensees in other media markets. Staff writers D. C. Fontana and John Meredyth Lucas reviewed drafts from novelists seeking approval from Desilu/Paramount, assessing their conformity to the quickly forming canon of *Trek* on television. Fontana once tore into author James Blish for consistently ignoring the institutions of Starfleet and the Federation as established on the series, but also criticized "too much development of Vulcan backstory," an intrusion into territory the television writers claimed for themselves.[18] Although the series' producers clearly held creative authority over licensees, their management did provide significant creator-to-creator interaction and exchange through the process of review (even if indirectly relayed through Desilu executives). In November 1967, Roddenberry wrote Desilu executive Ed Perlstein to stress the importance of his staff's review of licensed work to protect the *Trek* property; he proposed appointing a producer to serve as a permanent and paid liaison to licensees, charged with "guiding the writer and publishers away from technical inaccuracy or outright bad taste."[19] Perlstein agreed and sought the producers' help in managing licensee creativity, but he simultaneously refused to pay for that work. "[It] should be their labor of love," Perlstein wrote.[20] The burden of creative liaison with licensees was thus firmly placed upon the television creators.

By the time of the spin-off series *Star Trek: The Next Generation* in the late 1980s, creative liaison with licensees had been largely displaced to bureaucratic sites within the institutions of Paramount (first a Gulf+Western subsidiary and later part of the Viacom conglomerate). Prior to the premiere of the series, Paramount executive Frank Mancuso created a Star Trek Office within the studio's licensing department to manage and coordinate with licensees on behalf of and in lieu of Gene Roddenberry and his writing staff. Assistants working in the office were responsible for summarizing licensed novels, comics, and other materials in memos that would in theory be forwarded to the *Trek* producers for comments to be relayed to the licensee (through the licensing department), all within ten working days. In practice, however, television creators were often removed from this process.[21] If the licensing department approved them on its own, projects would receive the go-ahead. Coordination between the television writers and the licensed creators was an unnecessary step in the franchise process. Moreover, because licensee use of the property was bureaucratically isolated from "official" use on television, the creative subordination of licensees intensified. Richard Arnold, an assistant working in the office, even denied the creative legitimacy of licensed novelists within the franchise: "They should not call themselves *Star Trek* writers because they're not. They are writers of fiction based on *Star Trek* for licensees."[22] The growing distance between television writers and licensed creators served to further delegitimate the creative work of the latter group.

Throughout the 1990s, executive management of licensing continued to isolate television creators from their licensed counterparts in the *Trek* franchise. While working on *Deep Space Nine* in 1997, Ronald D. Moore explained that he and his colleagues "haven't read the comics and . . . have no contact with anyone at [licensee] Marvel. I assume they have a contact at Paramount in Merchandising and Licensing who tracks the continuity. On the show, we only consider *filmed* material to be canon."[23] The creative production cultures within the *Trek* franchise had become entirely isolated, with licensing bureaucrats, rather than creative forces, serving as the only interface for collaboration with the franchise.

Given this lack of creative control, consistency, and unity, the idea of authorship was managed through the imposition of a creative hierarchy and, within it, the diminution of some creative acts compared with others. Canonization made creative authority imaginable in an instance where multiple, discrete nodes of production had otherwise made it institutionally difficult to conceive or achieve. The canonized creative hierarchies that historically marginalized licensed work have also been reproduced by licensees themselves—an understandable development, given their need for approval from the licensing offices endorsing these hierarchies. In 2007, Andrew Harris of IDW Comics explained his intention as a licensee to honor *Trek*'s onscreen history of more than six hundred television episodes, but not any previous iteration from books or comics. He asked fans incredulously, "Are our writers and editors really going to read every ST books [*sic*] out there, and then meticulously conform our stories to the contents of every single published page?"[24] Though himself a licensee, Harris embraced the hierarchies that delegitimated other licensed work.

Given the burden historically placed on licensees to conform to privileged television creativity, contemporary licensees have developed several sets of strategies to negotiate their position in relation to that authority. Glen Dahlgren, design director for *Star Trek Online* in 2006, explained that licensed video game producers face "two fundamental (although sometimes conflicting) goals." First, they must design games to be fun in their own right; simultaneously, however, the games must "fulfill the popular fantasies," assuming players would want to "visit the places they know" rather than those they did not.[25] In this case, the creative process demanded that licensees reproduce existing experiences rather than create entirely new ones. As one video game developer explained, the "world building part is sort of taken care of for you" when working with the *Star Wars* license. "You have an immediate visceral reaction to seeing an AT-AT marching down the plains of Hoth . . . and that's what we get with the license. We can focus on making the gameplay, the

FIG. 1-1 The official Xbox Live Arcade *Battlestar Galactica* game features a simple, top-down view with basic two-dimensional directional control of a Viper fighter. (Frame grab)

controls, the visuals really great as opposed to worrying spending time also on building the world right."[26] In the case of the *Battlestar* game for Xbox Live Arcade in 2007, the play experience re-created a number of battle sequences, including the original Cylon attack on the colonies, already familiar to viewers of the television series. But even the reproduction of familiar experiences can bring creative opportunities. As the *Star Wars* licensee explained, "There's a lot of challenges back and forth to make that experience evocative of the movie and also fun for the game player."[27]

Similarly, comic book publishers licensed to use Hollywood properties seek out writers whose passion for the existing universe allows them to come to the project extremely well versed in the continuity of that world. IDW editor Andrew Harris, for example, praised *24* comic book author Beau Smith, explaining that his detailed, internalized knowledge of the "continuity, character traits, and tone" of the television series eased the process of gaining Fox's approval: "so unless we're accidentally treading on something that Fox has planned for [television], we've already given Fox the kind of story that it's looking for."[28] Needing licensor approval, Harris preferred authors with the ability to master other creators' work as authorial canon. Similarly, Brandon Jerwa, hired to write *Battlestar* comics books for licensee Dynamite Entertainment, claimed that "they don't just randomly recruit writers for their licensed properties; you have to be 'in it to win it' and have that real passion for the subject matter. All of the *Galactica* project writers are knee-deep in the continuity and we all keep very close communication between ourselves."[29]

With this internalization of continuity, licensees could identify pockets of the franchise safe for elaboration without preempting their development on television. Dynamite, for example, pursued writer Kevin Fahey because of his experience as a *Battlestar* writers' assistant (and later staff writer) on the television series. With this "direct connection" to the production of the series, Dynamite editors claimed to gain a much better vantage point from which to identify "areas that they're not exploring."[30] As Fahey recounted, "I spent some time brainstorming what *Battlestar* stories would be unique as a comic series, what stories I knew the show wouldn't be able to tell."[31] Significantly, the Dynamite editors did not attempt to coordinate with the more senior television producers of the franchise, but sought only to identify from the low-ranking writer the pockets of storytelling possibility that held little or no interest for those producers. Most often, these pockets have been found in the diegetic past, in ground already covered and passed over by the television series. The majority of Dynamite's mini-series and ongoing titles have looked to a time period even before the start of the television series, resulting in premises like *Battlestar Galactica: Season Zero* (set one year prior to the television mini-series), and an *Origins* title that explored the lives of characters before the Cylon attack. As *Comic Shop News* suggested in its preview of *Zarek*, a 2006 mini-series that provided a backstory for revolutionary Tom Zarek (but not the contradictory backstory later developed by the television series), licensed extensions for *Battlestar* were designed to function as nonlinear storytelling devices.[32] Exploration of the narrative past, or "continuity mining," presented a viable strategy for licensees looking to elaborate a franchise without upsetting creative hierarchy in the process.[33] Through these flashbacks, licensed creators could defer to central creativity authority while at the same time situating their original work within that privileged narrative continuity. Licensed work remained a creative enterprise, but it participated in a hierarchical collaborative partnership with television producers from a deferential position of creative subordination.

For their part, privileged creators and intellectual property holders have also developed new strategies to impose coordination among various contributors to the larger franchise. Until their dismissal from the *Heroes* writing staff in 2007 and 2008, respectively, television producers like Mark Warshaw and Jesse Alexander labored to return the site of license management to the creative arena. As a producer who had formerly managed the extension of the *Smallville* television series into webisodes and other online venues, Warshaw proposed to NBC in 2007 a three-pronged "transmedia team" to manage all narrative extensions of the *Heroes* world as part of a single, consistent story. One arm would manage merchandising, a second would coordinate all narrative mobilizations of the property across comics, the Internet, and the like, and a third would work with the stars of the series to secure their participation in promotions and content made for these new media. By extending television authorship to licensed markets, Warshaw's plan would allow for direct creative collaboration within the licensing unit.[34] While Warshaw's model proposed an internal infrastructure for managing licensed creativity, a second, parallel strategy enforced unity across networked production cultures by shifting management to an external site. Companies like the New York–based Starlight Runner Entertainment began to develop franchise plans that promised ongoing external creative management for the licenses owned by its clients. Starlight Runner promoted itself as an "international clearinghouse" willing to guide licensees through the creative process on behalf of the intellectual property holders.[35] Both of these top-down strategies held that licensed creativity could best be managed by adopting a coherent, collective, but hierarchical authorship that brought unity to a network of different production cultures. Nevertheless, these strategies were initially limited in their deployment or considered disposable by entertainment executives. In the case of *Heroes*, Warshaw and Alexander were dismissed, and the struggling series was encouraged to dial back its ambitious transmedia storytelling designs starting in 2008.

Battlestar, despite its many innovations as a television series, remained rather traditional at the level of licensing and industrialized multiplication of creative labor, essentially replicating Ronald D. Moore's prior experiences with *Star Trek*. Moore seemed to expend little effort in managing the down-market creativity of licensees. These licensed production cultures deferred to a creative authority from which they remained largely disconnected, subordinated, and perhaps even alienated as only one part of a heavily bureaucratic intellectual property management entity. Moore's archived production papers at the University of Southern California underscore this break between television and licensed production, and offer little evidence of his participation in licensed creativity beyond a draft of the mini-series novelization—and even that lacks Moore's annotations. Although the lack of documents cannot be deemed conclusive evidence of Moore's total uninterest in collaborating with licensees, it stands in contrast to the thoroughness with which he has archived story outlines, studio notes, and edit decision lists related to the television series proper. His detailed engagement with these other activities suggests that his many creative responsibilities as showrunner were largely limited to the television text. Even the "Resistance" webisodes produced for online distribution in 2006 by members of the television production staff did not generate archival traces of Moore's participation. This archived construction of Moore's roles and responsibilities as a television author obscures the complex power differentials operating among the various industrialized creative cultures linked by the contemporary media franchise.

Licensing Users

If *Battlestar* sustained historically hierarchical and rigid structures when it came to industrial licensing, its owners proved far more proactive about incorporating non-industrial and amateur production cultures into the franchised multiplication of creative labor. In the context of the digital information economy, the *Battlestar* license extended to nonprofessional production operating outside the bounds of industry. *Battlestar*'s online video offerings and social gameplay experiences depended largely upon the activities of these extra-industrial, de facto licensees.

Theorists like Yochai Benkler have identified in digital technology the social capacity to support decentralized, cooperative networks of peer production that operate outside the logic of the market and threaten to displace industrial modes of production.[36] Axel Bruns makes similar claims in conceptualizing "produsage" as a non-industrial process wherein modes of production geared to generate discrete products give way to collaborative ventures based on open participation and ongoing revision of shared cultural properties by users. According to this view, the power of networking offers an alternative to the well-defined, contractually prescribed collaboration afforded by licensing and its exchange of payment for the right to produce culture. Bruns, however, argues not for a strict opposition between production and produsage, but identifies a continuum where market-based and nonproprietary forms of production interweave and recombine in an ongoing process of creation.[37] Bruns sees a potential disruption of proprietary culture in this continuum; however, he simultaneously tempers that utopianism with the recognition that the most successful corporations will be those that turn the collaborative and open qualities of produsage to their advantage. The new possibilities afford by the digital economy have not gone unnoticed or undisrupted by industrial institutions, and, as Tizianna Terranova argues, corporations that colonize digital media have profited from the labor of networked users. Instead of hiring professional producers to fill corporately owned virtual spaces with content, the culture industries can depend on consumers to regenerate those spaces constantly by posting, updating, uploading, and otherwise using them. In this turn toward user-generated content, the open, participatory, and processual qualities of digital networks put would-be produsage in service to industry. As Terranova explains, the culture industries can exploit the productive use of culture in order to create "monetary value of knowledge/culture/affect."[38]

The *Battlestar* franchise incorporated this logic of produsage and absorbed the labor of consumers into the existing model of licensed production. In a real legal and economic sense, consumers became licensees/franchisees of NBC-Universal, taking up positions within the decentralized network of labor surrounding the intellectual property. A first instance of this "enfranchisement" of the *Battlestar* audience occurred in summer 2007, when NBC-Universal's Sci-Fi Network sought to sustain viewers' interest in the off-season by offering them access to the BSG Videomaker Toolkit, which contained a collection of establishing shots, visual effects, sound effects, and other stock cutaway images typically used on the series. Fans were invited to combine these professional materials with their own original footage to generate four-minute episodes that could be uploaded to the official Sci-Fi website and made accessible to all, with the promise that a few lucky contributors would see their work air on television during the upcoming season. In aggregate, hundreds of these videos helped animate and refresh *Battlestar*'s online profile, given that the constant stream of homemade narratives encouraged frequent, repeated visits to the site over the summer, despite the lack of new professionally produced content. A consumer base was thus transformed overnight into a production base of virtual employees willing to work without pay to sustain a beloved franchise. Instead of policing the use of its intellectual property, NBC-Universal saw the economic advantages of licensing consumers.

"Join the Fight," the BSG social networking game introduced in 2008 (http://jointhefight.scifi.com), similarly illustrated NBC-Universal's efforts to extend licensed production to consumers of *Battlestar*. Asking "Are You a Human or a Cylon?" the game prompted players to swear allegiance to the human Colonials or their cybernetic antagonists and then to engage in team-based competition. To compete, players earned "cubits" (the money of the *Battlestar* universe) to pay for ongoing upgrades to the engines and armaments in their personal fleets of space fighters. Of course, these virtual fleets are never completed, resulting in an arms race that required players to work endlessly. To earn cubits, players completed "tasks" that called attention to the role of labor in this virtual mode of consumption. Weekly quizzes, for example, skewed attentive viewership from a pleasurable act into a type of labor to be rewarded with cubits. Play was literally coded as work within a capitalist framework. Other modes of engagement that rewarded players included acts of creative labor such as designing team logos and writing team mottos, and promotional activities such as recruiting more players. Within the context of a game, viewers performed ongoing creative and promotional labors that ultimately served the needs of industry by expanding both the content and the reach of the franchise.

The status of this work as specifically licensed labor was driven home by the "terms of service" contract to which fans had to consent before participating in this social network. To gain access to the *Battlestar* franchise, NBC-Universal asked its producers to agree that "such Material is licensed to you by Join the Fight and Join the Fight does not transfer title to any such Material to you."[39] This contract demanded that nonprofessional producers submit to the industrial licensing model in two additional respects. First, producers must possess "all the rights, licenses, permissions and consents necessary to submit the Content" before they could add to the *Battlestar* content pool. In other words, no user-generated creation was recognized outside the boundaries of the licensing agreements made in the professional, industrialized world; users were permitted to play with the *Battlestar* license, but not the *Star Trek* license. Second, would-be users "are granting Join the Fight including without limitation NBC Universal, Inc., its licensees, successors and assigns, the perpetual and irrevocable, non-exclusive right" to anything their labors might produce in the context of the game. Like any licensee bound by the licensor's contract (or, for that matter, any for-hire worker), these producers forfeited any claims of authorship or ownership over their labor. The licensing contract constituted by the terms of service, therefore, defused and reoriented the threat of produsage and networked information economies to permission-based modes of decentralized creative collaboration based in hierarchy, authority, and licensing.

Produsing Outside of the Industry

Not all digitally networked creative collaborations concerning *Battlestar* have been contained by licensing contracts. Some, like the unofficial *Battlestar* game *Disapora*, the development of which began during production of the series in 2008 but continued even beyond the first release of a playable build in 2012, operated in direct opposition to the license-based restrictions imposed by contracts like the "Join the Fight" terms of service. As an unauthorized game mod or "total conversion," *Diaspora* modified the 1999 flight simulator *FreeSpace 2* and transformed it into a *Battlestar* game. Even though Volition, the *FreeSpace* developer, released the game's source code for consumer modification in 2002, the collaborative alteration of that code to produce a *Battlestar* game still defied the licensing arrangements at the heart of that franchise. The professional game publisher Sierra Entertainment paid for the right to make a game based on the *Battlestar* property in 2007, for example, and *Diaspora* therefore violated that and similar licenses governing professional production. Yet unauthorized projects like *Diaspora* ultimately helped to add value to the franchise by engaging unpaid producers in the work of creative production and demonstrated the compatibility of decentralized creative collaboration with the perceived authorship of those at the top of the industrial hierarchy.

The origins of *Diaspora* can be traced to another unauthorized *Battlestar* game called *Beyond the Red Line* (2007). Though incomplete, the *Beyond the Red Line* demo offered a fully playable version of *FreeSpace 2* modified to reflect the look and feel of piloting one of *Battlestar*'s Viper fighters. This demo generated a remarkable amount of attention, not just in the gaming press, but also in wider publications like *Popular Science*.[40] The game was later voted 2007's "Mod of the Year" by users of Mod Database.[41] In his unflattering

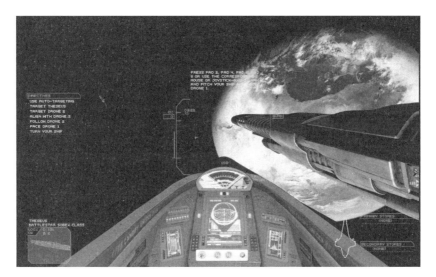

FIG. 1-2 The fan-made *Disapora* positions players within the cockpit in three-dimensional space in control of pitch, thrusters, afterburners, and an array of complex flight systems for more of a simulator experience. (Frame grab)

review of Sierra Entertainment's officially licensed *Battlestar* game that same year, AtomicGamer.com critic Jeff Buckland recommended that players try the unofficial *Beyond the Red Line* instead: "If you're really itching for something a little more authentic, then take a look at Beyond the Red Line.... Even the demo is better than the full version of this game."[42] In other words, this review undermined the industrial economics of franchising, dismissing the licensed game in favor of an unauthorized counterpart. Because of this and other potential challenges, the team responsible for *Beyond the Red Line* actually sought to reframe their labor as commensurate with the industry's licensing logic. Preempting any attempt by NBC-Universal to shut down the project, team member Chris Hager wrote in 2007: "Copyright infringement is a very thin line, and we have done our best to stay on the legal side of that line as much as possible. We use originally-built models and textures, our own writing, our own music compositions and audio effects; aside from model likenesses, theme, and core plot points, we have taken nothing directly from the show. You may notice that the name of our game is absent the name of the TV Show; this was done for legal reasons, as well as aesthetic."[43] While this statement danced perilously around the terms of licensing used by NBC-Universal (trademarked "model likenesses, themes, and core plot points" are hardly "nothing"), the team's aim was to position its game as a part of the promotional apparatus for *Battlestar* by arguing that it was a non-competing entity that might serve an industrial utility as "free publicity" for the television series.⁴⁴ These amateur developers merely extended the legal line in the sand put forward by NBC-Universal when it granted nonprofessionals the right to certain trademarked elements of the franchise in contracts like the "Join the Fight" terms of service agreement.

By positioning their unauthorized game mod within proprietary industrial models, however, these upstarts brought into the collaborative networks of produsage the potential for proprietary struggles over creative authority and control. In October 2008, a schism within the *Beyond the Red Line* team produced a second, competing *Battlestar/FreeSpace 2* mod called *Diaspora*. Hassan "Karajorma" Kazmi, who jumped ship to become *Diaspora*'s project lead, explained that "a mixture of creative differences and disagreements about release schedules" drove the split.⁴⁵ Although neither team would disclose the details of the conflicts to this author, these differences foreground problems inherent in both authorized and unauthorized game production. Open-source architectures and collaborative networks do not preclude creative difference any more than authority-driven industrialized labor networks do. The production of *Beyond the Red Line* ground to a halt as much of its talent and resources moved to *Diaspora*. In one detailed blog entry, Kazmi explained how custody of *Beyond the Red Line* assets was determined after the split; modelers on both sides claimed ownership of the files they had created and demanded exclusive access to the use of them. As Kazmi put it, "Diaspora was originally founded on the idea that we wouldn't be using anything from BtRL that wasn't made 100% by members of Diaspora or which we weren't given permission for by the creator."⁴⁶ Even though the creators of both mods rejected corporate proprietary control of culture, they paradoxically insisted upon maintaining ownership over their own private resources. Thus, a de facto system of authorship, ownership, and licensing emerged to structure and limit open collaboration.

While this system supported proprietary claims at a micro level—where discrete 3-D models, textures, and music cues could be controlled by single contributors who claim ownership of them—the organization of labor at a macro level within *Diaspora* nevertheless emphasized openness, collaboration, and shared authorship of the entire work-in-progress. According to the project website, the *Diaspora* team was composed of more than two dozen volunteers, each with specialized talents, including but not limited to: 3-D modeling and texturing; coding and scripting to modify the capabilities of *FreeSpace 2*; music composition; and "fredding" (the design of mission scenarios with the *FreeSpace 2* editor, or "FRED").⁴⁷ Despite these different skill sets and backgrounds, production operated according to a relatively democratic process. Kazmi explained he was "not interested in setting up any kind of complex management structure for the team."⁴⁸ Creative decisions for *Diaspora* were made largely by collective vote, quite unlike the authority-driven management guiding the industrial production of *Battlestar* in its television and officially licensed incarnations. Moreover, since the choice of a storyline, for example, would drive production at all levels (determining what models, sounds, and gameplay mechanics would be needed), Kazmi and the other developers decided against leaving story decisions to a dedicated and authoritative writer staff. Modelers, coders, and fredders alike would be invited to develop the narrative collectively: "We were allowed to pick apart each other's ideas and say what we liked and didn't like about them as well as what would and wouldn't work."[49] After collective deliberation, the group decided to craft a new storyline called "Shattered Armistice" that would take place concurrently with the events seen in the 2003 television mini-series, but from the perspective of the Battlestar *Theseus*—newly imagined by the *Diaspora* team.

In this desire to steer clear of using the *Galactica*, however, the central creative authority of Ronald D. Moore continued to figure into the organization of collaborative creative labor within the *Diaspora* team. In contributing to and widening the narrative history of the franchise, the *Diaspora* team risked encroaching on the authorial turf perceived to belong to Moore. Though not official licensees, *Diaspora* developers frequently demonstrated deference to the privileged television showrunner in a manner strikingly similar to subordinated industrial producers, particularly where the notion of canon was involved. Kazmi acknowledged that Moore "created the reimagined BSG universe so obviously we have the greatest respect for him. Our general philosophy is to try to follow the show's universe as closely as possible while still making a game that is fun to play. That means that while we aren't only to show what has already been seen in the show, everything we make should hopefully feel like something you could have seen in the show."[50] Thus, the collective decision to avoid using the familiar *Galactica* in "Shattered Armistice" can be seen less as an attempt to diverge creatively from what the privileged author had done with the property, and more as a strategic move to stake out a less contested ground and respect the showrunner's creative monopoly on central events and characters. With development of the game happening at the same time that Moore and his television team were completing the fourth and final *Battlestar* television season in 2008–2009, Kazmi reasoned that the first release of *Diaspora* would be "set during the fall of the Colonies precisely because it made it harder for Season 4 to make the entire thing contradict canon. In addition it is a great setting with a lot going on to make stories about but very little actually seen or even talked about on camera."[51] So, despite its own internal rejection of a centralized, industrial monopoly on the use of creative resources, *Diaspora* ultimately respected the authorial power of Moore and the expression of singular, central authority through notions of canonicity.

The privilege accorded the showrunner figure by the *Diaspora* team did not extend to authorized licensees or other de facto producer-licensees, however. Kazmi explained that "when it comes to everyone else, well we have enough trouble not falling down [our] own plot holes without dumping everyone else's on us." Much as licensees for properties like *Star Trek* have reinforced their own creative subordination, the producers of *Diaspora* privileged the perceived center of creativity in the *Battlestar* franchise over the wider and decentralized network in which they themselves participated. Kazmi added, however, that disregard for professional and amateur licensees alike "doesn't mean we don't respect the other BSG games and mods out there. They're trying to do the same thing we are after all."[52] The developers' deference to centralized authorship should not be seen as a blind loyalty to the author figure, but perhaps more significantly as the recognition of competition and opposing claims to creative collaboration within the *Battlestar* franchise. With the creative labor of these decentralized production cultures offering so many competing narrative possibilities in a context that privileges canonicity, embracing the figure of the centralized creative authority, and throwing one's lot in with him, helped support claims to creative legitimacy.

In summary, although the production of *Diaspora* was arguably an illegal appropriation of the *Battlestar* license, the produsage-based labor processes animating it can be conceptualized in relation to and as part of the industrial base of the *Battlestar* franchise in at least four ways. First, the *Disapora* team sought to position its creative endeavors as free labor with an industrial utility, as part of the promotional apparatus surrounding the series. Second, despite boldly challenging NBC-Universal's official licensing partnerships by making *Battlestar* into a collective, communal resource, *Diaspora* embraced a proprietary model of authorship in which the use of game assets could be at least in some cases owned, approved, and controlled. Third, while embracing a democratic mode of creative collaboration, the team deferred to centralized creative authority by accepting the notion of a singular canon. In doing so, fourth, the team reinforced the creative subordination of a decentralized network of licensees and producers working in relation to the franchise, privileging power at the center. In all these ways, *Diaspora*—and the networked organization of creative labor it illustrates—should not be considered a revolutionary mode of production in opposition to industry, but rather as a part of the licensed franchise in which creative authority and cultural power continue to be negotiated.

As this essay has argued, projects like *Diaspora* call attention to the continuous relationship between industrial strategy and the networked organization of culture emerging in the context of convergence culture. The open, collaborative, processual production networks made possible in the digital economy do not always stand in opposition to traditional industrial models and have frequently been incorporated within corporate strategy—not just as free labor, but through the persistence of proprietary ownership and centralized authority over culture within new socially networked labor patterns. Produsage and wikinomics are not breaks from earlier models, but recombined extensions of them. Of course, one could certainly argue that in their compatibility with the industrial nodes in the franchise network, projects like *Diaspora* are not, strictly speaking, true evidence of produsage; they slide too far down Bruns's continuum toward production and exist more as vehicles for industrial publicity. If that is the case, however, and the collaborative, open creative networks of game modding do not embody the possibilities of which scholars like Bruns, Benkler, and others write, then we might want to consider the degree to which those possibilities are utopian chimera, never to be actually achieved in a cultural economy still driven by capital, ownership, and authority.

To explore this tension between old and new systems of creativity within convergent capitalism, this essay has explored the industrial management of creativity as media franchises like *Battlestar Galactica* become licensed across a decentralized set of production cultures. In the industrial model, showrunners like Ronald D. Moore have been afforded a creative and authorial privilege, despite the multiplicity of creative labor encompassed by the franchise and uncontained by this author figure, and licensed creativity has been marginalized and made deferent to the seat of power at the center of the franchise. At the same time, the new possibilities embodied by projects like *Diaspora* required non-authorized licensees to negotiate these older corporate models of cultural production based in centralized creativity and intellectual property ownership. So, while creativity and authorship remain up for grabs in the new economy, old claims on them have not been entirely displaced. Producers, like licensees, continue to create in relation to institutionally empowered producers, even as the latter look for new ways to capitalize on these collective forms of labor.

Notes

1. See Michael Z. Newman, "From Beats to Arcs: Toward a Poetics of Television Narrative," *Velvet Light Trap* 58 (Fall 2006): 16–28; James L. Longworth Jr., *TV Creators: Conversations with America's Top Producers of Television Drama* (Syracuse, NY: Syracuse University Press, 2002).
2. Derek Kompare, "More 'Moments of Television': Online Cult Television Authorship," in *Flow TV: Television in the Age of Media Convergence*, ed. Michael Kackman et al. (New York: Routledge, 2010).
3. "The Captain's Hand," narrated by Ronald D. Moore, *Scifi*, http://en.battlestarwiki.org/wiki/Podcast:The_Captain%27s_Hand.
4. Quoted in William McCarty, "Brandon Jerwa on New Battlestar Galactica Season Zero," *Newsarama.com*, May 22, 2007, http://forum.newsarama.com/showthread.php?t=113605.
5. Ibid.
6. Don Tapscott and Anthony D. Williams, *Wikinomics: How Mass Collaboration Changes Everything* (New York: Portfolio, 2006), 20–30.

7   Tiziana Terranova, "Producing Culture for the Digital Economy," *Social Text* 63, 18 (2000): 33–58.
8   Axel Bruns, *Blogs, Wikipedia, Second Life, and Beyond: From Production to Produsage* (New York: Peter Lang, 2008), 24–30, 397.
9   Quoted in Vaneta Rogers and Cliff Biggers, "Bauer's Playing It Cool," *Comic Shop News* 1074 (January 2008): 1.
10 Interview with the author, Los Angeles, August 24, 2007. The subject's name has been omitted for privacy.
11 Interview with Danny Bilson (game writer), Los Angeles, January 17, 2008.
12 See Avi Santo, *Transmedia Brand Licensing Prior to Conglomeration: George Trendle and the Lone Ranger and Green Hornet Brands, 1933–1966* (Austin: University of Texas Press, 2006); Simone Murray, "Brand Loyalties: Rethinking Content within Global Corporate Media," *Media, Culture, and Society* 27, 3 (2005): 415–435; M. J. Clarke, "The Strict Maze of Media Tie-in Novels," *Communication, Culture & Critique* 2, 4 (2009): 434–456.
13 Quoted in Paul Dini and Chip Kidd, *Batman: Animated* (New York: HarperCollins, 1998), n.p.
14 Gene Roddenberry, memo to Pato Guzman, "Star Trek Emblem," August 10, 1964, Gene Roddenberry Star Trek Television Series Collection, 1966–1969 (hereafter Roddenberry Collection), box 27, folder 4, Arts Library Special Collections, Young Research Library, University of California, Los Angeles..
15 Gene Roddenberry, memo to Matt Jeffries, "Phaser Weapons," April 26, 1966, Roddenberry Collection, box 30, folder 2.
16 Gene Roddenberry, memo to Ed Perlstein, "Star Trek Rifle-Toy Development Arrangement," July 2, 1965, Roddenberry Collection, box 31, folder 10.
17 Matt Jeffries, memo to Gene Roddenberry, "Art Direction," April 12, 1966; Ed Perlstein, letter to Don J. Beebe, AMT, September 14, 1966, Roddenberry Collection, box 5, folder 4.
18 D. C. Fontana, memo to Gene Roddenberry, "Star Trek #2 Paperback Manuscript by James Blish," August 22, 1967, Roddenberry Collection, box 30, folder 3.
19 Gene Roddenberry, memo to Ed Perlstein, "Star Trek Novelization by Mack Reynolds," November 14, 1967, Roddenberry Collection, box 30, folder 3.
20 Ed Perlstein, note to Gene Roddenberry, November 16, 1967, Roddenberry Collection, box 29, folder 4.
21 Timothy W. Lynch, "An Interview with Richard Arnold," posted to Compuserve, 1991. In addition to interviewing Arnold about his experiences in the Star Trek Office, Lynch followed up with research at Paramount's licensing department to get a glimpse of the approval process.
22 Ibid.
23 Ronald D. Moore, post to America Online Experts Forum, "Answers," July 28, 1997.
24 Quoted in Gustavo Leao, "Interview: IDW Editor Andrew Steven Harris Talks Future of *Star Trek* Comics," *Trekweb.com*, September 3, 2007, http://trekweb.com/stories.php?aid=46dccc613517b.
25 Quoted in John Callaham, "*Star Trek Online* Interview," *Firing Squad*, September 8, 2006, www.firingsquad.com.
26 Interview with the author, Los Angeles, August 24, 2007. The subject's name has been omitted to protect his identity.
27 Ibid.
28 Quoted in Rogers and Biggers, "Bauer's Playing It Cool," 1.

29 Quoted in McCarty, "Brandon Jerwa on New Battlestar Galactica Season Zero."
30 Quoted in Matt Brady, "Inter-Galactica Odyssey," *Comic Shop News* 1064 (2007): 2–3.
31 Ibid., 3.
32 Cliff Biggers, "Dynamite Explores the Revolution Within," *Comic Shop News* 1005 (2006): 1. Notably, the *Battlestar* character flashbacks in the comics function much like those built into the contemporary television series *Lost*. The *Battlestar* comics were conceptualized as a means of exploring character and revealing "how earlier events helped to bring them to the focal point of the [television] series."
33 Matt Brady, "Jerwa: The Ghosts of *Battlestar Galactica*," *Newsarama.com*, July 15, 2008, http://www.newsarama.com/comics/080715-BSGGhosts.html. Writers like Jerwa have also created comic book series like *Battlestar Galactica: Ghosts* that introduce entirely new casts of characters and situations and place them within the already explored history of the franchise universe. As Jerwa notes, "it's probably pretty easy for [the licensor] to let me run free with this one, because it doesn't have to tie into a million different things and really isn't subject to the same specific character scrutiny that the regular books are."
34 Mark Warshaw (personal interview), Los Angeles, June 24, 2008. Though Warshaw's initiative would be accepted, NBC scaled back his plan, compressing the three prongs proposed into a single position.
35 "The Process," *StarlightRunner.com*, January 1, 2009, http://www.starlightrunner.com/process.html.
36 Yochai Benkler, *The Wealth of Networks: How Social Production Transforms Markets and Freedom* (New Haven: Yale University Press, 2006), 3-5, 56.
37 Bruns, *Blogs, Wikipedia, Second Life, and Beyond*, 397.
38 Terranova, "Producing Culture for the Digital Economy," 38.
39 "Terms of Service," *Battlestar Galactica: "Join the Fight,"* January 28, 2009. The site no longer exists.
40 Chuck Cage, "Have a Blast," *Popular Science* 272, 4 (April 2008).
41 "2007 Mod of the Year Awards," *Mod Database*, January 20, 2009, http://www.moddb.com/events/2007-mod-of-the-year-awards.
42 Jeff Buckland, "*Battlestar Galactica* PC Review," *AtomicGamer.com*, November 4, 2007, http://www.atomicgamer.com/article.php?id=488.
43 Quoted in Nick Breckon, "Mod Is Dead: Beyond the Red Line," *Shacknews.com*, July 31, 2007, http://www.shacknews.com/featuredarticle.x?id=489.
44 Ibid.
45 Hassan Kazmi (aka Karajorma), "Inaugural Address," *Diaspora Development Blog*, June 7, 2008, http://www.hard-light.net/forums/index.php?PHPSESSID=br7idj37rp6t17jngrs10h2494&topic=56784.0.
46 Hassan Kazmi (aka Karajorma), "Our Rag Tag Fleet," *Diaspora Development Blog*, December 8, 2008, http://www.hard-light.net/forums/index.php?PHPSESSID=br7idj37rp6t17jngrs10h2494&topic=56784.0.
47 "Team," *Diaspora-Game.com*, January 28, 2009, http://www.diaspora-game.com/team.html.
48 Hassan Kazmi (aka Karajorma), e-mail interview with author, December 23, 2008. Perhaps referring to the split with *Beyond the Red Line*, he added, "We've tried that and it caused more issues than it solves. *Diaspora* is a hobby and hobbies should be fun, not a second job you don't get paid for. Instead, we've adopted an informal method of leadership.... Decisions are taken either on the spot after short

discussions with whoever is on the channel at the time (when minor), or by forum vote after IRC discussions between the team (for major ones)."
49 Hassan Kazmi (aka Karajorma), e-mail interview with author, December 29, 2008.
50 Kazmi (aka Karajorma), e-mail interview with author, December 23, 2008.
51 Kazmi (aka Karajorma), e-mail interview with author, December 29, 2008.
52 Ibid.

2
In the Game

• • • • • • • • • • • • • • • • • • • •

The Creative and Textual Constraints of Licensed Video Games

JONATHAN GRAY

When the video game industry's blockbuster hits pull in huge earnings, as for instance when *Call of Duty: Modern Warfare 2* grossed $550 million in its opening week, the press is fond of comparing these profits to those from film or television hits, as if the different media were locked in mortal combat.[1] But far from competing with each other, Hollywood and the video game industry continue to converge, buying out, buying into, and/or working with one another. On the one hand, Hollywood often mines the game industry for ideas, both adapting films into video games and mimicking the visual style of video games.[2] On the other hand, every year sees the rollout of multiple new games based on popular film or television franchises, and profits have clearly been good enough to keep Hollywood licensing its best intellectual property to game designers. More than a hundred *Star Wars* video games and multiple *Simpsons* games exist, and a whole spate of other famous and not-so-famous films and television shows have found themselves pixilated for joystick- or mouse-wielding fans.[3] Some of these have proven critically successful; for instance, the Nintendo 64 *Goldeneye 007* and the PC *X-Wing* are often noted as breakout hits. Yet overall, and despite monetary success, the process of adapting film and television into games has been a fraught enterprise. Arguably, the most reviled video game in the medium's short history is the 1982 Atari cartridge, *E.T. The Extra-Terrestrial*; industry lore highlights the game's central role in bringing about the collapse of Atari and the home console market in 1983 and 1984, leaving thousands of the unsold cartridges to be buried in a landfill in New Mexico.[4] Of course, the industry recovered, but many licensed games continue to languish in fan ratings, becoming objects of scorn and derision. If the film that "isn't as good as the book" has become one cliché of adaptation, the other is that of the game that "isn't as good as the film or television show."

Given the oft-contested nature of video game expansions of film and television story worlds, the licensed game offers a fascinating riddle for anyone interested in the processes of adaptation, convergence, and franchising. Although licensed games have become a frequent butt of gamer jokes and symbols for the worst byproducts of an increasingly convergent industry, what might this reputation tell us about the process of transmedia storytelling in the digital era? Why have so many games been seen to fall short of the mark, and what does this failure tell us about this fabled mark?

In an effort to address these questions as they relate to "wired television," I immersed myself in published reviews of licensed video games. Ideally, the more free-flowing, responsive nature of the games' and/or shows' discussion forums would have offered a fuller sense of how gamers and television fans negotiate meanings surrounding the games, but a preliminary examination revealed that many such sites offer little more than walk-through–style hints and cheats or quick judgments, with scant elaboration upon the actual virtues or sins of various games. By contrast, using the review aggregator site Metacritic (www.metacritic.com), I was able to access a huge slate of reviews of games based on television properties. Reviews are by no means a perfect sounding board for audience reactions, so I advance the following comments tentatively as a way of identifying rifts that might exist in the licensed game business more generally. I am aware that game reviewers stand in for neither gaming nor fan populations in total. Nevertheless, with no real tradition of game reviews in highbrow venues such as newspapers or high-end magazines, the tone and diction of these reviews is often relaxed and colloquial. The barriers to entry for reviewers are frequently much lower than in highbrow venues, and quite simply the proliferation of voices and styles produces a wide variety of review types and modes. Thus, reviews still offer the researcher a promising site for initial discussions of and varied reactions to licensed games.

Narrowing the field to a manageable number of games for this chapter, I examined all sixty-one extant reviews listed at Metacritic for *24: The Game* (developed by SCE Studio Cambridge, published by 2K Games and Sony Computer Entertainment); *Lost: Via Domus* (developed and published by

Ubisoft); *The Simpsons Hit and Run Game* (developed by Radical Entertainment, published by Vivendi Universal Games); *CSI: New York: The Game* (developed by Legacy, published by Ubisoft); *Desperate Housewives: The Game* (developed by Liquid Entertainment, published by Buena Vista Games); and *Little Britain: The Video Game* (developed by Revolution Studios, published by Blast! Entertainment Ltd. and Mastertronic). These games are based on a variety of television genres featuring fictional narratives, ranging from an action-espionage drama to a serial pseudo–sci-fi drama, an animated sitcom, a detective procedural, a parodic evening soap, and a comedy sketch show. As might be expected, the games' genres are similarly diverse: *24* works predominantly in a "run-and-gun" mode; *Via Domus* is an "adventure game" with a few puzzles thrown in; *Hit and Run* is a *Grand Theft Auto*–style driving game; *CSI: New York* is a "point-and-click"–style detective game; *Desperate Housewives* is a *Sims*-like game with a mystery to solve; and *Little Britain* represents a composite of several games. With only one exception, all sixty-one critiques were written by different reviewers.

My key interest lies not in reporting the specific reception of each game, but rather in exploring the evaluative criteria used by the reviewers. What, in other words, made for a "good" or "bad" adaptation of a television show? Were there trends in what was seen to work or not work? What prospect do video games have as a viable site for the telling and expansion of television narratives? After first surveying some of the literature on game adaptations, I will use the reviews to answer such questions. The literature and reviews indicate that rushed production processes and widespread disrespect from Hollywood for the creativity and resources (both cash and original production personnel) required to design a licensed game are common sources of perceived failure. However, I will argue that we must go beyond simply blaming the production process, since the reviews also point to particular problems with the process of adapting the *experience* of a television show, suggesting that game adaptations require as much attention to translating temporality as to imbuing spatiality. Merely exploring the spaces of a diegetic world is clearly not enough for many gamers, who also seek—and rarely find—the temporal experience of the show.

Console-ing Film and Television Narratives

To date, little has been written academically on the process of adapting film or television to games. In a 2009 "In Focus" essay section of *Cinema Journal* on film, television, gaming, and convergence, Judd Ruggill offered a brief history of the industries' collusion; at six pages, it was five pages longer than most histories.[5] That said, when licensed games are discussed, the conclusion most often reached is that such projects are sabotaged by a perilously short development and production period that predetermines quality. This is Nick

Montfort and Ian Bogost's moral to the story of *E.T.*'s failure, for instance.[6] As noted by Derek Johnson (in this collection) and by Avi Santo, the licensing arrangement traditionally involves selling limited rights to develop merchandise or properties such as games to independent subcontractors.[7] The key word in that sentence is "limited": game developers are rarely given full access to the creative team behind the licensed entity and are usually restricted from doing certain things with characters or locations. "Day-and-date" assignments are also common, whereby the developers are expected to have a game in hand for a specific release date that coordinates with the film or show's publicity schedule. Thus, licensed game design regularly occurs under significant restraints that may leave the designers no more inspired to create a game that meaningfully contributes to the licensed property than is a clothing company making a line of T-shirts emblazoned with the property's logo.

Trevor Elkington notes these and other problems with what he describes as the "self-defeating" adaptation process of creating licensed games, attributing most of the problems to "incompatible production practices between film studios and game studios."[8] He observes that games tend to require more time to develop than do films; forcing game designers to speed up to meet a day-and-date release schedule will inevitably have dire consequences for the finished product. Video game design is too often regarded as an afterthought by the property holder;[9] designers are given insufficient time and money to succeed and/or are micro-managed by too many overseers, given no room to innovate. He quotes a developer of one licensed game: "We were like taffy, being pulled between licensor, sub-licensor, and publisher, all of whom wanted something different. We still got the game done, but at a heavy price. Everyone felt like they were forced to make a game they didn't believe in."[10] By contrast, Kristin Thompson's chapter on the development of the *Lord of the Rings* video game suggests the rule for licensed games by examining an exception; her numerous quotations from the game's design team note a rare allowance of time, properties from the movies, and respect from director Peter Jackson.[11]

Adaptation, therefore, is a process that frequently disturbs the fault lines between apparently convergent industries and media. All too often, licensed video games do little more than slap several character and place-names from a film or television story world onto an established game genre. The *Star Wars Episode I: Racer* game, for example, is little more than a standard racing game dipped lightly in the *Star Wars* universe. Based predominantly on one scene from the movie, the game "contains almost no story" and "only a suggestion that makes the player frame her actions within the game as part of this specific fictional universe."[12]

Such instances may seem to offer the critic about as little to discuss in terms of textual adaptation as do Bart Simpson key chains or *Star Trek* linens. They also clearly point to the need for, as Elkington proposes, greater central management,[13] along with greater respect, time, and resources given by property holders to the process of game design as a whole. But it would be naïve to think that time, money, and respect alone will solve the widely perceived problem of bad licensed games. Rather, game adaptation poses a *textual* challenge as well as a challenge to *production cultures*. Elkington notes, after all, that adaptations are rarely even given a chance by many reviewers,[14] which raises broader issues of how adaptation works in general and what obstacles its fraught history presents for the would-be licensed game designer.

In *A Theory of Adaptation*, Linda Hutcheon begins by noting how frequently adaptations of any sort are panned and resented by critics: "If we know [the] prior text, we always feel its presence shadowing the one we are experiencing directly."[15] But if purists often regard that shadow as a felt presence, they relate to the adaptation as an absence. Purists will criticize not only an adaptation's inability to replicate the original, but also the affront to the original text that is supposedly suggested by the act of re-creating it. Citing Kamilla Elliott, Hutcheon notes a type of "heresy" that most adaptations commit by suggesting that form and content can be separated.[16] Adaptation's threat to the individual fan is a sloppy rendering that damages the original, but adaptation's threat to the tradition of textual analysis is more profound, suggesting that the soul of a work might be successfully transferred to another body and that this soul (the text itself) might exist independently of multiple elements and details of the work. To this end, Hutcheon steps away from discourses of purism and fidelity, noting that "recognition and remembrance are part of the pleasure (and risk) of experiencing an adaptation; [but] so too is change." She writes of adaptations as combining "the comfort of ritual" with "the piquancy of surprise."[17]

The trickiness, though, lies in sorting through precisely which parts of a text are replicated and which are changed, what is ritually repeated and what surprises us. Such a dilemma confronts most licensed video game creators. On one hand, as Elkington notes, those games that try slavishly to re-create the plot of a film often stumble, cursed by cut sequences that rob the player of agency. An examination of numerous critics' comments on film adaptations leads Elkington to conclude that "what video game consumers seek from adaptations is not a simple, interactive rehearsal of film events but in fact further expansion of a narrative world via an engaged relationship with an interactive medium."[18] On the other hand, even those games that aim to "expand" or "extend" a narrative world require a high-stakes gamble from their developers on precisely which elements of the original narrative should be transferred, which can be discarded, and what can be added without damaging the audience's beloved text.

In a brief introduction to the above-mentioned "In Focus," I used that most clichéd of video game moves, the "double jump," to explain this process of textual expansion. The double jump is "an apt metaphor for understanding what happens to the text—as economic, aesthetic, and participatory unit—as it attempts to move from media platform to media platform. Convergence is not movement as usual—it is a special move that requires something special of text, industry, and audience, and a move that engages its own laws of physics."[19] In writing that, I was particularly inspired by the work of Tanya Krzywinska in the collection *ScreenPlay: Cinema/Videogames/Interfaces*, one of the few extensive examinations of film-game convergence to date.[20] Krzywinska examines not a specific adaptation of a horror film, but rather the adaptation of the genre to the realm of games. She argues that games offer a particularly intriguing spin on the usual experience of watching a horror film and cringing as characters make stupid decisions that will get them killed. We need the characters to behave that way for the story to continue, and thus their behavior is pleasurable at the same time that it inspires fear and horror at what will happen. This tension, she argues, helps to produce one of horror's signature characteristics, namely, a loss of control and the subjugation of agency to a (usually tortuous) sense of destiny. However, when we as gamers are responsible for moving a character around a blind corner or for engaging in any other act that is bound to invite trouble, the loss of control is heightened by a seeming promise of control (of the joystick), thereby amplifying a sense of our fated encounter with the terrifying. Krzywinska's argument suggests the degree to which games can be active contributors to the creation of a new type of story world, perhaps even playing an important role in accentuating the pleasures of a genre or text when adapted and in creating new pleasures beyond those experienced in a film or television show . . . if the designers understand and can replicate or accentuate the temporal experience(s) that are linked to the text.

Matthew Weise and Henry Jenkins further explore this dynamic in their study of how the "affective mechanics" of the *Aliens Versus Predator 2* game differ from those of *Aliens* the film. Whereas the film offers multiple perspectives, the game "might better be described as ego-centric, restricted to what can be known and experienced by a single character who can count on nobody to come to his or her assistance. What happens happens to us. We are immersed fully in the action; our choices determine the outcome."[21] A simplistic reading of both Krzywinska's and Weise and Jenkins's work might produce the gloss that games offer more immersive experiences than do film or television shows. Certainly, games open up immersive possibilities, though the utopian rhetoric of immersion often forgets the strict rules and limitations of games that work against immersion. But both Krzywinska and Weise and Jenkins also point to the *experiential* element of narrative. A film or television story world is not just a combination of plot, characters, and a setting; it also has a tone and a timbre, and audiences are given an experience. Adaptation thus is not simply about throwing viewers into a story and locale and letting them meet the characters;

a certain experience of the film or television show must seemingly also be replicated or amplified.

Louisa Stein's contribution to the "In Focus," a piece on the *Gossip Girl Second Life* world, is telling here.[22] She studied the degree to which the adaptation, and fans' interactions with it, can reveal the nature of fans' relationships to the overarching, transmediated text. As fans try to situate themselves in relation to gender, materialism, and technology with *GGSL*, we as analysts see clear traces of how they situate themselves in relation to such issues with *Gossip Girl* itself. Thus, *GGSL*, especially in its open-endedness, allows us to see the experiential core of *Gossip Girl*'s story world for at least those fans in *GGSL*.

It is this "experiential core," I will argue, that many game designers bypass. Talk of transmedia storytelling has often focused inordinately on narrative and plot. Thus, for instance, when Henry Jenkins writes of the Wachowski brothers' grand experiment in transmedia storytelling, creating the *Matrix* sequels across various platforms, including the console game *Enter the Matrix*, his attention is on how a continuing plot was parsed into various media.[23] Jenkins cites a particular challenge for transmedia storytelling: how can writers craft plots that offer rewards to those players who track them from platform to platform but are still intelligible to those who do not follow the transmedia thread? Yet, transmedia storytelling is about much more than plot or characters alone, and thus the challenges mount when designers must also transfer tone, style, pace, and timbre across different media that are variously enabled or disabled to offer up other elements of the experiential core—all within the constraints of licensed games. Hutcheon argues that some media are better equipped for experiences of telling, some for showing, and some—among which she counts video games—for interacting.[24] So we must ask not only how a story can be told across media, but also how tone, timbre, pace, and experience can encompass various modes and media. What experiences do they capture, which do they amplify, what is added, and what is the effect on players' overall evaluation of the licensed games? Functionally, many licensed games fail because of a rushed production process; but if they had extra time and resources, on what aspects specifically could and should designers be focusing their labors?

Immersion for Beginners: Authentic Plots, Voices, and Visuals

Before examining the critiques of individual games, it is worth noting the reviewers' oft-repeated criticism of the entire category of licensed games. For instance, Neville Nicholson writes of the "incessantly mounting pile of hideous licensed games,"[25] while DJ Dinobot notes the large pool of "crappy exploitation" within which most television licenses "languish."[26] Kristan Reed, writing specifically of the *Simpsons* franchise, opens her review with a longstanding complaint: "'Why can't anyone make a decent *Simpsons* game?' has been the plaintive wail of millions of gamers for over a decade now, tired of being fed licensed pap year upon year. With such a fantastically rich property to draw upon, it's almost criminal that various developers over the years have been able to get away with this systematic trashing of the brand."[27] Many other reviewers similarly begin with the presumption that most licensed games are bad and that audiences expect them to be bad in advance of playing them. They clearly see it as their task either to argue for the exceptional quality of a particular game (even if "exceptional" only meant that it was passable or mediocre) or to reaffirm the received wisdom that licensed games are always poor. Part and parcel of this judgment is the specific complaint that licensed games have inflicted damage upon the original property, as is evident in Reed's suggestion that the developers' "trashing" of *The Simpsons* is a "criminal" act. Daniel Sayre echoes this view when he opines that most developers have "put their hopes on the license, instead of the gameplay, to move copies."[28] Jim Powell asserts that "publishers in most cases spend most of a game's budget on the license, then more often than not throw together some half-arsed game that vaguely resembles the movie/cartoon/TV show and then flog it to the unsuspecting public. It doesn't need to be a good game because by the time people buy it and get it home it's already too late, and the fat, cigar-chomping producer has just earned himself another swimming pool."[29] While acknowledging the financial success of licensed games, few reviewers regard them as good.

Within this framework, then, it becomes important to scan the reviews for the reasons so many licensed games are judged to have failed. A common opening for a review conveys the reviewer's status as a fan and hence his or her investment in the televisual text. Many reviewers clearly want the game at hand to be good. Consider this opening from a review by Chris Roper: "Let me begin by saying that I'm an enormous fan of *24*. I've been watching the show since it first premiered, and on the one occasion that I thought I'd missed an episode, I was ready to swear off TV entirely for 8 months until the DVDs hit so that I wouldn't accidentally catch spoilers in a commercial. . . . So, it should go without saying that I would expect a lot from a game based on *24*."[30] Other reviews begin with an explicit warning to those who are not already fans of the television show to stay away from the game, and even to skip the review. Patrick Klepek, for example, begins one review: "If you aren't a hardcore fan of ABC's serial sci-fi mindbender *Lost*, you can go ahead and skip this one."[31] Hilary Goldstein starts with *Lost*'s cursed number sequence, 4–8–15–16–23–42, after which she notes that those to whom the numbers have no meaning may wish to turn away now. "But," she continues, "you're a *Lost* fan, right? So of course you will soldier on."[32] Time and time again, reviewers assume that only fans will bother reading these reviews, since fans have significant interest in the games and high hopes. Beyond the reviewers' own claim to fan status, they often invoke legions of hypothetical fans who are imagined to be desperately awaiting the game's arrival (or the arrival of *any* paratext, for that matter).

Key to satisfying expectations, at bare minimum, is the inclusion of the original cast. David Simpson, for example, notes how much more excited he was about the *Little Britain* game after hearing that David Walliams and Matt Lucas had signed on: "This wouldn't be the awful *The Godfather* without Al Pacino as Michael; this would be an authentic Lou and Andry under my control."[33] "Authentic" as an adjective appears disproportionately often when reviewers discuss voiceover work, as when most reviewers marvel at the *24*, *Hit and Run*, and *CSI: New York* games' inclusion of all cast members or, alternately, when missing cast members become central to critical judgments. Patrick Klepek faults *Via Domus* because "your ability to immerse yourself in the game's island world comes to a screeching halt when Locke—one of the show's (and game's) principal characters—begins speaking, sounding like someone faking your grandpa's voice . . . badly. This pattern continues for all the stand-ins [. . . so that] you're forced to just grit your teeth and move on."[34] By contrast, reviewers who find the impressions more convincing are quicker to praise a game on the whole. Interestingly, for instance, the *Desperate Housewives* game features impressions of all characters except the show's narrator, yet most reviewers found the impressions well done and appreciated the "authenticity" of the authorial voiceover provided by Brenda Strong. Clearly, then, voice work matters a great deal, forming, as Klepek suggests, an important barrier or entry point to seamless immersion in the story world.

The look is important too, and reviewers of the six games directed numerous comments toward the rendering of various characters. Amusingly, many critics noted with some concern that *The Simpsons* characters' three-dimensionality does not match the show's two-dimensionality, but all consoled themselves with the fannish knowledge that the rendering does at least invoke Homer's voyage into the third dimension in "Homer3," a story from the episode "Treehouse of Horror VI."

Authenticity of plot also matters. Indeed, *Via Domus*, *24*, *CSI: New York*, *Desperate Housewives*, and *Hit and Run* all received their most positive feedback and commentary for their plots (*Little Britain* is a sketch show). Reviewers of *Desperate Housewives* were united in feeling that the game captures the soapy, yet quirky and wryly camp tone of the television show. As Gabe Boker noted, "the story at hand is your standard *DH* fare. You are the newest ('desperate') housewife on Wisteria Lane, a small suburban area that appears innocent enough on the outside, but is filled to the brim with gossiping, back-stabbing and pretentious residents . . . as you cheat, lie and scandal your way to becoming the most popular (and hated) mistress on the block."[35] The reviewer known as Luke similarly observed that "the story is standard fare and about what you would expect from a *Desperate Housewives* game."[36] Numerous reviews noted that the game was penned by one of the show's writers, as did reviews for *Hit and Run*. Kristan Reed praised the latter for a "stupendously excellent script" typical of the show,[37] while Jeff Haynes thought that the plot "sounds like something you'd actually see in the show."[38] Most reviewers were united in finding it funny in a quintessentially Simpsonian manner. *Via Domus* evoked slightly more mixed responses, but many reviewers found its style true to the original. Matt Cabral liked the presence of *Lost*'s "slick storytelling style" and its "edge of the seat ending,"[39] and Gus Mastrapa thought the game had adeptly re-created the "vibe" of the show, so that playing it felt like playing an episode.[40] In a poor review of *CSI: New York*, Mark Arnold nevertheless found the stories to be "one redeeming factor": "They are of high enough quality to be pulled from the show itself."[41]

24's reviewers positively raved about the game's plot. Garnett Lee was particularly effusive: "This is season 2.5 in the form of a game, and it shows. In fact, the story line of this *24* is so strong that it will almost certainly wind up among your favorites; you have to wonder how they didn't decide to hold onto it for traditional production. . . . Suffice it to say that the script is a classic, packed with the unexpected twists and turns that makes the series so addictive."[42] John Powell was excited to find that the plot "does rival anything on the series itself."[43] Among the twenty-seven reviews of *24*, there was near consensus that the plot is excellent. Just as many reviewers found the division of *Desperate Housewives* into twelve distinct "episodes" cute, and the *Via Domus* reviewers enjoyed its similar division into episodes, complete with "previously on *Lost*" sequences at the beginning and with (playable) flashbacks. *24*'s reviewers heaped further praise on the game for capturing the show's visual style, dividing the screen into various scenes, and for "filming" cut sequences in a suitably and richly cinematic mode. Tom Orry summed up: "The way the game mimics the show's style is impressive, with all the trademark split-screen action, on-screen clock, tension mounting blips and all the rest. You even get the introduction to each hour and most end on a revelation or key event of some sort."[44]

Based only on the review excerpts above, each of these games may at first appear to have been judged as a success. The aspects singled out for praise also illustrate how studios have started to take games more seriously, making "above the line" talent available to game designers. *24* in particular is celebrated for its staff writers' involvement, trademark visual style, inclusion of all cast members, and plot that honors the show, fitting in between two seasons, no less, and thus slotting conveniently into the canonical narrative. Having writers and cast on board clearly matters a great deal to gamers. But plot, sound, and visuals are only three components of a televisual narrative, and no reviewer regarded them as reason enough to warrant high recommendations for a particular game. The production processes of some of these games suggest a greater willingness on the part of the property holders to commit resources to the process of game design, but asset availability is clearly not sufficient. Rather, though many reviewers found much to compliment in these six games, and though the games tended to be judged as better than standard licensed game fare (*Little Britain* excepted), the critics perceived the lack of certain *experiential* elements from the televisual original. *24* in particular may look and sound authentic, but it was lambasted by reviewers who, like Lee, Powell, and Orry above, often began with compliments but soon dismissed the game for its inability to deliver an authentic *24* experience.

Immersion for Pros: Capturing the Experience

"Immersion" in the televisual story world is often central to a game's pull on viewers. "Authenticity" of plot, voice, and visuals can help, but many reviews suggest that the *spatial* immersion that surrounds a player with recognizable characters and locations must also be accompanied by *temporal* immersion within the pace, feel, and phenomenological experience of a text. As Mikhail Bakhtin noted in his famous discussion of the novel, all narratives are characterized by their "chronotopes"—a spatio-temporal matrix that is as much about time as it is about space.[45] Game adaptations, perhaps, are mostly topos, with not enough chronos.

On one hand, game time allows players to slow down and explore a world, thereby aiding spatial immersion. Such a privilege was of no consequence to reviewers of *Little Britain*, *24*, and *CSI: New York*, suggesting the degree to which the geographies of such worlds may be relatively unimportant (even when, ironically, two of the titles are geographical by nature!). Reviewers of *Via Domus*, *Hit and Run*, and *Desperate Housewives*, on the other hand, found the ability to explore, respectively, "the island," Springfield, and Wisteria Lane quite titillating. The "sandbox" style of the latter two games was praised, with frequent favorable comparisons in the case of *Hit and Run* to *Grand Theft Auto*'s much-lauded style. "What you've got here," wrote a reviewer in *Gamers' Temple*, "is *Grand Theft Auto III* (*GTA3*) without the hookers and bloodshed. Instead of running around Liberty City, you get the chance to cruise the streets of Springfield, complete with traffic, wandering characters from the series, well-known locations such as Moe's Tavern, and enough references to *Simpsons* episodes to keep even the Comic Book Guy happy."[46] The games' ability to capture minutiae of their respective worlds and allow exploration and interaction appealed to many critics. *Via Domus*, too, was complimented for offering a beautiful rendering of the island, complete with such fan-recognizable items as Locke's wheelchair and small details in Dharma stations. As Matt Cabral enthused, "Fans are able to explore lesser seen areas like the other side of the

FIG. 2-1 The *Lost: Via Domus* (Disney-ABC) video game replicates locations from the series. (Frame grab)

magnetic wall and even the Swan's bathroom—c'mon what *Lost* fan wouldn't want a closer peek?"[47]

On the other hand, all the games suffered from limitations of temporality, most notably endless repetition. Numerous *Desperate Housewives* reviewers complained about characters who continually reveal the same secret, offering nothing new with future conversations, a complaint echoed by *Via Domus* reviewers. *Hit and Run* reviewers criticized the repetition of specific phrases by the game's characters, especially the player's avatars. Repetition was also seen as standing in the way of otherwise strong narratives. *Via Domus* reviewers universally criticized the clumsiness of player navigation in the jungle. Players can all too easily get stuck, requiring them to repeat scenes after being forced to watch an unskippable cut sequence for one's (momentary) death at the hands of the island's "smoke monster." All games include mini-games that most reviewers found to be incredibly boring impediments to play. *Little Britain* was especially criticized for becoming decidedly unfunny while repeating jokes and catchphrases ad infinitum. Self-professed fan of the show David Simpson exclaimed: "After playing the Vicky Pollard level for ten minutes I couldn't bear to hear another 'yeah but no but' speech ever again! I think what really grates at me is that these larger-than-life characters are for the most part soulless hateful entities and whereas I can take a three-minute sketch of Maggie being racist or homophobic, playing as her for any length of time I began to realise just how horrible and wrong that character is."[48] When players are forced into cycles of repetitive banality, the experience of immersion risks feeling like being stuck in quicksand.

Temporal problems of other kinds haunt the games too. Grant Holzhauer, for instance, complained about the pace of the game experience of *Via Domus*: "[*Lost*] hooks its audience with the slow paced, methodical release of information, carefully building the protagonists' stories and unraveling the mysteries piece by piece. *Via Domus* . . . blows through the story so fast that those unfamiliar with the show will be left confused, while fans will wish they could have deeper experiences of these events."[49] Miguel Lopez also faulted the game's length, noting that *Lost* fans have proven themselves remarkably adept at dealing with long, hard puzzles, in contrast to the petty, simple, and quick challenges of *Via Domus*.[50] Predictably, several *24* reviewers bemoaned the game's relative lack of a hard and fast timeline, a key feature of the television show that is difficult to translate into a game in which players start, stop, and repeat levels after "dying" in them. Although some of these games may allow players the opportunity to experience the spaces of the television story worlds, numerous reviews found them incapable of replicating the temporality and the phenomenological experience of watching the television show, thereby strictly limiting the possibility of immersing players in such worlds.

Reviewers of both the *CSI: New York* and the *24* games missed other core experiences from the shows. Poorly done action scenes were especially criticized. The *CSI: New York* game, for example, asks players to solve murders but includes none of the gunplay of the show, leading all reviewers to find it decidedly unexciting. In the case of *24*, much of the game involves various forms of action, from run-and-gun to driving and sniper missions. Yet the reviewers universally found the gameplay lacking, with iffy controls, a poor targeting

FIG. 2-2 The *Lost: Via Domus* (Disney-ABC) video game replicates characters and actions from the series. (Frame grab)

system, and villains displaying ludicrously simple artificial intelligence. Thus, despite marveling at how much the plot is worthy of the show (see above quotations), Garnett Lee warned that "you'll be reaching for the TiVo remote to fast-forward past it all, as if it was the interruption of commercial breaks in the broadcast show. But you can't, so you'll have to suffer through it." The strong script notwithstanding, he eventually concluded that "so much of what makes the TV series hasn't found its way into the game."[51] In general, *24* reviewers agreed that the game looks like the series (split-screen camerawork and impressive cinematography), sounds like it (original actors reprise their roles), and reads like it (outstanding plot) but that it failed at *feeling* like *24* because it isn't exciting.

Here we must pause to examine what "excitement" is as an experience and in terms of gameplay. The *24* game puts the player in the place of the antiterrorist superhero Jack Bauer and his team of Counter Terrorist Unit (CTU) colleagues as they take down the bad guys. This device may seem to offer ample excitement to an audience member, who is usually a spectator rather than a participant. But clearly, the excitement of *24* as a television series stems as much, if not more, from its ability to pace its gun battles and to create intrigue, suspense, and action with precision timing. In the game world, even with good visuals, voice work, and plot, the show's skill at managing time is no longer present, resulting in a bad to mediocre experience. Neither *Little Britain* nor *The Simpsons* received similar complaints, since neither relies heavily upon suspense or action; still, the repetitive progress of the games slows the shows' rapid-fire pacing of jokes. All six games, in one way or another, were judged negatively because they create a new temporality for the televisual story worlds and because they move players through these worlds at speeds that lose the experience of watching the shows.

Another aspect of the games that was mentioned repeatedly as a hindrance to the reviewers' ability to engage with what they saw as the core experiences of the television shows was the proclivity to mimic other established games or game genres. Since designing licensed games is frequently a rushed, thankless job, designers regularly borrow from established games rather than attempt to break new ground. Consequently, numerous reviews noted that *Via Domus* looks better than *Uncharted: Drake's Fortune* and hence showcases the abilities of next-generation gaming consoles; but such observations usually set up further discussion of how superior *Uncharted* is as a game. Similarly, all nine *Desperate Housewives* reviews referred to the game as a lesser version of *The Sims 2*, and all thirteen *Hit and Run* reviews noted its attempt to replicate *Grand Theft Auto III*. Daniel Weissenberger even framed his review around charges of plagiarism: "For anyone who has played *Grand Theft Auto III* (*GTA3*), the gameplay will be incredibly familiar." He concluded by broadening the critique to encompass the entire *Simpsons* game franchise: "Rather than trying to create an original game that is born logically from the characters and premise of the show, they seem satisfied to make *Simpsons* versions of already popular games (the last three *Simpsons* games, *The Simpsons: Road Rage*, *The Simpsons Wrestling*, and *The Simpsons Skateboarding* were, respectively, the *Simpsons* versions of *Crazy Taxi*, and two cynical attempts to cash in on hot genres). What used to be a name that represented cutting-edge satire is now a sign of intellectual theft."[52] In a postscript he noted that Sega had filed suit against Fox Interactive and Radical Entertainment over the *Road Rage/Crazy Taxi* similarities, and he hoped that "this lawsuit might finally serve as impetus for the industry to finally shift away from 'me too' titles." *CSI: New York*, meanwhile, was described as a not very good throwback to point-and-click adventure games, and *24* was often compared unfavorably to multiple first-person shooters. Only *Little Britain* escaped comparison with other games, perhaps because its reviewers found it so uniquely bad that it occupies a class of its own.

Admittedly, Weissenberger was the only reviewer to see a problem with *Hit and Run*'s appropriation of *GTA3*. Other reviewers enjoyed the intertextual links, often framing them in terms of the show's parodic urge to copy the world around it, thus immersing players in the self-reflexive ethos of Springfield. But reviewers of the other games found the allusions to established game properties to be tired and unoriginal, markers of their inferiority. Moreover, each game's similarities to other, better games occasionally dominated reviews, and in this respect Weissenberger's attitude was not uncommon in my sample. If game reviewers often struggle to see the world of the television show as they know and experience it, a game's similarities, intentional or not, to other games clearly risk intertextually coding it as, for instance, a poor *Sims* copy rather than a *Desperate Housewives* game. Ironically, the *show*'s core experiences may be missing or watered down, but the substance and looks of other games are both present and relatively dominant.

Conclusion

The combined weight of the reviews surveyed here suggests that the path forward for licensed games, or for any transmedia, is a rocky one, requiring not only shifts in convergence production cultures but also greater attention to the adaptive process in general. Developers have at times shown themselves equal to the task of creating the spaces of beloved television narratives, whether the suburban cul-de-sac of *Desperate Housewives* or the lush South Pacific island of *Lost*. They have also shown themselves capable of enlisting a show's voice talent and writing talent, resulting in "authentic" characters with whom players can interact in plots that, at their best, are on a par with anything the television show offers. Here we see some signs of the genre improving. However, the promises to "immerse" fans in their favorite story worlds are still unrealized, beset on one hand by many games' close adherence to other video game properties and genres and, on the other, by issues of *temporality*.

Beyond the challenge of replicating the characters and plots of television lies the harder task for transmedia of capturing the (temporal) experience of the story world, a quintessential part of any narrative wherein resides much of its tone, timbre, and personality. Certainly, licensed games can still be great fun. As long as key chains, slippers, and other mundane merchandise branded with beloved characters continue to sell, a market for all licensed games, good or bad, will continue to exist. Also, we must not discount the significant pleasures of exploring one's favorite spaces from within televisual worlds. But to create stronger transmedia connections from show to game, more attention must be paid to replicating and/or playing off the core experiences of the television series, many of which may be encoded in the series' mastery of pace, time, and timing. Texts, in other words, are not situated just in space. They *happen*, and adaptation needs to attend to the phenomenology of a text if it is to approximate that text's ontology. With all the spatial metaphors used to discuss transmedia—starting with the prefix "trans" and extending to "convergence," "overflow," "multi-platforming," "expansion," "world building," and so forth—and with talk of "transporting" a reader/viewer/player into another story world, perhaps as much attention needs now to be paid to examining the temporal elements of transmediation.

As I write this chapter, the trade press and gaming magazines have grown fond of reporting that ties between Hollywood and game designers are improving. Warner Bros., for instance, has recognized the need to bring its game publishing, and increasingly design, in-house,[53] and, as previously mentioned, Peter Jackson realized the importance of allowing a free flow of information to the *Lord of the Rings* designers.[54] In a personal interview, game designer Matt Wolf also described how his job on *The Simpsons Hit and Run*—by far the best rated of the games discussed above—consisted predominantly of ensuring that the game remained "true" to *The Simpsons*, thus suggesting the industry's increased realization of the need to broker the process of adaptation actively, not simply to let it happen in a slipshod manner. Wolf subsequently worked on *The Bourne Conspiracy* and noted the importance to that project of working without the imposition of a delivery date.[55] Perhaps, then, we are beginning to see signs of Hollywood appreciating the need to give designers time, information, and the freedom to create. If so, however, this study suggests that developers of licensed games must pay considerably more attention to issues of temporal, phenomenological transference, situating the player not only in the familiar spaces of the television show but also in its experiential core. They must find ways to incorporate more of that which truly matters about a show and its story world for fans "in the game."

Notes

1. Ben Fritz, "*Call of Duty: Modern Warfare 2* Video Game Gets Hollywood-Scale Launch," *Los Angeles Times*, November 18, 2009, http://www.latimes.com/business/la-fi-ct-duty18-2009nov18,0,5238209.story.
2. Will Brooker, "Camera-Eye, CG-Eye: Videogames and the 'Cinematic,'" *Cinema Journal* 48, 3 (2009): 122–128.
3. Nick Montfort and Ian Bogost, *Racing the Beam: The Atari Video Computer System* (Cambridge, MA: MIT Press, 2009), 119.
4. Ibid., 127.
5. Judd Ethan Ruggill, "Convergence: Always Already, Already," *Cinema Journal* 48, 3 (2009): 105–110.
6. Montfort and Bogost, *Racing the Beam*, 134.
7. Avi Santo, "Batman versus The Green Hornet: The Merchandisable TV Text and the Paradox of Licensing in the Classical Network Era," *Cinema Journal* 49, 2 (2010): 63–85.
8. Trevor Elkington, "Too Many Cooks: Media Convergence and Self-Defeating Adaptations," in *The Video Game Theory Reader 2*, ed. Bernard Perron and Mark J. P. Wolf (New York: Routledge, 2009), 214.
9. Ibid., 222.
10. Ibid., 224.
11. Kristin Thompson, *The Frodo Franchise: The Lord of the Rings and Modern Hollywood* (Berkeley: University of California Press, 2007), 224–253.
12. Simon Egenfeldt-Nielsen, Jonas Heide Smith, and Susana Pajares Tosca, *Understanding Video Games: The Essential Introduction* (New York: Routledge, 2008), 171.
13. Elkington, "Too Many Cooks," 229.
14. Ibid., 223.
15. Linda Hutcheon, *A Theory of Adaptation* (New York: Routledge, 2006), 6.
16. Kamilla Elliott, *Rethinking the Novel/Film Debate* (Cambridge: Cambridge University Press, 2003); Hutcheon, *A Theory of Adaptation*, 9.
17. Hutcheon, *A Theory of Adaptation*, 4.
18. Elkington, "Too Many Cooks," 219.
19. Jonathan Gray, "Introduction. Moving Between Platforms: Film, Television, Gaming, and Convergence," *Cinema Journal* 48, 3 (2009): 104.
20. Tanya Krzywinska, "Hands-On Horror," in *ScreenPlay: Cinema/Videogames/Interfaces*, ed. Geoff King and Tanya Krzywinska (London: Wallflower, 2002), 206–224.
21. Matthew Weise and Henry Jenkins, "Short Controlled Bursts: Affect and Aliens," *Cinema Journal* 48, 3 (2009), 115–116.
22. Louisa Stein, "Playing Dress-Up: Digital Fashion and Gamic Extensions of Televisual Experience in *Gossip Girl's Second Life*," *Cinema Journal* 48, 3 (2009): 116–122.
23. Henry Jenkins, *Convergence Culture: Where Old and New Media Collide* (New York: New York University Press, 2006), 95–134.
24. Hutcheon, *A Theory of Adaptation*.
25. Neville Nicholson, "Yeah, but no, but . . . NO," *PALGN*, February 25, 2007, http://palgn.com.au/playstation-2/6680/little-britain-review/.
26. DJ Dinobot, "*The Simpsons: Hit and Run*," *GamePro*, September 16, 2003, http://www.gamepro.com/article/reviews/31080/the-simpsons-hit-run/.

27. Kristan Reed, "*The Simpsons Hit and Run* Review," *Eurogamer*, October 30, 2003, http://www.eurogamer.net/articles/r_simpsonshitandrun_ps2.
28. Daniel Sayre, "*The Simpsons Hit and Run*," *Game Chronicles*, October 8, 2003, http://www.gamechronicles.com/reviews/ps2/simpsons/hitandrun.htm.
29. Jim Powell, "*24: The Game*," *AceGamez*, n.d., http://www.acegamez.co.uk/reviews_playstation2/24_The_Game_PS2.htm.
30. Chris Roper, "Jack Bauer's Worst Day Yet," *IGN*, March 2, 2006, http://ps2.ign.com/articles/692/692726p1.html.
31. Patrick Klepek, "*Lost: Via Domus* (PS3)," *1UP*, March 4, 2008, http://www.1up.com/do/reviewPage?cId=3166799&p=37.
32. Hilary Goldstein, "When Is a Game Not a Game? When It's *Lost*, Of Course," *IGN*, February 28, 2008, http://ps3.ign.com/articles/855/855831p1.html.
33. David Simpson, "*Little Britain: The Video Game*," *AceGamez*, n.d., http://www.acegamez.co.uk/reviews_playstation2/Little_Britain_The_Video_Game_PS2.htm.
34. Klepek, "*Lost: Via Domus* (PS3)."
35. Gabe Boker, "*Desperate Housewives: The Game Review*," *GameZone*, November 1, 2006, http://pc.gamezone.com/gzreviews/r29413.htm.
36. Luke, "A Guilty Pleasure," *PALGN*, November 23, 2006, http://palgn.com.au/pc-gaming/5866/desperate-housewives-review/.
37. Reed, "*The Simpsons Hit and Run* Review."
38. Jeff Haynes, "*The Simpsons Hit and Run*," *Game Over Online*, November 3, 2003, http://www.game-over.net/reviews.php?page=ps2reviews&id=156.
39. Matt Cabral, "*Lost: Via Domus* Review," *Cheat Code Central*, n.d., http://cheatcc.com/ps3/rev/lostviadomusreview.html.
40. Gus Mastrapa, "*Lost: Via Domus* Review," March 3, 2008, http://g4tv.com/games/xbox-360/38195/Lost-Via-Domus/review/.
41. Mark Arnold, "*CSI: NY: The Game*," *Impulse Gamer*, February 2009, http://www.impulsegamer.com/pccsinythegame.html.
42. Garnet Lee, "Think There'll Be a DVD Version?" *1UP*, March 1, 2006, http://www.1up.com/do/reviewPage?cId=3148512&did=1.
43. John Powell, "*24* Flawed but Fun," *Wham! Gaming*, March 13, 2006, http://wham.canoe.ca/ps2/2006/03/13/1485663.html.
44. Tom Orry, "*24:* The Game Review," *VideoGamer*, March 20, 2006, http://www.videogamer.com/ps2/24_the_game/review.html.
45. Mikhail M. Bakhtin, *The Dialogic Imagination*, ed. Michael Holquist, trans. Caryl Emerson and Michael Holquist (Austin: University of Texas Press, 1981).
46. "*The Simpsons Hit and Run* Review," *Gamer's Temple*, n.d., http://www.gamerstemple.com/vg/games5/000771/000771r01.asp.
47. Cabral, "*Lost: Via Domus* Review."
48. Simpson, "*Little Britain: The Video Game*."
49. Grant Holzhauer, "Some Things Should Remain *Lost*," *GameDaily*, March 4, 2008, http://www.gamedaily.com/games/lost-via-domus/playstation-3/game-reviews/review/6800/1967/.
50. Miguel Lopez, "*Lost: Via Domus*," *GameSpy*, February 28, 2008, http://ps3.gamespy.com/playstation-3/lost/855841p1.html.
51. Lee, "Think There'll Be a DVD Version?"
52. Daniel Weissenberger, "*The Simpsons Hit and Run*," *GameCritics*, December 24, 2003, http://www.gamecritics.com/review/simpsonshitrun/main.php.
53. See James Brightman, "Warner Bros. Buying Stake in SCi/Eidos," *Game Daily*,

December 15, 2006, http://biz.gamedaily.com/industry/feature/?id=14755; Marc Graser, "Hollywood Revises Gaming Strategy," *Variety*, May 29, 2009, http://www.variety.com/article/VR1118004317.html.
54 Thompson, *Frodo Franchise*.
55 Matt Wolf (game designer), personal interview, Burbank, CA, August 7, 2008.

3

Going Pro

●●●●●●●●●●●●●●●●●●●

Gendered Responses to the
Incorporation of Fan Labor as
User-Generated Content

WILL BROOKER

On January 15, 2009, a soaring female vocal cut through the recorded announcements at London's Liverpool Street station. Passengers on the busy concourse stopped and smiled, recognising Lulu's "Shout." And then two of them started to dance. And then three. The music stuttered through a mix into another upbeat pop track, "The Only Way Is Up." By now there was a cluster of dancers in the center of the concourse, executing a routine too tight and coherent to be improvised, too spontaneous and fluid to seem rehearsed. Like a virus, the choreography spread. A woman casually watching suddenly switched perfectly into the rhythm as it segued into the Pussycat Dolls, picking up the steps while the people around her stared. Observers transformed into participants, hearing their cue and becoming part of the act. By the ninety-second mark, as the stately waltz of the "Blue Danube" shuffled and scratched into "Get Down On It," more than three hundred dancers had claimed the center of the concourse, their slick routine echoed by the more ragged, genuinely improvised, and unrehearsed audience participation. Young mums bounced and punched the air, white-haired couples jigged together, a guy with a hi-top fade grinned as he sacrificed his cool to copy the professionals.

The magic of this moment comes from alchemy, from clever chemistry; from the discovery of that perfect balance between improvisation and professionalism, between a grassroots flash mob and an official advertising campaign. Because, of course, that's what it was: a two-and-a-half-minute commercial for the phone company T-Mobile, painstakingly planned and executed to look like a genuinely spontaneous event. The advertisement incorporated the techniques of the flash mob—a culture-jamming social performance whereby crowds gather via digital messaging to act out a "silent disco" at a transport hub or stage a pillow fight in a city center—but T-Mobile's resources (dance auditions, rehearsals on-site, elaborate placement of cameras around the concourse) made it slicker and tighter than a grassroots event. But not too slick, not too tight. The cameras deliberately captured movement glimpsed through the crowd; the dancers stumble, slip momentarily out of sync, negotiate unpredictable space, and interact with passengers who aren't in on the joke. The latter were the people—catching the magic so quickly, cheering generously, and inventing their own quirky dances—who really made the ad work as an expression of joy rather than as a cynical exercise. "Life's for Sharing," says the slogan as the dancers split up and become passengers again, and those who witnessed the event are left giggling and clapping, texting and calling their friends to report on what they just saw.

That alchemy—that giddy, ecstatic meeting between the shiny resources of producers and the raw unpredictability of audiences—is hard to achieve. The T-Mobile ad finds a sweet spot on a continuum between the slick product of professionals and the rough creativity of amateurs. Somewhere else on that continuum are the fan films like Kate Madison's *Born of Hope* (2009), a labor of love almost indistinguishable in parts from the *Lord of the Rings* movies to which it pays tribute.[1] At another point are weird artifacts like NBC's *Zeroes* (2009), a parody of its own *Heroes*, designed to look like a cheap but imaginative fan work.[2] This oddly contrasting pair of texts—the fan film that tries to look like official product and the official product that tries to look like fan film—captures something of the contemporary relationship between official media producers and their active audiences. Fans are commonly driven by the impulse to produce the most "professional" work possible—that is, as similar to the admired source text as possible—while the media industry increasingly seeks to engage fan creativity, drawing it into official platforms and harnessing that ground-level energy. This engagement has numerous potential benefits for producers. It offers fans a feeling of investment and ownership, encouraging longer-term loyalty to a brand. It taps into free and enthusiastic labor, whether in the production of content or in support of viral marketing campaigns such as the McDonald's-sponsored "Avatarise Yourself" campaign (2010) that transformed Facebook thumbnails into blue-skinned Na'vi, which in turn replaced the *Mad Men* mannequins of 2009 and *The Simpsons*–style portraits of 2007. It also channels fan activity from the uncontrolled wilderness of the broader Internet into a walled garden where amateur producers can be tempted by cool toys and tools and a public stage for their work, then kept within strict guidelines in terms of content, approach, and copyright.

This chapter focuses on a specific case study, the *Battlestar Galactica* Videomaker contest of 2007. Like the T-Mobile ad, the Videomaker promotion aimed to hit the right spot between professional slickness and raw energy, inviting fans to produce short tribute films that incorporated official *BSG* clips and sounds. The films were constrained within firm rules, and the best would be screened on television. As such, the contest serves as a perfect example of the industry's attempts to Pied Piper fan creativity into its own playground, co-opting fan labor with the promise of official recognition.

Unlike the T-Mobile ad, the contest never sparked magic. The chemistry went sour; the balance tipped. In fan terms, the project was a flop. This chapter asks why it went wrong. It sketches the precedent for the Videomaker campaign through previous, similar cases and compares it specifically to the attempt of FanLib, a commercially funded online community, to capture the energy and output of fan fiction writers, also in 2007.

Like Videomaker, FanLib failed; but it failed differently. It was also rejected by the community it tried to embrace and incorporate; but it was rejected for an entirely different set of reasons. I will examine these different forms of failure and their relationship to gendered expectations among the distinct communities of fan fiction authors (predominantly female) and fan filmmakers (predominantly male).

Inevitably, this gendered distinction is a generalization, but, unfortunately, one that I will have to perpetuate throughout this chapter to avoid slowing the discussion to a crawl. So a disclaimer is due here. Statistics to back up the dramatic contrast between male and female fan activity are rare, and the evidence is often anecdotal; but studies from the early 1990s onward strongly suggest that fan fiction is an overwhelmingly female-dominated pursuit. In 1991, for instance, Henry Jenkins agreed with Camille Bacon-Smith's assertion that "media fan writers in general" were 90 percent female. My research into *Star Wars* followers, conducted between 2000 and 2002, drew an even more dramatic finding from one fan fiction website: its visitors and contributors were 95.7 percent female.[3] Jenkins restated in 2006 that fan fiction "is almost entirely produced by women."[4] By contrast, but as further confirmation of the gender divide, Julie Russo's study of the 2007 Videomaker project calculated that its fan film community was approximately 80 percent male, and Francesca Coppa asserted in 2009 that "fan filmmaking remains male dominated."[5]

For the most part, fan fiction explores the emotional relationships among characters; one of its best-known forms, slash, foregrounds the homoerotic connections between characters who, in the original, are never explicitly coded as gay. The male fans of my *Star Wars* study were either indifferent or hostile toward "the idea of gay Jedi" and described fan fiction as, for the most part, "pure unadulterated horse crap . . . tacky, tasteless." "I do know a few gay men who write slash," one of my correspondents told me. "I don't know any straight men that do though. I think that's mostly because straight men don't like to even imagine homosexual situations between males." The male-dominated culture of fan filmmaking was, in my experience, more concerned with perfecting special effects and producing professional action sequences than with character, performance, or even dialogue. As a telling indication, my research found that the *Star Wars* fan filmmaking tutorials at *TheForce.Net Theatre* extensively covered computer-generated modeling, fight choreography, and title crawls, with only a handful addressing issues such as story structure.[6]

Vidding, a long-established fan practice that creates montages from film and television clips matched up to popular music, may seem similar to filmmaking. However, it is less involved with creating new footage and more concerned with manipulating existing, found material to draw out emotional subtexts or alternate narratives through creative editing, combined with songs that echo or add new associations to each scene or shot.[7] As such, vidding has far more in common with fanfic. Though its tools are digital video rather than prose, vidding also tends to rework existing stories in terms of character relationships, rather than focusing on effects and action; and it is practiced, Coppa confirms, "overwhelmingly by women."[8] The two forms are distinct but closely related, and are part of a network of often overlapping communities and practices within female-dominated media fandom, including costume-making, songwriting, and visual art. The term "fanfic community" does not adequately cover or contain the diverse and multiple ways in which women celebrate media texts; but in the absence of a convenient catch-all term, it is used throughout this chapter as necessary shorthand.

It should also be noted that, while fan filmmaking can certainly be considered predominantly male and fan fiction predominantly female, there are exceptions to the rule. Even if the visitors to the "Master and Apprentice" slash fiction site were overwhelmingly female, 4.3 percent of them were male. Similarly, Russo's calculation that around 80 percent of the Videomaker participants were male clearly suggests a significant group who were not. Like the "few gay men who write slash," and indeed *Born of Hope* director Kate Madison, these creators going against the gender trend should not be forgotten, even if, once again, they cannot be mentioned at every turn. Finally, we should remember that to categorize fans as male or female is to impose a gender binary that some would find problematic. Russo is careful to count only the "typically male names" within her 80 percent—the others are either female or "indeterminate"—reminding us that fans, particularly online, may choose to resist either category and consciously identify outside that binary as trans* or genderqueer.

Ultimately, this chapter suggests that, as campaigns based around user-generated content become more embedded within popular advertising, consumers are more readily accepting the implied contract between themselves and mass media corporations. Videomaker failed in 2007 because it misjudged its audience in terms of the specific rules and rewards, not because the basic premise was flawed. Similar campaigns in the following years have been more successful, as media producers learned from the mistakes of early exercises and adapted in response, offering fans fairer terms of engagement, less complex demands, and more persuasive returns on their efforts while at the same time insisting on tighter controls. There is also a broader training in evidence. Younger people are recruited into the relationship, which is presented to them as a game. Casual audiences, outside the culture of fandom and lacking a sense of its historical push-pull with official producers, are invited by advertisers to create content and rewarded with a sense of belonging to a corporate-created brand community.

There will be fewer failures now that a large cross-section of the mass audience is growing accustomed to the concept. However, the largely female practice of fanfic remains resolutely separate from this new dynamic of producer-controlled fan content. If submitting to official sites and engaging creatively, donating free labor in return for professional tools, credits, and a public platform, is to become a dominant model within the audience/producer relationship, what will become of the fanfic communities that reject not just a specific set of rules, but even the whole premise of "going pro"?

FanLib and Other Stories

Videomaker's contest was not the first corporate attempt to draw fan creativity into an official playground and onto an official platform. Similar exercises have been documented since the early 2000s. Some were sensitive in their approach and successful in their engagement with fan activity, while others merely alienated the communities they hoped to embrace. A rare example of the former is "Dawson's Desktop," an online overflow from the WB teen television show, which expanded the narrative between its weekly installments by offering access to the main characters' in-box and deleted files. The concept was inspired by fan fiction that imagined backstory, predicted future plotlines, and filled in gaps, and a select team of creators was recruited from fandom to consult with producers and guide the website's development. "We're in touch with them all the time," claimed project leader Andrew Schneider. "We wanted to make sure the fans were getting what they wanted. They helped us design the interface and they told us what they liked and did not like."⁹ Although "Dawson's Desktop" made no attempt to incorporate existing fan fiction, it encouraged fans to write to the show's protagonists in character (and replied to them in the same vein), inviting them into the Capeside story world in a mutually respectful dynamic of creative play-acting. "In that way," commented Henry Jenkins in his report on the exercise, "the producers integrated the creative energy of the fan community into developing new content, which, in turn, would sustain fan interest."[10]

Such thoughtful collaboration is unusual in the brief history of user-generated media content. Lucasfilm met with resistance to two distinct but equally ill-judged initiatives in the early 2000s, both of which disguised containment as empowerment. First, the offer of free webspace in the backyard of the official *Star Wars* site—initially received by fans as a supportive gesture—seemed less appealing in the light of a clause that signed over all the content within this playground to Lucasfilm, giving the producers power both to eliminate anything that infringed on its vision of the saga (such as slash) and to poach any fan work it admired. As I noted in *Using the Force*, "Lucasfilm could seek out and destroy objectionable fan sites with cease and desist letters, but this tactic risks making it look like a tyrannical father; the offer of official webspace puts the company in the role of a kindly uncle."[11] Meanwhile, *Star Wars* fan filmmakers were being invited to participate in a similar official platform, lured to the AtomFilms website by the offer of professional sound effects, a broader audience, regular prizes, and a share in any profit, but they were warned that "no attempts to expand on the *Star Wars* universe will be accepted." Only documentaries and parodies would be endorsed. This generic limitation reinforced Lucas's role as the only provider of original content and would have impoverished the releases of fan filmmakers, if they had taken up the deal rather than sticking to unofficial but less restrictive sites such as TheForce.Net.[12]

The aims of FanLib, its intentions, and the almost comically inept misjudgment of its potential market, which led to its rapid demise, have been covered extensively by other media scholars, most notably Suzanne Scott and Karen Hellekson in separate articles from 2009 and Henry Jenkins in a May 2007 blog entry that links to and collates various other fan commentaries.[13] Hellekson summarizes it as an "attempt of (male) venture capitalists to profit financially from (female-generated) fan fiction":

> FanLib, founded by industry insiders ... launched in May 2007 with $3 million in funds, sought to commodify fan fiction at a newly created fanfic archive site. Although outreach included targeting and e-mailing fanfic writers and encouraging them to upload fic to the site in exchange for prizes, participation in contests leading to e-publication, and attention from the producers of TV

shows like *The L Word* and *Ghost Whisperer*, FanLib's persistent misreading of the situation alienated fans, as did the draconian terms of service. One fan closes her analysis of FanLib's terms of service by noting, "It's perfectly clear—they get the bucks and we get the lawsuits." The site closed down in August 2008.[14]

The case has similarities with previous attempts to incorporate, contain, and (ideally) monetize or otherwise profit from fan energies. Like the "Dawson's Desktop" team, FanLib contacted a select crew of fans prior to rolling out its official website, offering them a flattering invitation to join the project. (In another twist, the "Naomi" who gushed to female fans, "My colleagues and I want it to be the ultimate place for talented writers like you," was suspected to be a sockpuppet for the all-male board of directors.) Like the official *Star Wars* webspace, it offered a more official centralized territory, gathering creativity in one place; like AtomFilms, it unveiled a platform where grassroots creativity would supposedly reach a broader audience, receive prizes for effort, and, at least potentially, be noticed by professionals.

FanLib failed for several reasons. As already suggested, its approach toward the community of predominantly female fanfic authors was a pantomime of slapstick clumsiness. Its primary advertising campaign was a Charles Atlas–style contrast between a skinny guy posing against a pink background with a muscular dude backlit in blue. This cliché of the effeminate fan geek becoming macho beefcake earned scorn, and then slash fiction, from the female community.[15]

Cheery Naomi, who sent out those personal e-mails, was never convincingly identified as a real person.[16] As with the *Star Wars* containment exercises, fans examined the terms of service and voiced their disillusionment with clauses that shifted ownership of their fiction to FanLib, but held the authors responsible in the case of lawsuits and protected media producers from charges of plagiarism should they in turn poach anything from fan creativity. The profit system was also skewed: fan authors were expected to be happy with T-shirts, prizes, and points for each story they uploaded, which slowly built up into $50 Amazon.com gift certificates.[17] Needless to say, the board of directors was aiming to earn real dollars rather than Monopoly money and store coupons from the venture.

So far, so familiar. Yet the most interesting reason behind FanLib's failure was specific to the community of predominantly female fanfiction writers. Put simply, they didn't want the prizes. They didn't want the platform. They didn't want the legitimacy. They didn't want the official status, the profits, or the professional recognition. Chris Williams, the hapless CEO who took most of the FanLib flack, protested in an online interview that the site offered fan authors the chance to be published in a HarperCollins e-book; that another event transformed fan stories into videos for distribution on the *Ellen*

DeGeneres Show; that his team had hosted the first collaboration between fans and a television creator, resulting in a joint-authored *L Word* episode.[18] Overwhelmingly, the fanfic authors weren't interested. They already had what they wanted. To quote just two of the responses to Williams's argument:

> In case you haven't heard, we already have a place to share our fanfiction, to get it on the radar, to get it read by people who matter—other fans. LiveJournal. Fanfiction.net. A vast number of archives and sites that range from small, private affairs to the behemoths like Fictionalley. And don't give us that crap about "wider audience"—if you crave fame in fandom, there are ways to get it, good and bad. And they are just as free to use, and don't come hand in hand with limits and editing by people I don't know from Adam.[19]

> Underlying all of FanLib's rhetoric about this exchange is the upsell—the push for fans to have their work acknowledged by someone higher along the hierarchy of creative production, by getting "closer" to the "talent behind" their favorite canons. The implication is that that hierarchy (imposed upon the fandom by outsiders) carries with it its own authority, its own implicit weighted system of value.
>
> But that's not how fans see their experience, and it's not how fans view fanfic. Mr. Williams, fans *already have that*. Only they have it within an autonomous arena where the act of posting a fanfic is its *own* best end—no upsell needed.
>
> It's true that many fans might be excited about the prospect of a creator reading their fic; but many more would not be, and in fact it would make them uncomfortable, because what makes fanfic unique, what makes it *fanfic*, is that it operates outside the sphere of the media production machine that created its canon counterpart. It has an authority all its own.[20]

As one fan author, Telesilla, concluded her Livejournal summation of the case: "We are valid and we are legitimate and we do not need FanLib."[21]

This was the crucial point in FanLib's failure. The terms of service could have been considered more thoughtfully and structured more fairly. The misguided recruitment and advertising campaign and the defensive protests from Chris Williams on behalf of his colleagues could have been handled with greater cool, sensitivity, and intelligence. The community would not have been so thoroughly alienated, and the ground between fanfic authors and Williams's venture not so irredeemably scorched. But this fundamental barrier would have remained. The female-dominated fanfic community exists within its own ecosystem, what Hellekson dubs a gift economy, a means to its own end.[22] Whatever the terms, whatever the deal, the majority of these authors do not see fanfic as a calling card to mainstream success or a stepping-stone to big-name recognition. By and large, fanfic does not want to "go pro."

Case Study: Videomaker

The concept and premise behind Videomaker are detailed by Derek Johnson in this volume and discussed at length by Julie Levin Russo. Russo sums up the state of play in August 2007:

> Videomaker Toolkit is a fan-driven promotion... heavily advertised on the official site. Its instructions invite us to "be a part of *Battlestar Galactica*" by creating a four-minute tribute film, the best of which will be selected to air on television. In order to "help give your videos the *Battlestar* look and sound," a menu of downloadable audio and video clips is provided, while the rules place a premium on an archaic "ex nihilo" model of originality by stipulating that the only additional material permitted is that which "you created." Moreover, these "tools" are limited to less than 40 short CGI-based establishment and action sequences (divided into "land" and "space" and including mostly ships, architecture, and explosions), plus a number of signature sound effects and only seven partial music tracks (also included is the show's logo image and a required ending clip plugging "new episodes of *Battlestar Galactica*" and Videomaker itself). That is, Videomaker's conception of sanctioned derivative filmmaking is extremely narrow, notably excluding the character-based dramatic scenes that make up the majority of the show.

Russo's account of Videomaker's restrictions are followed by two further, key observations: first, that more than a hundred submissions had been accepted at the time of her report, so the contradictions inherent in the project "don't seem to be crippling"; and second, as noted above, that "81 of the authors listed for the first 100 submissions have typically male names."[23]

The gender distinction in terms of fan expectations and approaches must already be obvious. The female fanfic authors cited above would scoff at any offer to create "tributes" to a beloved show that excluded character-based drama and focused on "ships, architecture, and explosions."[24] As we saw, the fanfic community remained unmoved by FanLib's offers of professional networking and a public platform. Russo comments skeptically, therefore, on Videomaker's assumption "that recognition by and on television is incentive enough to channel this artistic labor out of the internet at large and into SciFi's walled garden." And yet, one hundred fans signed up. They signed up because frameworks that allow certain forms of fan expression (such as action-packed, effects-based tribute) and forbid others (such as expansion of the story world and character exploration) are, in Jenkins's words, "anything but gender neutral."[25] They signed up because, overwhelmingly, they were men and because, overwhelmingly, male fans have an entirely different agenda.

Soon after Videomaker's launch, NBC's Sci-Fi channel (now Syfy) opened

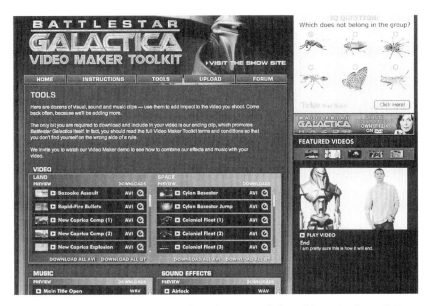

FIG. 3-1 The SyFy.com Videomaker Toolkit online contest helps publicize *Battlestar Gallactica*. (Photo courtesy of NBC-Universal publicity).

an online discussion forum for community support and critique, announcing the project with a burst of exclamation marks and bold text.²⁶ "Download all the music, sounds, and visual effects clips you'll need to make your own *Battlestar Galactica* videos, then post them on SCIFI.COM for a chance to see them on SCI FI! Plus, executive producer David Eick will name one lucky fan the BSG Videomaker of the Year!"²⁷

In dramatic contrast to the passionate rejections and scathing rejoinders that greeted FanLib's offer of professional recognition, the response from Sci-Fi's community included an ecstatic "sooo coool!" and "this is a glorious day."²⁸ The few quibbles concerned its details, such as deadlines, format, and file quality, but not the premise. These fans did not object to the offer of action clips; they wanted more. "I'd like to second Katchin's request for a planet shot of Caprica and add one for more shots of Centurions like those from 'The Eye of Jupiter Part II.'"²⁹ They didn't remark on, or perhaps didn't even notice, the exclusion of character-based drama; instead, they voiced mild disappointment about the exclusion (for legal reasons) of European- and Canadian-based fans from the project. "Nuts, I guess that counts me out then. You Americans are lucky, have fun with it."³⁰

Female authors on LiveJournal had scorned FanLib's model of "a coloring book [where] players must 'stay within the lines'" and its promise to business partners of a mass audience happy to "collaborate democratically in a fun online game that *you* control."³¹ Again in marked contrast, the predominantly male fans who set to work on Videomaker were happy to stay within the lines and even welcomed restrictions as a spur to creativity. The prospective filmmaker who asked for an establishing shot of Caprica added, "I know the challenge is to work around it and I'll try." The overriding attitude on the forum was that amateurs had been handed a brief from above, were grateful for the contract, and, while that contract was upheld, would prove their skills and enthusiasm by doing the job without complaint. An offer had been made, a reward had been promised. The deal seemed more than fair. They jumped at the chance.

During the competition, from February to June 2007, the forum rang out with the banter and shared advice of a busy workshop crowded with specialists and keen amateurs in friendly competition. Members discussed streaming compression, interlacing, fonts, codecs, anti-aliasing, and, to a lesser extent, acting. The focus was on creativity, but channeled in a direction different from that of a fan fiction community. Whereas fanfic often celebrates imaginative, yet convincing detours from the original text, here much of the effort was devoted to accuracy and fidelity, whether in terms of building a complete and detailed Raptor cockpit "realistic right down to the dashboard,"[32] designing sound effects from scratch, or duplicating the title sequence. "To get things to look like they do on the show," one contestant "took an episode I bought from iTunes and went over it frame by frame to see exactly how they did their titles. Then I just figured out how to recreate that look in AfterEffects."[33] The achievement for another player "is that the sound designer was able to completely re-create the sound of a Cylon eye osculating [sic]. How accurate do you think it was?"[34]

These fans had no problem with the chance of "public recognition by and on television" as a return on their labor. Although the pleasures of collaborative creativity, sharing films, gaining feedback, and improving technical skills were part of the reward, the prospect of seeing their films hosted on SyFy.com and possibly more broadly was clearly a driving incentive. "I am so chuffed at the moment. Our film is featured on the front page of www.scifi.com as the Sci-Fi Pulse Featured video. This was quite a surprise to see. Earlier on today we saw that our video was featured in the newsletter, but then to see it on the front page of the site is awesome! Got a screen capture to savour our 5 minutes of fame."[35]

Intriguingly, the sole voice of radical dissent, from a member called GirlwGuns, echoed the views of the fanfic community: "DON'T DO IT PEOPLE! DON'T HELP THEM WRITE THE SHOW! STOP NOW WHILE YOU CAN!"[36] One poster raised his eyebrows: "Chill pills. Get your chill pills here." The other community members ignored the protest and began jointly composing a battle chant or work song ("The honor! The power! Hail and kill! The rain! The thrill!"). GirlwGuns was dismissed, seemingly through the stereotype of the hysterical woman. Her warning, at the time at least, was simply not relevant to the rest of the group. As another member commented flatly, "It's the recognition people are after."[37]

So, what went wrong, and when, and why? By June, the forum members were expressing mild regret and polite disappointment, though still couched in an absolute deference to producers (specifically, the show's executive producer and the competition judge) that again contrasts starkly with the self-contained, self-sufficient nature of the fanfic community. "It's too bad the website has been rather inconsistent in listing all the entries," one contestant wrote, "but fortunately, what we all think about them isn't part of the judging process . . . it's up to Mr Eick."[38] Gradually, the fan filmmakers began complaining about the delay between submission and appearance on the website, then about the lack of any response or feedback from the site staff at all. They knew only one person could win the grand prize, but the "recognition" that people were after worked on several levels; having a fan film showcased on Sci-Fi.com for public feedback was also an incentive. When the competition organizers disappointed on this score, failing to reply and delaying uploads to the main website, the energy on the boards turned sour.

> I really wanted to give them the benefit of the doubt, but obviously something is wrong. You don't create a promotional campaign to stir up interest in your product or program, then drop the ball and leave the people you sucked into the campaign hanging in the wind. I've never participated in anything like this before, but I did it because I like the show and wanted something creative and fun to do. I really expected much more from the organizers of this. I grow more and more disappointed every day. And hey—YOU—the people that came up with this idea and thought this would be a great marketing campaign . . . it's backfiring. All this is doing is making SciFi look incompetent, thoughtless and totally unorganized. Hopefully you'll get your act together and start treating the people you sucked into this with a little respect.[39]

There remained a plucky determination to make the best of the exercise, with forum members commenting on how much they had learned and on the pleasures of simply sharing and evaluating each other's work—pleasures similar to those within the fanfic community. Despite a remarkable display of patience, tolerance, and respect for an institutional hierarchy within which fans accepted their place near the bottom ("they'll update the page when they're good and ready, no sooner"[40]) and a continued excitement about being recognized ("I made it! Holy shit!"[41]), a feeling of dissent grew in the ranks when the competition seemed to have been forgotten by the producers of the website, let alone by the show itself. "It really is a slap in the face to fans," one wrote, "when they set up a promotion like this, then ignore people's hard work and keep them in the dark about what's going on."⁴² Resignation and resentment took over: "I think we all need to face the facts . . . the show has been cancelled, they've forgotten about us or just don't care anymore. Either way I'm beginning to think this whole thing was a huge waste of time and effort. This whole thing stinks of, 'Let's just forget the whole thing and hope no one really notices.' "⁴³

By August, with most fans having deserted the project and forum traffic dwindling to a few posts, the contest was over. The winner, never officially announced on the forums, was another slap in the face: the DAVE (Digital Animation and Visual Effects) School at Universal Studios, Florida, walked it with a CGI animation. Understandably, the remaining fans on the Sci-Fi board were disillusioned with the result: the DAVE School entry was not "homemade," and the institution had obvious links to the show's producers. "Did any of you think there would be an entry of this calibre, from a world class school?" Tumbleweeder asked the few diehards still reading. "This isn't the fan base, I was imagining, to enter the contest . . . shame on me, I guess . . . I truly believe the contest was over before it began."⁴⁴ Tumbleweeder had, in fact, summed up the key issue two months beforehand, in June: "I don't believe they have kept their end of the bargain."⁴⁵ His comment captures the fundamental difference between the female authors' reaction to FanLib and the predominantly male fans' disillusionment with Videomaker. The fanfic authors didn't just reject the deal; they were barely able to engage with it. It was, in cultural terms, proposed in another language, from a community of male venture capitalists that didn't understand them (one commentator compared the approach to "predatory colonialism"⁴⁶). The filmmakers embraced the deal; but the deal was broken. The distinction lies in the two groups' contrasting attitudes toward mainstream hierarchies, official corporate culture, and the concept of "going pro."

Going Pro

The Videomaker fans, it should be clear, saw their engagement with the project as a form of professional contract. They were happy to keep to the rules and color within the lines in exchange for public recognition and professional attention. Their willingness to accept this deal demonstrates a notion deeply embedded among male fans—and largely absent in the female fanfic community—that their amateur creativity is a means to an end, a calling card or show reel, a passport to the big league.⁴⁷

The ranks of fan filmmaking are sufficiently crowded with examples of amateurs who have gone pro to establish a precedent and to suggest that the odds of being spotted and crossing over are reasonable, albeit slim. Kevin Rubio, who created the fan film *Troops* in 1997, was recruited by Lucasfilm to a professional gig writing *Star Wars* comics in the early 2000s and is currently working on the *Clone Wars* television series. *George Lucas in Love*, directed by Joe Nussbaum in 1999, earned a fan letter from George Lucas and, on its commercial release, landed on top of Amazon's sales chart.[48] As Henry Jenkins observes, "Lucas and his movie brat cronies clearly identified more closely with the young digital filmmakers who were making 'calling card' movies to try to break into the film industry than they did with female fan writers sharing their erotic fantasies."[49] It's interesting to note that this hierarchy of amateur and pro—and the particularly male trajectory of progression up through the ranks—is reinforced in the fan taxonomy of two (male) academics. Nicholas Abercrombie and Brian Longhurst's breakdown of fans into the categories of consumer, fan, cultist, enthusiast, and "petty producer" is informed by a previous study of American hot-rod enthusiasts, who are divided into "two main groups: professionals and amateurs." "Individuals," they note, "can move from one sector to another over time."[50] However, Abercrombie and Longhurst do not give the impression of a flat playing field, with each category equally valid; rather, the sense is of a promotional ladder. The continuum from consumer—strongly coded as feminized through association with "*follower*-like" young mothers, "distractedly abreast with the series while cooking tea for their children"—to petty producer, or professional, "may represent a possible career path." This trajectory is further valorized in terms of a "more systematic engagement with 'serious leisure,'" a development leading to the choice between an amateur or a professional career path and "the decision to go commercial."[51] Consumers with "relatively little background knowledge" with which to judge the quality of the media they enjoy—again, the term is feminized through association with daytime soap operas—are contrasted with petty producers through the criteria valued by male fandom: "it is possible to see an increase in technical skills across the continuum."[52]

The sense of fan practice as a steady progression to professional status is not unique to amateur filmmaking (or hot-rodding). Matt Hills notes that the BBC Wales reboot of *Doctor Who* was built on the loyalty and energy of fans-turned-professionals. "The cultural productivity of a generation of fans was, it seemed, fully integrated into the UK's creative industries, and into the programme itself. Executive producer Russell T. Davies and producer Phil Collinson were card-carrying fans; the actor playing the tenth Doctor was a fan; writers such as Paul Cornell, Mark Gatiss, Steven Moffat, Gareth Roberts and Rob Shearman were all fans; even *Radio Times* writer Nick Griffiths had grown up as a *Doctor Who* fan."[53]

And not a woman is named among them. Sheenagh Pugh observes in her study of fan fiction that many professional *Doctor Who* writers came from the "grass-roots level," but that "they all seem, typically for that fandom though very atypically for fanfic as a whole, to be men."[54] Quotations from her interview with *Doctor Who* novelist Paul Magrs are stunning for their air of self-confidence: "[Question:] *What made you first want to write a Doctor Who novel?* [Magrs:] I was writing them when I was ten. There was a bit of a gap for a while, I wrote some other novels and published them. I'd been at school with Mark Gatiss and did my MA in creative writing with Paul Cornell, so I thought, 'Hang on—shouldn't I be writing one as well?' I knew it would be great fun."[55] Contrast this absolute sense of entitlement with the self-deprecatory shyness of Kel, a female author and, Pugh suggests, "a publishable writer if ever I read one." She would "of course dearly like to do this (fiction) professionally. However, I tend to regard that as Mittyist in the extreme. Kel the Novelist ranks right up there with Kel the Replacement Bassist in Metallica [giggling] it just ain't gonna happen."[56]

Pugh admits she was initially "baffled and astounded . . . that many fanfic writers chose never to try the profic waters."[57] She suggests fear as a possible reason—the unwillingness of many female writers to face the harsher public world of rejection. As she notes, "There is a cultural expectation that women will be more diffident about their abilities than men, and less inclined to competition." And so it follows that "the collaborative, non-competitive environment of fanfic might give some female writers a confidence to write and share their writing that they would not have outside it. If that's so, then the alternative for these writers probably wouldn't be to write profic . . . but not to write for an audience at all."[58]

Pugh adds, though, that it would be an arrogant error to assume that "the vague chance of everlasting fame" was something "all writers would pursue at any cost." As suggested by the responses to FanLib, most female fanfic authors do not see their practice as a stepping-stone or calling card. Neither do they hold back from sending their work to professional agents merely because of fear of rejection. As Aja pointed out in her reply to Chris Williams, cited above, a community like LiveJournal is "an autonomous arena where the act of posting a fanfic is its *own* best end—no upsell needed."[59]

This autonomous arena offers its own pleasures, self-contained, unique, and even exclusionary. Karen Hellekson suggests that "the off-putting jargon and the unspoken rules [mean] that only *this* group of *that* people can negotiate the terrain."[60] Part of fanfic's appeal, as Pugh notes, is the open-access democracy; other points in its favor, which "going pro" cannot match, are the lack of competition, the guaranteed audience, the instant feedback, and the support of beta-readers. "To lose all this in exchange for a minuscule amount of money and an equally minuscule chance of fame," Pugh concludes, "isn't necessarily such a good bargain."[61] But an online fanfic community like LiveJournal has qualities that go even deeper, down to the formal level.

First, as Karen Hellekson discusses, female fanfic writing exists in a gift economy. Hellekson's edited volume with Kristina Busse opens with a demonstration of the friendly to-and-fro. "It starts like this. Somewhere in cyberspace, someone complains: 'I had a lousy day! Need some cheering up.' Soon after, a friend posts a story dedicating the piece: 'This is for you hon—your favourite pairing and lots of schmoopy sex. Hope it'll cheer you up!' "[62]

In her later article Hellekson develops the point, again noting the community's exclusionary aspects, but drawing out another key dimension of LiveJournal and similar spaces.

> To the uninitiated outsider, media fandom as it's currently practiced online in blog spaces such as LiveJournal makes little sense: strange jargon with unclear acronyms and lots of punctuation sits next to YouTube or Imeem video embeddings. Perhaps a post announces part 18 of a long piece of fan fiction. In the comments someone has left the writer a gift: a manipulated image of her two favourite characters cleverly sized so she can upload it into the blog software interface.[63]

It's not just the support, the feedback, the cozy greetings after a lousy day, and the homemade gifts that would be lacking in the commercial sphere of professional fiction. The print-based professional novel or short story collection could not, on a technical level, hope to duplicate the multimedia collage of links, embedded video, and animated images that characterizes, and is an integral part of, online fanfic.

Second, as the essays in Busse and Hellekson's collection make particularly clear, female online fanfic is not based around the traditional model, still the basis of professional fiction, of an author presenting finished work to her largely silent readership. Of course, an online forum allows for instant feedback, which may shape the ongoing work. But more radically, online storytelling is a type of performance,[64] consisting not just of sole-authored fiction and response, but also of collaborative writing, of romance epics constructed through dialogue between two or more individuals against a constant commentary and chorus from the audience,[65] and, perhaps most interesting, of a physical dimension through exclamations that work as stage directions. Professional fiction excludes the possibility for "squeeing" at another author or the option of appraising writers with the textual action of a "*lick*," "*snog*," or "*hump*," let alone "::squishes your boobies::."[66] The pleasures of collaboration and joint authorship—coupled with a tradition of "real person slash" in which the appeal of celebrities is not their unattainable distance but their endearing insecurities and flaws—contribute to problematize the conventional ideal of individual success and radically complicate the male trajectory of a straightforward progression from fan to professional. Female fans who publish profic sometimes do it in secret, keeping their amateur and professional identities apart,[67] while Big Name Fans, the fanfic equivalent to Abercrombie and Longhurst's petty producers, are regarded ambivalently "as something to be aspired to (or derided)."[68]

Fan Labor and User-Generated Content

It would be unwise, within this short space, to generalize about such a broad and complex topic as gendered expectations of work and payment, reward, and currency. However, on the level of these specific case studies, we can again see a striking contrast between male and female responses to offers of possible mainstream success and professional recognition—the promise of "going pro"—in return for fan labor. Hellekson's article on the fanfic community as a gift economy stresses that the items exchanged, gestures of time and skill, "have no value outside their fannish context." "Gifting is the goal. Money is presented less as a payment than as a token of enjoyment."[69] Telesilla, on her LiveJournal blog about FanLib, offers one reason for this noncommercial system: "One of the two underlying 'rules' of fanfic has always been that we fly under the radar and we don't make a profit from fanfic. These two rules are linked tightly, because, as long as we don't shove fanfic in the copyright holder's faces, they can pretend we don't exist."[70] However, as Hellekson notes, the reason goes beyond mere self-preservation. "Online media fandom is a gift culture in the symbolic realm in which fan gift exchange is performed in complex, even exclusionary symbolic ways that create a stable nexus of giving, receiving and reciprocity that results in a community occupied with theorising its own genderedness."[71] Hellekson explains that the fannish process of gift-giving, in rejecting the male-gendered monetary model, foregrounds "the female-gendered task of maintaining social ties" and privileges the female-coded social sphere over the traditionally male system of economic competition. Further, by giving of their time, effort, skills, and energy, female fans make themselves into a gift that they can donate to other women—a reclaiming of agency that overturns the traditional patriarchal model of woman as currency.

No wonder, then, that female authors responded to FanLib's male-dominated, capitalist initiative with hostility. "Among the feminists," Icarus-ancalian concluded, "an attempt to make money off of women fanfiction writers with no compensation went over like a lead balloon."[72] Commenting on another LiveJournal blog, Valarltd saw "a bunch of men going 'cooollll, something else to sell, and we don't even have to produce it! The women will do all the work for us!'"[73] The FanLib project tried to impose a male, commercial paradigm onto a female community that had established its own self-contained rules and currency; very much like a colonial army attempting to win the hearts and minds of an entirely different culture and mining its wealth in exchange for trinkets.

Yet the directors of FanLib surely didn't think they were exploiting the fanfic community. It seems very likely that they viewed the offer as a fair deal, hence Chris Williams's anguished spam on multiple LiveJournal pages, protesting, "We're good people here and you make us sound like we're an evil corporation or the govt. sending your kids to war or something."[74] Based on the behavior of male fan communities offered the same bargain, Williams could hardly be blamed. While Valarltd was mocking FanLib as a "cooolll" money-making racket for its male producers, SharonMustLive was exclaiming without irony that Videomaker was "SOOO COOOL!" As demonstrated above, the male fan filmmaking community saw the promise of official recognition, possible fame, and contact with the show's producers as a perfectly fair exchange for their labor. It was only when the contract was broken that they finally gave up trying to defend the project, and even then they expressed dismay about the way it had been handled rather than about the concept itself.

This tendency of male fans in particular to accept willingly such terms of exchange, offering free labor in return for a token reward from the producers, has been repeated in a range of promotional campaigns and online initiatives since Videomaker's demise in 2007. In this volume, Derek Johnson details the *Battlestar Galactica* "Join the Fight" social networking initiative that followed in 2008. Players choose their loyalty to either the human or the Cylon army and enter competition with other fans in order to earn "cubits," which in turn enable them to upgrade "engines and armaments in their personal fleets of space fighters." Close attention to the primary text of the show pays off in competitive quizzes, again rewarded by cubits, which enable further tooling up of the players' military capabilities. "The prize is more than bragging rights," promises the website copy; cubits also provide "the currency you'll need to 'pimp' your very own virtual Viper or Raider."

This game is based on barracks-room bragging, on pimping up military vehicles, on performing "ongoing creative and promotional labors that ultimately served the needs of industry by expanding both the content and the reach of the franchise."[75] And the reward is imaginary money. The female fanfic community would laugh the game out of town. Yet the first post on SciFi.com's "Join the Fight" forum, from CylonJefferson, exclaims "So cool!"[76] It is impossible accurately to judge the gender of an online community, but the avatar of a goateed guy appraising the game as "killer sweet," the trash talk about recruiting a team to "kick some fakin' toaster tail," and the wisecracking about waiting "for Starbuck's boobs to stop drooping" give the sense of a male group happily adopting the personae of a military squad.[77] More crucially, they seem content to be recruited into a hierarchy, to begin at the bottom, and to earn their way up through the ranks, even through imaginary currency and pretend structures of power.

The same dynamic—taken to a more absurd degree—can be seen at the online platform of the NBC comedy series *The Office,* "*Dunder Mifflin Infinity.*" Here, instead of joining the fight as a rookie, visitors are invited to submit an application for a lowly job in a fictional paper company. Successful applicants get to join a branch and engage in Internet-based tasks, such as writing a song or poem, answering a quiz, or submitting a St. Patrick's Day photograph. "Diligent and hard-working employees will be rewarded in SchruteBucks, which can be spent on decorating their Virtual Desks," the site explains. Successful earners have the opportunity to apply for a branch transfer or a promotion to regional manager. Where *Battlestar Galactica* offered players participation in a war and succeeded in generating content for the franchise in return for an invented currency, *The Office* invites them to join a mundane white-collar environment and provide the website with creative labor in return for ShruteBucks and the chance of promotion. The concept takes the self-contained gift economy, where fan effort is meaningless outside the specific community, and puts it in the service of major television corporations. That it might work seems far-fetched, but the forums, launched in early 2008, remained popular in 2010.

This premise of recruiting visitors into the ground level of a fictional organization's ranks and rewarding them with a drip-feed of token rewards and public recognition in return for creative labor lies at the heart of various recent marketing projects. Some are hugely ambitious. *The Dark Knight* alternate reality game, "Why So Serious?" (2008), successfully persuaded hundreds of young men from around the world to submit photos of themselves and their friends in Joker make-up in exchange for a minor role in the online story as part of the arch-villain's army of henchmen.[78] The same engine also powers smaller-scale campaigns, such as Kellogg's 2009 advertising promotion for Krave cereal. "Open a Choc Exchange account," the site promises, "and you'll be credited with 10,000 Krave Choc Chunks."[79] Choc Chunks are, it should be noted, not a physical foodstuff but another imaginary currency. Players can "earn extra chunks by performing all sorts of tasks online and on Facebook." Again, the rewards for players are dubious, while the benefits for the company are obvious. Judging from the cereal's Facebook page, this "Join the Fight" dynamic, whereby players sign up as brand loyalists and then recruit others in order to earn points for themselves, is extremely successful: Krave had 39,528 followers at the time of writing.

This cross-gendered, probably cross-generational following on Facebook's mainstream social network, inspiring loyalty and commitment from an audience group beyond conventional media fandom, is perhaps the most striking aspect of Krave's campaign. Emma Jayne Evans posts, "iv hav them for breakfast and I dnt eat breakfast in years but eat these or a mc donolds breakfast well nice my fav [*sic*]." Diana Williamson reminds the group that Krave is "STILL only a pound in iceland," and Linda Mapplebeck happily confesses that Krave has "tempted me from a 5 year run of Weetabix for breakfast every day, absolutely delicious!"[80] They represent a level of publicity that money can't buy, and it seems unlikely that they and the other 39,500 followers were trained in online devotion and demonstration through the kind of fandom that academia still tends to study.

The model of user-generated content employed by Videomaker has expanded beyond niche genre fandom to reach a mass market, albeit in a modified form. What we see here is, I suggest, not the straightforward poaching of a user-generated content strategy from a relatively specialist minority (*Battlestar Galactica* followers) to a broad consumer base (Kellogg's cereal purchasers), but a convergence of strategies. Kellogg's is effectively borrowing from and simplifying the dynamics of fan communities in order to create an instant sense of brand belonging and engagement.[81] However, as will be seen below, the producers of cult television shows are now adopting very similar approaches, drawing on both the Videomaker model and its simpler breakfast-cereal equivalent to create campaigns that encourage easy, unskilled labor rather than extensive, original creative work. A model of casual, undemanding fan engagement—the invitation to create content from a quick game with ready-made toys and to join a community with little commitment or effort beyond signing up and performing online tasks—is increasingly becoming the dominant strategy.

The minority appeal of "Join the Fight" and the pleasures of belonging to an exclusive fan group are sustained through a project like Lost University,[82] another fictional institution that, like Dunder Mifflin Infinity, entices players to apply for a second life of tasks and rewards on top of their real-life responsibilities. A student taking real-world classes in Jewish studies and law at the University of Wisconsin–Madison, for instance, could spend her evenings studying PHI 101 ("I'm Lost Therefore I Am") or interpreting Egyptian hieroglyphics at the virtual Lost University, earning credits toward an imaginary "graduation." Enrollment for LU will presumably remain low; Lostpedia reports that "the course content is accessible to those who own the Season 5 Blu-Ray set."[83] However, mundane games of the same order, which immerse the player not in fantastic alternate existences but in simulations of second jobs, have also overwhelmed Facebook's far broader audience. Farmville, in which players tend land in exchange for the virtual currency of "farm coins," had 83 million players in April 2010. Other copycat games are catching up: Hotel City, for instance, had 13 million active players and Mafia Wars had 24 million at the time of writing. A model that encourages users to offer their participation in exchange for negligible forms of "currency" is therefore the central engine both of overflow simulations like Lost University, which reward a dedicated fan mentality, and mainstream Facebook games that attract millions of players internationally.

Further examples demonstrate the prevalence of this simple user-generated content model both in mass marketing and in promotional campaigns that attempt to create a cult following. During 2009–2010, ABC's *Flashforward* successfully persuaded viewers to record video confessions and submit textual testimonies of their future visions for its "Mosaic Collective" website, fleshing out the show's story world with the contributions of "real people."[84] User-generated content gives its website the fresh, unpredictable quality that a team of professional scriptwriters could never create, and encourages contributors to invest greater commitment in the show by becoming part of its mythos. At the mainstream end of the market, Oxo's 2009 "Oxo Factor" campaign relied on the free labor of British families who gamely acted out a script for the chance to become stars of an amateur advertisement "shown in the commercial break during the finale of the latest *X Factor*."[85] Released in spring 2010, both the ultraviolent comic book adaptation *Kick-Ass* and the family adventure *How to Train Your Dragon* invited website visitors to enter the diegesis, train with its characters through online games, and earn points with no currency outside the high-score table. Introducing the young viewers of a Dreamworks movie to this model of user-generated content and its negligible rewards for free labor is a form of training in another sense,[86] echoed by the Cartoon Network's integration of a "game creator" in its online supplement to the *Batman: The Brave and the Bold* cartoon.[87] The kids who build a Batman/Green Arrow platform game, supplying the website with content in return for a place on its leaderboard, are being caught and taught young.

The mistakes of the Videomaker campaign have been avoided, or learned from, through this less sophisticated model of user-generated content. ABC's *Lost* was, at the time of writing, encouraging viewers to vote for their favorite fan-created videos submitted to its new competition. "The winning finalist's *Lost* finale promo will be shown *on air* during *Lost: The Final Season*."[88] However, rather than complicating the issue by inviting original digital video, this "Ultimate *Lost* Fan Promo Contest" reassures participants that no technical skills are needed. "Show your passion for the show by creating a 35-second *Lost* finale promo video and you could see your work on TV! No need to be a video maven, use either the mash-up tools on ABC.com or your own editing software to put your 5+ years of *Lost* knowledge to good use!"[89] The shortlist of videos consists of clips arranged into various orders, each stamped firmly with the warning: "PROPERTY OF ABC STUDIOS." The invitation reaches beyond the niche culture of fan filmmakers (or "video mavens") who invested significant time, effort, and budget into the Videomaker project and expected a fair return. By extending the scope of "fan" to a broader, more casual viewer and redefining "promo video" to mean a knocked-together mash-up of official clips rather than a sustained original film, ABC kept a tighter control on the project and took less of a risk.

Even the British government and its opposition adopted user-generated content models in the months leading up to the 2010 general election: Labour used "the first political poster created by a member of the public" in its billboard campaign, while the Conservatives relied on the free labor of a public Twitter feed for their "Cash Gordon" website, offering points and the public recognition of a leaderboard in return for contributors' efforts (and therefore employing exactly the same model of reward as the *Kick-Ass* promotion).[90] Meanwhile, T-Mobile's 2009–2010 campaign enabled freelance musician Josh Ward to achieve his dream of a leading a super-group by inviting him to recruit musicians across the United Kingdom through social networking and texts, and then recording a commercial track.[91] While not every commentator was convinced—Charlie Brooker sneered at "simpering middle-class mop" Josh, his "utterly spontaneous and not-at-all-stage-managed musical quest," and the trend for "loser-generated content"—T-Mobile was still, quite clearly, aiming at that alchemy where corporate resources meet grassroots enthusiasm and spark into a kind of magic.[92]

Where, then, does this trend leave fan fiction? In the years since the very different failures of FanLib and Videomaker, the mediascape has shifted considerably. Producers increasingly seek user-generated content to build their brands, and fans—in the broad sense of the 83 million Farmville fans and the 39,500 fans of a new cereal—give so readily of their labor that we can hardly call it theft.[93] Academia must address this mainstreaming and redefinition of "fandom," where commitment and loyalty to a brand mean little more than clicking a "Become a Fan" button and working up through the structured ranks of a website game. But what of those individuals who remain fans in the old-school sense, whose engagement with a text involves a different league of effort, energy, skill, and emotional investment? Emma Jayne Evans, who announced that Krave was "well nice my fav," is unlikely to write stories or Photoshop avatars around her new choice of cereal; her attachment to Krave is, we have to assume, a minor aspect of her everyday life, and it may well fade when Iceland's special offer ends. Compare Evans's offhand Facebook praise with the deeply felt passion of Telesilla explaining her resentment toward FanLib:

> When someone comes from outside the community, makes a horribly bad pretence of giving a damn about us while showing no real understanding of the community, I get angry. This is our space, this is where we do what we do and where we meet the people we love and care about. You do not get to come into our space and tell us that we need what you can give us when what you want to give us is so clearly not in our interests and is not what a large number of us want.[94]

The fan community's sense of itself as a confidently self-contained space repelled FanLib's poachers in 2007. It continues to operate independently, self-sufficient and defiant, like a rogue green planet in an increasingly colonized system of shiny corporate worlds winking with advertising satellites and buzzing with traffic. And yet, after a year's operation, FanLib—despite the resistance to its project—had recruited 15,000 members. FanLib was mismanaged by a clumsy crew, but it served as a scout ship. There will, undoubtedly, be slicker, leaner, more intelligent approaches during the next decade. The fan fiction community now has an urgent choice to make: whether its separate sphere should tool up and monetize its practice before someone else does;[95] or whether it should stubbornly float free, refusing to engage with mainstream structures, sticking to its own currency, and maintaining an older notion of "fandom" at a time when the word's meaning may have shifted beyond its previous definitions. The dynamic whereby audiences generate free content in exchange for tokens and the promise of public or professional recognition is now becoming the dominant model. Fanfic authors may have to join the mainstream game on their own terms or watch hopelessly from the sidelines as a new crew of venture capitalists moves in and takes charge, playing to win this time.

Notes

Thanks to Denise Mann, Alexis Lothian, and Julie Levin Russo for their insightful comments on earlier drafts of this chapter.

1   Kate Madison, *Born of Hope*, http://www.bornofhope.com/Welcome.html.
2   Avi Santo, "From 'Heroes' to 'Zeroes,'" *In Media Res*, February 20, 2009, http://mediacommons.futureofthebook.org/imr/2009/02/20/heroes-zeroes-producing-fan-vids-without-fans.
3   Henry Jenkins, "*Star Trek* Rerun, Reread, Rewritten," in *Close Encounters: Film, Feminism and Science Fiction*, ed. Constancy Penley et al. (Minneapolis: University of Minnesota Press, 1991), 178. Will Brooker, *Using the Force: Creativity, Community, and Star Wars Fans* (London: Continuum, 2002), 134.
4   Henry Jenkins, *Convergence Culture: Where Old and New Media Collide* (New York: New York University Press, 2006), 155.
5   Julie Levin Russo, "Indiscrete Media: Television/Digital Convergence and Economies of Online Lesbian Fan Communities" (PhD diss., Brown University, 2010), http://01cyb.org/diss. Francesca Coppa, "A Fannish Taxonomy of Hotness," *Cinema Journal* 48, 4 (Summer 2009): 107.
6   Brooker, *Using the Force*, 130–131, 134, 174–175.
7   See Coppa, "A Fannish Taxonomy of Hotness"; Alexis Lothian, "Living in a Den of Thieves: Fan Video and Digital Challenges to Ownership," *Cinema Journal* 48, 4 (Summer 2009): 130–136.
8   Coppa, "A Fannish Taxonomy of Hotness," 107.
9   Quoted in Jenkins, *Convergence Culture*, 118. See also Will Brooker, "Living on

Dawson's Creek: Teen Viewers, Cultural Convergence, and Television Overflow," *International Journal of Cultural Studies* 4, 4 (December 2001): 456–472.
10 Jenkins, *Convergence Culture*, 118
11 Brooker, *Using the Force*, 169. See also Jenkins, *Convergence Culture*, 152–153.
12 Brooker, *Using the Force*, 177–178, 196–197. See also Jenkins, *Convergence Culture*, 154.
13 Suzanne Scott, "Repackaging Fan Culture: The Regifting Economy of Ancillary Content Models," *Transformative Works and Cultures* 3 (2009), http://journal.transformativeworks.org/index.php/twc/article/view/150/122; Henry Jenkins, "Transforming Fan Culture into User-Generated Content: The Case of FanLib," *Confessions of an Aca-Fan*, May 22, 2007, http://www.henryjenkins.org/2007/05/transforming_fan_culture_into.html.
14 Karen Hellekson, "A Fannish Field of Value: Online Fan Gift Culture," *Cinema Journal* 48, 4 (Summer 2009): 177.
15 See "FanLib," *FanHistory.com*, http://www.fanhistory.com/wiki/FanLib.
16 "I am not going to publicly discuss personal details about our employees," was CEO Chris Williams's final word on the subject. See Henry Jenkins, "Chris Williams Responds to Our Questions on FanLib," *Confessions of an Aca-Fan*, May 27, 2007, http://www.henryjenkins.org/2007/05/chris_williams_respond_to_our.html.
17 See "Article Summing up FanLib," May 20, 2007, http://icarusancalion.livejournal.com/626928.html.
18 Jenkins, "Chris Williams Responds."
19 E. M. Pink, response to Jenkins, "Chris Williams Responds."
20 Aja, response to Jenkins, "Chris Williams Responds."
21 Telesilla, "'You Come in Here with Your Guns and Your Brush Cuts . . .' (Some Personal Thoughts on FanLib)," May 22, 2007, http://telesilla.livejournal.com/558793.html (accessed April 2010).
22 Hellekson, "A Fannish Field of Value," 114.
23 Russo, "Indiscrete Media."
24 See also Suzanne Scott's comment that "importantly, the raw material offered to fans was primarily composed of clips of gun battles, Centurion robots, and ships careening through space—fodder that certainly targeted male fan filmmakers over members of the (predominantly female) vidding community." Scott, "Repackaging Fan Culture," para 3.3.
25 Jenkins, *Convergence Culture*, 155. Russo ("Indiscrete Media") confirms that "the genres that enjoy legal and corporate sanction," like effects-based tribute, documentary, and parody, "are disproportionately produced by men, while creative works that explore relationships between characters and 'expand the universe' are the almost exclusive preserve of women."
26 The new name was adopted in July 2009. The forums, accessed in April 2010, now use a "Syfy" URL.
27 Post by SyfyDigital, *Syfy Forums*, February 21, 2007, http://forums.syfy.com/index.php?showtopic=2265116.
28 Posts by SharonMustLive and LawnmowerFace, *Syfy Forums*, March 5–6, 2007.
29 Post by Derangedmilk, *Syfy Forums*, March 9, 2007.
30 Post by Roslin_is_Gorgeous, *Syfy Forums*, March 6, 2007.
31 Icarusancalion, "Article Summing up FanLib," *LiveJournal*, May 20, 2007, http://icarusancalion.livejournal.com/626928.html.
32 Post by Chrisdfw, *Syfy Forums*, July 6, 2007.

33 Post by PainThatImaUsedTo, *SyFy Forums*, June 2, 2007.
34 Post by Indiana Smith, *Syfy Forums*, June 22, 2007.
35 Post by Indiana Smith, *Syfy Forums*, June 27, 2007.
36 Post by GirlwGuns, *Syfy Forums*, March 14, 2007.
37 Post by Blackmoon21, *Syfy Forums*, March 28, 2007.
38 Post by PainThatImaUsedTo, *Syfy Forums*, June 2, 2007
39 Post by Markinetic, *Syfy Forums*, June 19, 2007.
40 Post by Tumbleweeder, *Syfy Forums*, June 13, 2007.
41 Post by TallyJC, *Syfy Forums*, June 20, 2007.
42 Post by JustASciFiFan, *SyFy Forums*, June 16, 2007.
43 Post by Dad, *SyFy Forums*, June 17, 2007.
44 Post by Tumbleweeder, *Syfy Forums*, August 23, 2007.
45 Post by Tumbleweeder, *Syfy Forums*, June 19, 2007.
46 Pine, response to Jenkins, "Chris Williams Responds."
47 See also Scott, "Repackaging Fan Culture," para. 2.10: "Male fans have historically sought professional status or financial compensation for their creative works more frequently than their female counterparts, and . . . fan practices deemed 'masculine' (game modding, fan filmmaking) are generally considered more viable as professional calling cards."
48 See Brooker, *Using the Force*, 175, for further examples.
49 Jenkins, *Convergence Culture*, 154.
50 Nicholas Abercrombie and Brian Longhurst, *Audiences: A Sociological Theory of Performance and Imagination* (London: Sage, 1998), 132, 133.
51 Ibid., 141, 142.
52 Ibid., 143, 145.
53 Matt Hills, *Triumph of a Time Lord: Regenerating Doctor Who in the Twenty-First Century* (London: I. B. Tauris, 2010), 56.
54 Sheenagh Pugh, *The Democratic Genre: Fan Fiction in a Literary Context* (Bridgend, Wales: Seren, 2005), 147.
55 Ibid. Note also the further implication of class and educational privilege here.
56 Ibid.
57 Ibid., 144.
58 Ibid., 148.
59 LiveJournal, once the center of female media fandom, has recently lost some of its traffic to Dreamwidth (since April 2009) and Archive of Our Own (since November 2009).
60 Hellekson, "A Fannish Field of Value," 113.
61 Pugh, *Democratic Genre*, 146.
62 Kristina Busse and Karen Hellekson, "Introduction: Work in Progress," in *Fan Communities in the Age of the Internet* (Jefferson, NC: McFarland & Company, 2006), 5.
63 Hellekson, "A Fannish Field of Value," 113.
64 See Francesca Coppa, "Writing Bodies in Space: Media Fan Fiction as Theatrical Performance," in *Fan Communities in the Age of the Internet*, ed. Busse and Hellekson.
65 See Eden Lackner, Barbara Lynn Lucas, and Robin Anne Reid, "Cunning Linguists: The Bisexual Erotics of Words/Silence/Flesh," in *Fan Communities in the Age of the Internet*, ed. Busse and Hellekson.
66 Kristina Busse, "My Life Is a WIP on My LJ: Slashing the Slasher and the Reality of

Celebrity Internet Performances," in *Fan Communities in the Age of the Internet*, ed. Busse and Hellekson, 208, 212.
67 See "Fans Turned Pro," *Fanlore*, http://fanlore.org/wiki/Fans_Turned_Pro. Of course, the fact that a significant strand of female fanfic involves queer pairings makes the transition from fan to pro more complex than is the case with straight action-adventure fan films.
68 Busse, "My Life Is a WIP on My LJ," 222.
69 Hellekson, "A Fannish Field of Value," 115.
70 Telesilla, "'You Come in Here'."
71 Hellekson, "A Fannish Field of Value," 114.
72 Icarusancalion, "Article Summing up FanLib."
73 Valarltd, post on *LiveJournal* blog of JustHuman, May 27, 2007, http://justhuman.livejournal.com/307483.html?thread=1420571#t1420571.
74 Chris Williams, post on LiveJournal blog of Telesilla (and elsewhere), May 17, 2007.
75 Derek Johnson, "Authorship Up for Grabs: Decentralized Labor, Licensing, and the Management of Collaborative Creativity." in this volume.
76 Post by CylonJefferson, *SyFy Forums*, April 4, 2008, http://forums.syfy.com/index.php?showtopic=2304555.
77 Posts by Baker, AmberLeo, and Spoonz, *Syfy Forums*, April 4–6, 2008.
78 See *Rent-a-Clown*, http://www.rent-a-clown.com/.
79 *Krave Choc Exchange*, http://www.krave.com/what.aspx.
80 Krave, *Facebook.com*, http://www.facebook.com/Krave?v=app_7146470109#!/Krave?v=wall.
81 On another level, its campaign is a new media version of traditional promotions, in which consumers would send in cereal box tops for gifts, badges, and club membership.
82 *Lost University*, http://lostuniversity.org/.
83 "Lost University" entry at *Lostpedia*, http://lostpedia.wikia.com/wiki/Lost_University.
84 *Mosaic Collective*, http://abc.go.com/shows/flash-forward/mosaiccollective.
85 *The Oxo Factor*, http://www.theoxofactor.com/.
86 I use the term deliberately to recall Walter Benjamin's phrase in *The Writer of Modern Life: Essays on Charles Baudelaire* (Cambridge, MA: Belknap Press, 2006), 191. See also my "Now You're Thinking with Portals: Media Training for a Digital World," *International Journal of Cultural Studies* 13, 6 (November 2010): 553–573.
87 *Batman: The Brave and the Bold Game Creator*, http://batmangamecreator.cartoonnetwork.com/.
88 *The Ultimate Lost Fan Promo Contest*, http://lost-promo-contest.abc.go.com/.
89 http://abc.go.com/shows/lost/fan-promo-contest.
90 See James Robinson, "David Cameron Depicted as Gene Hunt in Labour Poster," April 2, 2010, http://www.guardian.co.uk/politics/2010/apr/02/david-cameron-gene-hunt-labour-poster. *Cash Gordon*, http://cash-gordon.com/.
91 See Josh Ward on *MySpace*, http://www.myspace.com/joshward84.
92 "Charlie Brooker's Screen Burn," *The Guardian*, December 5, 2009, http://www.guardian.co.uk/tv-and-radio/2009/dec/05/charlie-brooker-screen-burn.
93 Compare Lothian, "Living in a Den of Thieves," 135.
94 Telesilla, "'You Come in Here'."
95 This is the suggestion made by Abigal De Kosnik in "Should Fan Fiction Be Free," *Cinema Journal* 48, 4 (2009): 123.

4
Labor of Love

• • • • • • • • • • • • • • • • • • • •

Charting *The L Word*

JULIE LEVIN RUSSO

The 2007 Writers Guild of America strike foregrounded the fact that labor, in both the institutional and the general sense, is an issue pivotal to current transformations in the entertainment industry. This dispute between screenwriters and executives illuminated the present-day predicament of mass media, which is hard pressed to keep up with a proliferation of content and platforms while squeezing ever greater efficiency out of its creative workers. These conditions have spurred not only the official exploitation of paid labor as expressed in the demands of the Alliance of Motion Picture and Television Producers (AMPTP) at the bargaining table, but also the industry's turn to a far more vast, dynamic, and affordable resource: the free labor of fans. Fan production has no doubt always held indirect economic value for corporations as a form of promotion and a stimulus to consumption, but until very recently, this phenomenon was rarely considered openly outside the science fiction niche. Now, as convergence puts pressure on television's obsolescing profit models, hit network shows like *Lost* (ABC, 2004–2010) and its derivatives are adopting cult media's tactics for attracting a loyal and engaged audience—in short, a fandom—as marketing's next frontier. In addition to aiding the presumptive value of active and insatiable consumers, the Internet's characteristics as a distributed, immediate, and continuous network make it practicable for the industry to mobilize fan labor directly as "user-generated content." At the same time, fans are able to expropriate media commodities directly, since television and movies, along with their multiplying complement of bonus features, can be downloaded at will to serve as the raw material for unauthorized creative work. Within this mainstreaming of the subcultural traditions of fandom, managing the production of queer readings, desires, and appropriations is a nexus of particular concern in the shift from broadcast's centralized and vertical model to the more decentralized and horizontal configuration of digital distribution.

The intimate relationship I propose between queer subjectivities and postindustrial capitalism is not arbitrary: as commodities become increasingly immaterial, the affective labor of desire, identification, and meaning-making accrues greater economic value. Paraphrasing a 1999 *Wired* article that boldly proclaimed the death of the "Old Web," Tiziana Terranova suggests that, with "new ways to make the audience work... television and the web converge in the one thing they have in common: their reliance on audience/users as providers of... cultural labour."[1] This labor, which is the productive force behind media convergence, exemplifies the architecture of the larger "digital economy," characterized by "a process of economic experimentation with the creation of monetary value out of knowledge/culture/affect."[2] Such relatively autonomous and freely conducted labor schemes—fan production included—break down the distinction between waged work and leisure; but this ambiguity does not place them outside of capitalist demands. In comparison to the sunny forecast for our much-vaunted "participatory culture," this view of convergence as incorporation may seem pessimistic: fandom is more commonly celebrated as a "gift economy" or alternative system of exchange that circumvents or even resists capitalism. Terranova argues that this outlook on free labor effaces the reality of its functional integration into the post-industrial economy. Her position does not, however, reduce fans and other digital enthusiasts to unwitting dupes of capitalism, colluding with the exploitation of their authentic practices by a monolithic machine. She emphasizes, by contrast, that "such processes are not created outside capital and then reappropriated by capital, but are the results of a complex history where the relation between labor and capital is mutually constitutive."[3] Given this interdependence, both the entertainment industry and its audiences have concrete demands and collective bargaining power in their immaterial labor negotiations.

These negotiations can take the form of punitive reactions in the guise of copyright enforcement and ideologies that devalue fan labor, but increasingly they also take the form of proactive enticements toward modes of participation that enrich the brand. Outside of cult and teen genres, one of the earliest forays into this terrain among television programs came from *The L Word* (Showtime, 2004–2009), the first American series to make lesbian romance its primary focus. In addition to thematizing issues of lesbian identity and representation onscreen, *The L Word* has innovated through online promotions that leverage its projected lesbian audience into an interactive fan community. At the intersection of lived subculture, virtual world, and marketing spectacle, the web-based tie-ins OurChart.com (a content portal and social networking site) and "You Write It!" (a platform for fan-written script contests) attempt to mobilize subjectivity as labor, exposing both the possibilities and the limits of such transmedia ventures. *The L Word* showrunner Ilene Chaiken has spoken of the push to dismantle television's fourth wall in the era of convergence: "In the beginning I said—and was given a very hard time for saying—'I don't listen, I write what I want to write.' But another way the world has changed since I started doing the show is that the internet has become a big part of our lives. Anybody who writes a TV show would be a fool not to interact with her audience. Our audience is particularly passionate and engaging, so I talk to them and I listen to them. I can't always do what they want to do, but there's an effect of hearing their voices and then deciding what stories to tell."[4]

Chaiken's growing willingness to listen and interact through the Internet is more than a nominal update to her job description. Implicit in her comments is the "L word" of her title: lesbian as a commodity that is produced as much by the "voices" of a "passionate" audience as by the program's own portrayals. There is thus another "L word" here, the one from my title: labor as an audience asset that the industry must now integrate. Both words—lesbian and labor—are taboo in the orbit of mainstream television, but, as rendered in the case of *The L Word*, both are central to key transformations in the mass media landscape. Here, I analyze the role of lesbianism as labor in *The L Word*'s commercial empire and, by extension, the role of subjectivity as labor in the emerging economy of convergence. My argument is that, even while more and more of fan production is subsumed into a capitalist topology, these conditions correspondingly intensify the underlying antagonism between audiences and corporations.

Immaterial Labor: Lesbians in Late Capitalism

An hour-long special created to air with the series finale of *The L Word* on Showtime (March 8, 2009) paid tribute to the program's heritage and legacy.[5] Producers and writers, cast members, minor celebrities, and an omniscient female narrator reflected on the significance of *The L Word* for lesbian representation—purportedly the culmination of years of broadcasting history, beginning with primetime television's first lesbian kiss on *L.A. Law* in 1991—and as a force for social change. Although the interviewees always returned to this refrain about the program's positive influence on gay equality at the level of the personal (by speaking to isolated or underprivileged youth) and the political (by portraying national issues like the military's "don't ask don't tell" policy and the lack of rights for same-sex parents), the special also reviewed some of *The L Word*'s more controversial and problematic narrative choices. Mixing contradictory discourses of inclusivity ("it's not about being gay, it's about being human," opined classical guitarist Sharon Isbin) and exclusivity (it's "a place of collective belonging" characterized by weekly viewing parties at lesbian homes and bars), this dialogue captured the dilemma of a niche show that must simultaneously appeal to a mainstream audience. Before *The L Word*, the fact that "lesbians on TV served more to titillate than to illustrate" was a common complaint; nonetheless, we should respect *The L Word* because it "unapologetically went 'all the way'" in its sex scenes to ensure that "straight people watched." By staking its very premise on the commercial viability of this overlap between the interests of gay and straight viewers, *The L Word*'s 2004 premiere heralded a moment when "lesbianism seemed poised for popularity." But according to the narrator, this alchemy did not come easily in the program's early seasons, as "its assumed audience... [of] lesbians felt the show had failed to deliver on its central promise: to represent the community in an accurate way."

The L Word's producers thus found themselves trapped between irreconcilable imperatives to be realistic and to be aspirational, to reflect lesbians authentically and to "break out of stereotypes" by rendering lesbians more conventionally appealing (with the latter leanings preferred due to the wider allure of glossy fantasy). One solution was to intervene in our cultural understanding of what constitutes "real" lesbianism. Amid criticism that the program portrayed only rich, beautiful, feminine women with no "Birkenstocks and flannel" in sight, for example, costume designer Cynthia Summers took it upon herself to "challenge the way lesbians think they should be looking or need to be looking to be able to be identified as 'a lesbian.'" These tensions—between normative and queer sexuality, between lesbian and mainstream audiences, between realistic and positive representations, and between portraying and fabricating a community—structured *The L Word*'s achievements and limitations throughout its six-season run.

My analysis of the ways that diegetic and metatextual labor served to support the program's commercial compromises is influenced by concepts from Autonomist Marxism, a poststructuralist hybrid with Italian origins (the movement's best-known proponent in the United States is Antonio Negri). This theory dovetails with axiomatic accounts of late capitalism by thinkers like Daniel Bell, David Harvey, and Fredric Jameson, but with particular attention to updating Marxist conceptions of labor relations and collective organizing for contemporary struggles.[6] Maurizio Lazzarato outlines the group's key ideas in an influential essay on "Immaterial Labor" that resonates with current conditions in the entertainment industry. According to Lazzarato's diagnosis, immaterial labor "seeks to involve even the worker's personality and subjectivity within the production of value," since, in a milieu that values intellectual property and brand image over the manufacture of material goods, subjectivity "becomes directly productive, because the goal of our postindustrial society is to construct the consumer/communicator."[7] As just one example, Negri writes: "Value exists wherever social locations of working cooperation are to be found and wherever accumulated and hidden labour is extracted from the turgid depths of society. This value is not reducible to a common standard. Rather, it is excessive ... [so] we must abandon the illusory notion of measurement."[8] Likewise, the Nielsen company's measurement of television ratings has been pushed toward an assortment of experimental metrics that aim to capture the excessive value of subjectivity and collectivity. Among them is *Hey! Nielsen* (launched in 2007), described by Susan MacDermid as "a new online social community, with ... features such as ratings (like Q Ratings), the ability to submit opinions and comments, to connect and to create a network of recommenders.... Its goal is to get fans rating, reviewing and blogging about their favorite shows, movies and stars."[9] By creating a social networking website in an attempt to mine qualitative data in communicative form, Nielsen acknowledged the unruly, unquantifiable character of late capitalism's immaterial commodities.[10] MacDermid quotes Nielsen executive Peter Blackshaw, who asserts that "understanding passion is the next frontier of market research ... we are paying very close attention to the root drivers and nuances around this level of emotion-charged consumer engagement."

"Passion" or "emotion"—colloquial cues for the desires that ground subjectivity—can become vital axes of the late capitalist economy because the most valuable labor and commodities are now immaterial. The ascent of the culture and information industries entails "the integration of the relationship between production and consumption, where in fact the consumer intervenes in an active way in the composition of the product," according to Lazzarato, rendering it "the result of a creative process that involves both the producer and the consumer."[11] Subjectivity and creativity are not as easily standardized and automated as widgets, though, and capitalism can only promote networked communication among workers and hope to superimpose on it some provisional mechanisms of control (like intellectual property law, for example). Thus Autonomism's pivotal argument is that labor, which is necessarily collective in organization and ubiquitous in scope, is not simply absorbed without resistance into the smooth space of capitalism, but rather negotiated through a process of struggle and antagonism with capitalism's perpetually insufficient procedures. By continuing to pry open the cracks in capitalism's containment of labor power, we can pressure it to innovate toward increasing accommodation of autonomous subjectivities. The concept of antagonism frames laborers, including fans, as a collectivity whose desires are not commensurate with those of a corporate system, and this recognition alone is a crucial corrective to the prevailing understanding of convergence culture.

For a melodrama driven by intimate relationships, the dimension of work may seem largely irrelevant to the narrative edifice of *The L Word*, a mere contrivance subordinated to its romantic intrigues. I argue that this apparent insignificance is in fact a symptom of the program's perfect rendition of the late capitalist transition to immaterial labor, wherein work is diffused throughout life and society. All of *The L Word*'s characters, insofar as their employment is represented onscreen, hold jobs in the services and cultural industries, the growth sectors in a post-industrial economy.[12] *The L Word*'s portrayal of their careers foregrounds immaterial skills that are becoming hegemonic under late capitalism: manipulating hierarchies of taste through hype and branding; leveraging personal connections and social networks; and communicating productively through various media channels. Moreover, in synergy with the genre of melodrama, the characters exemplify the interdependence of work and personal life, as their intimate relationships provide the material and the occasions for their professional advancement. To take this analysis even further, we might say that the characters on *The L Word* exemplify the importance of subjectivity itself as labor. It is their work on themselves (Jenny's identity crisis; Dana's coming-out process; Shane's ineffable style) and on their communicative capacities (Bette's taste-making; Tina's movie-making; Kit's community building) that makes them successful at their titular jobs. And for the purposes of *The L Word*, this labor is all concentrated in the production of "lesbian" as an economically meaningful category. Despite their occasional lip service against ghettoization, it is ultimately as lesbian professionals that the characters thrive in their careers, and thus they model lesbianism as an occupation. Of course, this portrayal is far from disinterested, since lesbian is also the category that works as *The L Word*'s eponymous brand. When Subaru hires Dana as a gay spokesperson for its "get out and stay out" ad campaign, the company is, in reality, paying Showtime for product placement that targets lesbian and gay-friendly viewers. Moreover, like Jenny's *Lez Girls* memoir and film, Chaiken's program banks on her experiences and credibility as a participant in lesbian communities. In a parallel that operates didactically, lesbianism is the *The L Word*'s privileged labor on both sides of the screen, as both its characters and its creators endeavor to render this identity lucrative in capitalist terms. Thus, if these characters are employed as lesbians textually, they are also employed as lesbians metatextually in that their job is to be spokeswomen for the program's trademark sexuality.

This representational strategy is more than a localized grab for a slice of the much-touted gay market, however; it is the larger regime of immaterial labor that makes it viable. With the general incorporation of subjectivity into post-industrial capitalism, cultural consumption (for example, watching television) is as important as the consumption of durable goods was in the industrial era. *The L Word*'s project to monetize a particular identity is one instance of today's new intensities of audience management, and the work of its characters or creators as lesbians echoes the work it asks of its viewers. Industrially, that is, what is productive for *The L Word* is not the willingness of its characters to take up the labor of lesbian identification and desire but the willingness of its viewers to do so. These viewers do not have to be lesbians, although that approximation is often convenient; but to want to watch (the desire that generates revenue for Showtime), they have to buy into the value of that position. The program promises various rewards that audience members might enjoy in exchange: the voyeuristic pleasure of seeing beautiful and often semi-nude women; the narrative pleasure of following a soap opera's relationship networks (posited as particularly characteristic of lesbian life); and the subcultural pleasure of participating in a recognizable community experience. But whatever their motivation, viewers must make a connection (however contingent or ambivalent) between themselves and *The L Word*'s manufactured lesbian subjectivity that sustains their involvement with the program. *The L Word*'s self-reflexive storytelling attempts to teach this vocation by example, through its object lessons in laboring to valorize lesbianism.

OurChart.com: The "Official Social Network" of *The L Word*

The L Word's most literal exemplar of a career in freelance lesbianism is Alice Pieszecki (Leisha Hailey), a bisexual-identified character who works throughout the series as a queer culture guru for media outlets, including *LA Magazine*, public radio station KCRW, and fictional television talk show *The Look*. Alice is certainly not the first queer woman to draw a diagram visualizing the complex web of hook-ups and break-ups that form the fabric of her community, but she is the first to make this graphic her trademark. The principle of her "chart" is introduced in the pilot episode when she plays a "six degrees of sexual separation" game with Dana, sketching out the serial couplings that connect the two of them with each other and with several friends. At the end of the scene, the camera tracks over their heads to frame a large bulletin board where Alice keeps a running tally of the links among her circle of acquaintances. But it becomes clear that the chart is more than a personal pastime for Alice. In the opening of the second episode, she pitches it to her editor as a marketable motif for an article: "The point is we are all connected, see? Through love, through loneliness, through one tiny lamentable lapse in judgment. All of us, in our isolation, we reach out from the darkness, from the alienation of modern life, to form these connections." Although her boss is unimpressed, Alice (or more properly, *The L Word*'s writing staff) here exhibits a savvy appreciation of the productivity of networked intimacy under late

FIG. 4-1 Alice Pieszecki's wall-sized chart (appearing here in *The L Word*'s opening credits) inspired virtual versions of this lesbian social network. (Frame grab)

capitalism. In a marked update from her initial pen-and-paper explanation, Alice demos the chart on her laptop using a graphics tablet. Only a few scenes later, she has implausibly launched a successful user-generated version online: "You know the chart? OK, I put it on the Internet. . . . This thing is growing. People are adding names, and it's growing exponentially." This vision of a web platform driven by relationships was prescient for its time (these episodes aired in January 2004, just before the inception of Facebook) and already signals the harmony between *The L Word*'s rendition of sexual community and the development of digital social networking.

While the network ethos of the chart is ever-present throughout the series, most notably in Alice's talk radio show based on the concept, the chart itself doesn't reappear until the beginning of season 4. In an eruption of metatextual instruction, Alice and Jenny introduce the character Helena to what is now a vibrant online community, telling her "it's so much fun, you don't know what you're missing! . . . It's like a social networking site—for lesbians." In Alice's opinion, the core feature of this diversified portal, now dubbed Our-Chart, is still its "hook-ups page," an interactive visualization of data on who has slept with whom. The graphics that represent this interface onscreen are artifacts of the program's technological imaginary, unrelated to any recognizable web browser or platform. Although Alice does describe in detail how to add a link by inviting someone to join, this scene's pedagogy is oriented more toward an ideology of transparency than tangible usage, hyping a fantasy of seamless equivalence between the sexual network and the online network. OurChart's discourse thus aligns perfectly with late capitalism's marriage of subjectivity and communication. The connectedness that Alice identifies as a hallmark of interpersonal relations in a sexual subculture is likewise a hallmark of the present-day organization of work, which depends increasingly on self-organizing cooperation facilitated by media and information technologies. *The L Word* capitalizes on these synergies by engineering a slippage between embodied participation in a web of love affairs and virtual participation in a digital archive of valuable network data. In this scene, for example, the mythology around promiscuous character Shane is transcoded from sexual into technical terms. As a "hub" of OurChart ("anyone who has slept with over 50 people," although in Shane's case the number is close to 1,000), she is instrumental in extending the site's pathways, much as her plentiful conquests link up more women in her community. Although the enterprise of representing this face-to-face social network online is not nearly so frictionless in practice as it appears here, the underlying logic equating sexual subjectivity with networked labor renders it coherent.

In contrast to season 1's more innocent reveries on the chart, the scene described above functions as an integrated promotion for the concurrent launch of the actual OurChart.com, itself a promotion for the *The L Word* in a sort of *mise en abyme* of transmedia branding. The tie-in website opened in January 2007, the same week as the season 4 premiere, but its interactive features were not up and running until several months later,[13] during which time the program's improbable vision hovered before fans as a self-fulfilling prophecy. Industry blogs reported that Chaiken, newly minted CEO of OurChart, confirmed that "the idea to migrate the chart to the Web grew out of a story line on the show. . . . Now, in the upcoming season, that character will realize that the chart has caught on. . . . At the same time, the real-world chart also will go live."[14] In the context of media convergence, with its intensifying interest in mobilizing viewership as immaterial labor, the harnessing of a real-world social network to work productively as an online social network is a predictable marketing strategy.[15] But OurChart.com, as portrayed within *The L Word*'s fictional Los Angeles, is symptomatic of the ideological payload conveyed in Showtime's move. The fantasy of an unmediated correspondence between subjective and digital layers can cover over the intercession of communication technologies and capitalist economies in the translation of the chart to the Internet. The site as rendered here is markedly unconstrained by funding or infrastructure; after Alice "put it on the internet," it just "caught on," with no apparent need for development, staff, advertising, or revenue. Moreover, beyond Alice's assurance that before someone can add one of her hook-ups to the chart the other party must opt in, the characters express no hesitancy over the alarming notion of porting intimate sexual histories into a searchable online database. These convenient erasures make OurChart.com formidable as a cutting-edge promotion precisely because it takes the *The L Word*'s economy of lesbianism as labor to its logical conclusion, enticing viewers-cum-users to work toward producing these values in more direct and centralized ways.

The implementation of the "chart" on OurChart.com materializes the many contradictions and insufficiencies that delineate *The L Word*'s ideology of commodity lesbianism. Much like the program itself, the website must find equilibrium between appealing to its niche fan base and to mainstream users and companies. But where the television series titillates to attract straight male viewers (among others), OurChart.com takes an opposing tack: it desexualizes its lesbian orientation in order to render it as a palatable assortment of consumer positions encompassing popular culture, chic style, and liberal politics. With unusual coyness for an *L Word* tie-in, the venture is billed as a "site where women can connect" ("About Us"), thus sidestepping queer sex by emphasizing an assumed gender stability that excludes male and transgender fans. Nowhere is the gap between OurChart.com's claims and its capabilities starker than in the failure of its hyperbolic promise to tell users who has hooked up with whom (which, according to the program's diegetic logic, has been the chart's primary impetus all along). Intuitive skepticism about the database project seems to prefigure its techno-cultural limitations, and these deficits are compounded by a conflict between the sexual archive concept and the site's move to advance a de-sexed brand of lesbianism. On *The L Word*, the imaginary OurChart's interface is portrayed as a navigable visualization of its entire user-generated record of intimate entanglements. On OurChart.com, by contrast, "friend" connections conveyed no more information than would be found on a typical online social network (send anyone a request, whatever your relationship may be, and the person can choose whether to approve it); furthermore, the Flash animation of the OurChart.com "chart" could display only about fifty of one user's friends in isolation.[16] In practice, it is not in the interest of Showtime or its target market to create a public record of a community's sexual history, and its equivalence between lesbian network and Internet network operates far better ideologically than technologically (at least as far as the "chart" graphics are concerned). But with the "hook-ups" feature diluted, it's more evident that the notion of the chart, a pivotal device in *The L Word*'s framing discourses, acts as an alibi: beneath the ideal of connection for the purpose of community building, Showtime is far more invested in the labor dimension of this diagram than the lesbian one.

Another fantasmatic equivalence in play at OurChart.com, beyond plotting a sexual network onto a technical network, is its conflation of onscreen and real-life communities. *The L Word*'s ultimate alibi is authenticity, and the website is a winning move in that rhetorical game: because real lesbians now chart their relationships just like the characters do, Alice and her friends evidently represent real lesbians. Thus OurChart.com not only advertises *The L Word* but buttresses its structuring ideology, leveraging user participation to heighten the verisimilitude of its portrayals. This device was not the program's first attempt to garner cultural credibility by layering behind-the-scenes narratives over its fictional soap opera. But in addition to amplifying the figurative parallel between production world and story world, OurChart.com provided a distribution channel for this ongoing stream of supplemental content. With regular submissions by Chaiken and actors including Beals, Hailey, and Moennig that promised fans insider access to *The L Word* empire and the opportunity to interact with its stars, OurChart.com enhanced the impression that the program engages an actually existing lesbian community (a role played here by the site's users). Blogs and videos by paid contributors augmented this prepackaged material and its subliminal creed of commodity lesbianism, with the implied assumption that, in order to appeal to *The L Word*'s audience, the website must be front-loaded for consumption.[17]

The corporate strategy underpinning OurChart.com follows a broader trend to position gays as a privileged marketing category. Pete Cashmore, founder and CEO of social network news portal Mashable, cites data suggesting that this assessment carries over to the Internet, where "gay, lesbian, and bisexual users are an extremely valuable demographic: social networks and blogs targeting this segment of the audience could perform well."[18] OurChart president Hilary Rosen parrots a similar doctrine, declaring that the site will "present marketers with a great opportunity to reach a consumer market that is targeted, financially independent and loyal"[19] and, later, that "the lesbian community is Internet-savvy and is twice as likely as heterosexual women to consider the Internet their prime source of entertainment."[20] Such mavens and, indeed, many of the analyses directed at the commodification of gay identity see this tendency in terms of an aptitude for consumption. The inference is that the web's primary innovation was to provide increased opportunities for advertising and sales. What the close relationship between *The L Word*'s onscreen representation and the online implementation of the chart demonstrates, however, is that the transition from broadcast to broadband enables an intensification of this dynamic that enlists what gay/lesbian demographics can produce as well as what they can consume, following the avenues that render identification and desire themselves productive in today's media economies.

And in fact, despite OurChart.com's heavy reliance on professional content to impose a consistent tone, its users did work. The site's social network was a lively one, with plenty of conversations, opinions, friendships, and no doubt hook-ups being forged beyond its "celesbian" encounters with *The L Word* stars. Because Showtime outsourced much of the labor of OurChart.com to its autonomous user base, the company could not guarantee that the subjectivities and discourses circulating there would conform to its intentions and interests. But if any such challenges occurred under the banner (and literally, the logo) of *The L Word*, could these unruly connections offer any significant disruption to the expropriation of users' work? Much like self-reflexive gestures incorporating fans' objections to the site's homogeneity into the program itself,[21] any unexpected, creative, critical, or even outright rebellious moments that erupted on OurChart.com played into the impression that the site was an authentic reflection of and platform for lesbian community. Simply by following the edict to "be subjects"—to desire, communicate, and invest immaterial commodities with meaning—fans are performing lesbianism as labor in accordance with *The L Word*'s teachings. The crucial fault line in this capitalist monolith, however, is that OurChart.com and similar user-generated online promotions do not capture the whole of this labor and its value: subjectivity is productive in excess of what a corporate schema can rationalize. Such negotiations between fan communities and the media industry are endemic to late capitalism, and given that both sides have their share of power in this milieu, the outcome of mediations between capital and fan laborers is far from a foregone conclusion.

FanLib's Challenge: "You Write It!"

The trend toward subsuming queer- and fan-oriented subcultures with relatively autonomous traditions into corporate regimes generates new antagonisms that demand delicate control. Beyond OurChart.com, *The L Word*'s most heavily engineered mobilization of fan production was a series of writing contests. Showtime launched this marketing campaign in 2006 with a scheme to prompt a complete "fanisode" (faux television script) and hired the company FanLib to design and run the web-based competition as one of the start-up's earliest projects.[22] For this initial contest, a member of *The L Word*'s writing team prepared a storyboard that filled in a diegetic gap of several months between the events of seasons 3 and 4, providing descriptions of the individual scenes that would make up an imaginary episode. Participants then voted for their favorite among the user submissions that realized each segment; and finally the winners were awarded prizes, and their scenes were assembled into a downloadable PDF version of the final script. This successful venture garnered a mention in the *Wall Street Journal*'s article about the transformation of fan fiction from a "fringe pursuit" to one that "helps unknown authors find mainstream success."[23] Because the fanisode wasn't intended for television production, we might speculate that it was organized in script format (as opposed to the more familiar prose fan fiction) precisely to appeal to aspiring screenwriters with polished skills.

Whether we read this move as nurturing or mercenary, it follows that certain expectations for a lesbian community of creative professionals are part of the impetus for *The L Word*'s FanLib promotions. In the introduction to the PDF e-zine that resulted from the fanisode, Chaiken celebrated *The L Word*'s fans, who "came at us enthusiastically with your reactions, your objections, your ideas, passions, preferences and opinions as to whether or not we were adequately and authentically representing the way that we live."[24] From the perspective of this politics of representation, which idealizes transparent portrayals of and by this categorical "we," it is necessary to the project of lesbian visibility that *The L Word*'s presumptively lesbian audience be encouraged to become involved with corporate media-making. However, as we have seen, the price of this brand of visibility is to render lesbian identity as a reified commodity that can be packaged and sold, not only by professionals but by each contest participant and each OurChart.com member. The feminist utopia of an "old girls' network," wherein mentorship leads to success within mainstream industries, here butts up against the converse heritage of fans' anti-commercial systems of value and recognition. Chaiken says that the writing competitions were inspired by the fact that "the fans of *The L Word* write a lot of fan fiction on their own,"[25] implying that submission of a scene in script form to a contest would have a comparable charm. But the majority of fan authors are not professional hopefuls like the winner interviewed in the *Wall Street Journal* (who was, incidentally, the only straight man to place in the fanisode competition). Chaiken's equivalence effaces the autonomous norms of fandom's gift economy, which cultivates alternative modes of sharing the characters and stories that originate in the corporate media. Meanwhile, it disavows the financial considerations underlying this opportunity to give advertisers "an exclusive shot at *The L Word* fans, since Showtime is ad-free ... [and] cut marketing costs ... [because] fans ... will write for next to nothing."[26] If, as the *Wall Street Journal* posits, "the rise of fan fiction is part of the spread of amateur-created content online ... on sites such as YouTube and MySpace,"[27] we should not expect ventures like FanLib's to negotiate the friction between capitalist mandates and "amateur" subcultures with any more consideration than these other commercial platforms.

Chaiken's statement is from a promotional video on Showtime's official website that presents a later FanLib installment (dubbed "You Write It!") featuring the lucky winner Molly as she claims her prize—a visit to the set to see her contribution filmed. "You Write It!" was structured like the fanisode competition, but its endgame made good on the promise that the victorious script would be included in an actual television episode (much to the delight of Molly, who was indeed a screenwriting student). It also had more open-ended instructions: "Choose a scene from *The L Word* seasons 1 or 2 to rewrite as a scene from 'Lez Girls,' Jenny's thinly-veiled, fictional account of *The L Word* characters' lives." While inviting fan-written scripts may imply a breakdown of the distinction between amateurs and professionals, this video's rhetoric emphatically reasserts the ideological gulf between fans and producers, quashing any intimation that fans' unpaid work could be afforded equal respect. The comments addressed to Molly, while well-meaning, are starkly condescending, apprising her of banal aspects of television production as if she didn't already have the knowledge to be a screenwriting success. The "You Write It!" contest was a perfect match with season 5's "Lez Girls," a movie-within-a-television-show that campily remixed *The L Word*'s early seasons. Molly's scene earned its winning vote tally by enhancing these self-reflexive layers with a *Charlie's Angels* mash-up, alluding to the history of lesbian viewing. In contrast to the discourses of "we" and "our" that characterize much of *The L Word*'s marketing, however, the turn to calling fans "you" highlights the restrictions on this openness to appropriation. Chaiken may profess an interest in "the way interactivity is taking over our lives," as borne out in *The L Word*'s cutting-edge online promotions, but this provocation extends only as far as fan labor channels value into the lesbian brand—because "you" work for free. Chaiken's outlook on the FanLib project both reflects and forwards this strategy, and like Jenny, Alice, or indeed Chaiken herself, Molly is an exemplar for fans' lessons in commodifying our passions.

In contrast to *The L Word*'s relatively harmonious deployment of OurChart.com and the FanLib contests as successful user-generated, fan-driven, for-profit corporate promotions, the infamous descent of the company FanLib stands as an object lesson in mishandling the exploitation of fan labor. Beginning in 2003, the start-up licensed custom software in order to run online fan writing contests for entertainment concerns (other than Showtime, HarperCollins Publishers was a notable customer). In addition to these commissioned projects, FanLib launched a commercial fan fiction archive in 2007, offering its industry partners the opportunity for "integrated customized marketing . . . capitalizing on existing communities around media."[28] To build interest in the site, the company issued flattering invitations to visible influencers and prolific writers in fandom; but as the fans they courted started investigating the business behind the e-mails, they were angered by the impression that it was instigated by outsiders and motivated by profit. Henry Jenkins summarized the facts that emerged in this grassroots probe, which sent FanLib's image and credibility among its target users into a downward spiral: "FanLib was emphatically not going to take any legal risks on behalf of the fans here . . . [only] providing a central portal where fans could go to read the 'best' fan fiction as evaluated by a board of male corporate executives . . . [who] talked about making fan fiction available to 'mainstream audiences.'"[29] In response to FanLib's mishandling of its appeal to the established fanfic community, this community organized to publicize its objections, reassert its values, and advocate for its interests. On LiveJournal, a group called "Life Without FanLib" was set up to track the issue and host the firestorm of discussion that followed. FanLib had promised to produce "consumer-generated media that is ready for the marketplace," one reporter wrote. "The result: More value for marketers, more manageability for producers."[30] The company's executives found that it was not as easy to commodify, monetize, and manage this surplus labor as they had speculated.

To FanLib's architects, the vast commons of freely exchanged fan works perhaps appeared to lack savvy businessmen who could repackage them as commercial promotions. But in fact, creative fandom has a rich tradition of conceptualizing the value of its labor by rejecting financial profit. For this reason, fan production is often understood as a women's "gift economy" or, in the words of fan scholar Karen Hellekson, a "gendered space that relies on the circulation of gifts . . . that deliberately repudiates a monetary model (because it is gendered male) . . . to permit performance of gendered, alternative, queered identity."[31] The problematic assumptions bound up with FanLib's emphasis on "mainstreaming"—which, as Hellekson suggests, map onto patriarchal and heteronormative coordinates of gender and sexuality—seemed to persist in the company's willful ignorance of its repugnance to many fans. One of FanLib's ads vividly illustrates the clash with the feminist and queer ethos that delineates the fan fiction subculture in question. The "Pink Guy/Blue Dude" image, which figured "Life without Fan Fiction" as a skinny, nerdy boy and "Fan Fiction at FanLib.com" as a muscular, shirtless man, implied that FanLib's corporate model masculinizes an activity that is otherwise markedly effeminate. This assumption offended a predominantly female community that nurtures alternative and perverse expressions of gender and sexuality and raised ire at the insinuation that FanLib's macho brand of commodification is the only legitimate way to envision fanfic. Fandom's response was to form, through grassroots mobilization online, a nonprofit organization with the mission of protecting the autonomy of creative fandom as an anticommercial, egalitarian commons. As for FanLib, its archive was shut down in advance of a buyout by Disney in 2008,[32] no doubt rendering the company a success in the owners' eyes, whether or not the site was able to recoup its $3 million in venture capital.[33] FanLib operated on the assumption that fans' labors of love have the same goals, motivations, standards, and economies as professional authorship—although in FanLib's business model, the corporation rather than the creators would reap the profits. Before fans either reject or embrace such terms for participation in the media economy, we should assess our structural position within this system as workers.[34]

The cover of *The L Word*'s fanisode e-zine features a photograph of the show's cast posed around a bed frame on a deserted beach, draped in satiny, revealing garments and staring vacantly out at their assumed audience. We could take this image as a metaphorical portrait of the network's vision of fan community: a neatly assembled, perfectly groomed, politically isolated demographic frozen in their consumer rictus. In its online promotions, *The L Word* constantly reasserts its own commodified portrayals as the outline of fan labor,

FIG. 4-2 Despite its fan-written content, the cover of the "You Write It!" e-zine pictures *The L Word*'s cast and is branded by Showtime and FanLib. (Publicity material)

demonstrating the limits of its gestures toward participatory engagement. But media fandom manifests alternative aspirations to queer female community that more concertedly oppose these capitalist contours. FanLib's gambit to harness fans' labor in a commercial archive foregrounded certain underlying constraints of online fandom, namely, its reliance on websites and infrastructure controlled by corporations. As a response, a watershed LiveJournal post by Astolat called for "An Archive of One's Own" that could materialize fandom's values of autonomy, openness, collectivity, and gifting in a platform owned and run by fans. Her manifesto catalyzed a grassroots campaign to lay the groundwork for this project, headquartered in the LiveJournal community "fanarchive" (later renamed "otw_news"). Barely a month after Astolat's provocation, a board of directors convened to plan the launch of a nonprofit, the Organization for Transformative Works (OTW), to advocate for the interests of fan producers. The Archive of Our Own (AO3) launched in October 2008 and reached open beta in November 2009. This mutiny coalesced because it had become essential for the community to react not only to FanLib but also to more widespread pressures on fandom's labor relations prompted by the industrial innovations of convergence. Companies' escalating interest in exploiting the work fans do via their creations, discussions, and desires has thus met with resistance—not necessarily to capitalism as a totality, but certainly to its unilateral imposition of new working conditions.

The conclusion to the saga of OurChart.com illustrates once again the vulnerability of fan communities when they rely on corporately controlled infrastructure, confirming the importance of efforts like the OTW's to advocate for the autonomy of fan labor. The site was shut down abruptly in January 2009, vaporizing the contributions and connections created by its active network of users. In Chaiken's farewell blog entry, which gave one week's notice of the closure, she wrote that "Showtime is not only OurChart's parent but one of Our Community's greatest champions . . . that's why in our final season of *The L Word*, we've decided to combine forces and host OurChart on sho.com."[35] This explanation was disingenuous, since hosting OurChart on Sho.com meant, in reality, that all the collectively generated content of the social network "chart" disappeared. Sho.com simply offered authorized tie-in content with token gestures of interactivity, such as "Q&A [with Chaiken] . . . behind-the-scenes podcasts . . . video specials . . . message boards . . . swag" and an "official" wiki. In a feeble attempt to continue a social media strategy, the star feature of Sho.com's OurChart page was a text box that allowed fans to post questions for Chaiken directly to an unmoderated Twitter account, perhaps an inadvisable move since it was immediately inundated with exclamations of outrage by OurChart.com members. Their outcry was in vain, however. Public information about why the site folded is slim, but it seems likely that, with *The L Word* entering its final season, the promotional value of OurChart.com was largely exhausted, and Showtime thus eliminated its funding.

The lesson for new media marketers is that, even though fan communities encompass a wealth of productive labor, very little of this labor can be monetized directly. Only this profitable surplus is of interest to corporations, but the subjective and collective desires in excess of this expropriation are what sustain the dynamic productivity of fandom. Autonomy is thus vital to the very processes of valorization that the industry is increasingly eager to exploit. The lesson for fans is that, if we depend on proprietary platforms like OurChart.com, our creativity and community will remain at risk until we fully conform to capitalist dictates. In this context, the relationship of queer fan production to media convergence is embroiled in double binds: would we prefer to end up marginalized or assimilated, unpaid or commercialized, subculture or target market? In addition to the stakes of defining the "our" that reverberates between OurChart and Our Archive, then, we might ask for whose interests "we" agitate from a Marxist, feminist, and/or queer perspective. Queer female fan practices embody an opportunity to galvanize antagonism within the industrial transformations in progress, and understanding, engaging, and defending the autonomy of these collectives will, I argue, contribute to everyone's freedom to labor queerly.

Notes

1. Tiziana Terranova, "Free Labor," in Terranova, *Network Culture: Politics for the Information Age* (London: Pluto Press, 2004), 95.
2. Ibid., 80.
3. Ibid., 94.
4. Neil Wilkes, "Q&A: 'L Word' Creator Talks Final Season," *Digital Spy: Tube Talk*, June 16, 2008, http://digitalspy.com/tv/tubetalk/a100754/qa-1-word-creator-talks-final-season.html.
5. Although the special was created to commemorate *The L Word*'s finale, it conspicuously avoided any discussion of the final season, an incoherent fiasco that was reviled by fans and critics. Apparently conceived more as an extended promo for Chaiken's unsuccessful spin-off series (a prison drama called *The Farm*) than as a consistent conclusion to the characters' narratives, season 6 melds a murder mystery into the program's soap-operatic format. The final episode withholds the promised resolution to this whodunit, however, retreating instead into maudlin reminiscences, complete with a diegetic tribute video that mirrors the extra-diegetic tribute special.
6. See Bell's *The Coming of Post-Industrial Society* (1973), Harvey's *The Condition of Postmodernity* (1989), and Jameson's *Postmodernism, or, The Cultural Logic of Late Capitalism* (1991).
7. Maurizio Lazzarato, "Immaterial Labor," in *Radical Thought in Italy: A Potential Politics*, ed. Paolo Virno and Michael Hardt (Minneapolis: University of Minnesota Press, 2006), 135, 142.
8. Antonio Negri, *The Politics of Subversion: A Manifesto for the Twenty-first Century* (Boston: Polity Press, 2005), 91–92.
9. Susan C. MacDermid, "Nielsen Moves from Measurement to Influence," *iMedia Connection*, January 7, 2008, http://imediaconnection.com/content/17833.asp.
10. Of course, the difficulties and contradictions of Nielsen's charge don't begin and end with the Web. In *Desperately Seeking the Audience* (New York: Routledge 1991), Ien Ang argues that the chaotic totality of audience characteristics and viewing practices will always exceed "desperate" attempts to fix them as quantifiable data.
11. Lazzarato, "Immaterial Labor," 141.
12. Notable examples include: Bette Porter (Jennifer Beals), art curator and administrator; Tina Kennard (Laurel Holloman), movie studio executive; Jenny Schecter (Mia Kirshner), memoirist; Shane McCutcheon (Katherine Moennig), hairstylist; and Dana Fairbanks (Erin Daniels), professional tennis player and Subaru spokesperson.
13. Pete Cashmore, "OurChart.com, Socially Networked," *Mashable*, March 30, 2007, http://mashable.com/2007/03/30/ourchart.

14. Wendy Davis, "'L Word' to Launch Social Networking Spin-Off," *Online Media Daily*, December 19, 2006, http://mediapost.com/publications/index.cfm?fa=Articles.showArticle&art_aid=52735.
15. See Candace Moore, "Liminal Places and Spaces: Public/Private Considerations," in *Production Studies: Cultural Studies of Media Industries*, ed. Vicki Mayer, Miranda Banks, and John Caldwell (New York: Routledge, 2009). Moore suggests that one motivation behind OurChart.com, like Nielsen's social network (as discussed above), is veiled market research. While "queer female cyber-identities are 'charted' (i.e., organized) on the site, and thus made ever more accessible to Viacom, the conglomerate that owns Showtime Networks, as a market demographic," it is equally true that identity is not so easily rationalized, since here "anyone can declare him- or herself a 'lesbian,' or indeed a 'friend of'" (134).
16. In a minor concession to the original idea, a second type of connection was added later, dubbed "friends plus." The site defined this modality in the vaguest possible terms, with no mandate that it involve a sexual entanglement: "We've created friends plus for everyone who's more-than-just-a-friend: exes, one-night stands, long-term partners, and any other players in your own personal dyke drama."
17. According to OurChart's blub on Facebook, the site's "blogging team delivers the latest on culture, sex, politics, lifestyle and entertainment. Our writers include Sirius radio host Diana Cage, Rigged fashion designer Parisa Parnian, playwright Lenelle Moise, 'My Husband Betty' author Helen Boyd, comedian Gloria Bigelow, comic book artist and *L Word* scribe Ariel Schrag, *New York Times* photojournalist Angela Jimenez, performance poet Staceyann Chin, and personal trainer to the *L Word* stars, Leah Beckingham." Among its video content, OurChart.com featured an original dramatic web series by Angela Robinson, *Girltrash*.
18. Pete Cashmore, "MySpace Users Slightly More Gay Than Facebook Users," *Mashable*, January 3, 2007, http://mashable.com/2007/01/03/myspace-users-slightly-more-gay-than-facebook-users.
19. Showtime Announcements, "*The L Word* Creator and Executive Producer, Ilene Chaiken, to Launch Venture to Extend the Well-Known Brand into Online Social Networking World," December 18, 2006, http://sho.com/site/announcements/121806ourchart.do.
20. Anne Becker, "OurChart to Preview *The L Word* One Week Before Showtime Debut," *Broadcasting & Cable*, December 5, 2007, http://broadcastingcable.com/article/111486-OurChart_to_Preview_The_L_Word_One_Week_Before_Showtime_Debut.php.
21. In the season 5 premiere, for example, transgender character Max confronts Alice (now an executive of the fictional OurChart) about her insistence that "OurChart is for lesbians," saying, "I thought OurChart is for everybody. It's *our* chart, doesn't that suggest it's inclusive?" When Alice grudgingly concedes that Max can write a featured blog about his transition, it appears as if OurChart simply offers a neutral forum where the lesbian community can air existing tensions rather than acknowledging how the site might aggravate those tensions.
22. FanLib's websites no longer exist, but an archived version of its *L Word* project can be found at http://web.archive.org/web/20060831222949/ http://lword.fanlib.com/.
23. John Jurgensen, "Rewriting the Rules of Fiction," *Wall Street Journal*, September 16, 2006, http://online.wsj.com/public/article/SB115836001321164886-CledovmXod4MomDQQvEU9VSfC6I_20070917.html.

24 Ilene Chaiken, "A Word from Ilene Chaiken," in "*The L Word*: A Fanisode," special commemorative e-zine (New York: FanLib, 2006), 4.
25 "Meet Molly," *Showtime Video*, http://sho.com/site/video/brightcove/series/title.do?bcpid=1304999811&bclid=1374480000.
26 Jon Fine, "Putting the Fans to Work," *BusinessWeek*, March 13, 2006, http://businessweek.com/magazine/content/06_11/b3975037.htm.
27 Jurgensen, "Rewriting the Rules of Fiction."
28 Kristen Nicole, "FanLib Emerges from Stealth Mode with $3M in Funding and Big Media Sponsors," *Mashable*, May 18, 2007, http://mashable.com/2007/05/18/fanlib/.
29 Henry Jenkins, "Transforming Fan Culture into User-Generated Content: The Case of FanLib," *Confessions of an Aca/Fan*, May 22, 2007, http://henryjenkins.org/2007/05/transforming_fan_culture_into.html.
30 Mary McNamara, "Internet Goes Nova over Showtime, Starz, Moonves Partnered FanLib.com," *Multichannel News*, May 28, 2007, http://multichannel.com/blog/TV_Crush/ 7482-Internet_Goes_Nova_Over_Showtime_Starz_Moonves_Partnered_FanLib_com.php.
31 Karen Hellekson, "A Fannish Field of Value: Online Fan Gift Culture," *Cinema Journal* 48, 4 (2009): 116.
32 Rafat Ali, "Disney's Buyout of FanLib Still On; Will Focus on Company Shows," *paidContent*, August 5, 2008, http://paidcontent.org/article/419-disneys-buyout-of-fanlib-still-on-will-focus-on-company-shows/.
33 Leva Cygnet, "On the Demise of Fanlib, and Why Fan-Run Sites Are More Likely to Succeed," *Firefox News*, July 24, 2008, http://firefox.org/news/articles/1677/1/On-the-Demise-of-Fanlib-and-Why-Fan-run-Sites-Are-More-Likely-to-Succeed/Page1.html.
34 Abigail De Kosnik has questioned the merits of a "gift economy" orientation, expressing legitimate concern that fans risk "waiting too long to decide to profit from their innovative art form, and allowing an interloper to package the genre in its first commercially viable format." "Should Fan Fiction Be Free?" *Cinema Journal* 48, 4 (2009): 120.
35 Renee Gannon, "OurChart.com Says Goodbye Forever," *Lesbiatopia*, January 19, 2009, http://lesbiatopia.com/2009/01/ourchartcom-says-goodbye-forever.html.

5

The Labor Behind the *Lost* ARG

• • • • • • • • • • • • • • • • • • • •

WGA's Tentative Foothold in the Digital Age

DENISE MANN

At a Society for Cinema and Media Studies (SCMS) workshop in March 2010, *Lost* showrunner Carlton Cuse observed that, given the rapid rate of change in the entertainment industry, the innovative series might already be part of network television's past. "We're like blacksmiths in the Internet era," he mused. "We're making a show that I'm not sure will ever be replicated given the tremendous resources we used. . . . So we are dinosaurs that are dying on May 23 [the date of the series' finale]."[1] The now completed *Lost* franchise (2004–2010) provides a useful means of examining the state of the post-network television industry in the digital age, given its creators' groundbreaking use of the web to craft digital expansions of its storylines as well as their use of blogs, alternate reality games (ARGs), and mobisodes to promote the series.[2] Furthermore, the complexity of the multi-platform, multi-genre, cross-promoted franchise prompted Disney-ABC's decision to use Apple iTunes, ABC.com, and other portals to stream the series digitally. Despite its arguably "prehistoric" status, the *Lost* franchise was influential on many fronts, and its trajectory sheds light on creative labor and the compensation and rights of production personnel who work on network television series and who are exploring transmedia storytelling. *Lost* showrunners Damon Lindelof and Carlton Cuse used their considerable leverage in the days and weeks before the Writers Guild of America (WGA) strike to engineer an agreement between the Alliance of Motion Picture and Television Producers (AMPTP) and the WGA.

This production study focuses on the insights ABC gleaned by engaging in creative partnerships with the *Lost* writers during the making of "The Lost Experience" ARG and speculates on the reasons the network later abandoned this type of intensive collaboration with creative personnel in the aftermath of the WGA strike. "The Lost Experience" experiment in interactive media marketing on a global stage began on May 3, 2006, in the United States (and a few days later in the United Kingdom and Australia) and ended on September 24, 2006; it was designed to keep fans invested during the summer hiatus between seasons 2 and 3. The production circumstances attending the making of "The Lost Experience" highlight the confusing overlap of job responsibilities that made it difficult for the WGA to secure payment for writers of derivative digital content. With so many corporate partners involved in making complex, demanding formats like the ARG, it is nearly impossible to discern what portion of the game is story-driven and what portion is promotional. To begin to tease out such issues, this essay relies on interviews and site visits with several members of the *Lost* writing team (Cuse, Lindelof, Javier Grillo-Marxuach, and Jordan Rosenberg), two members of the WGA (Chuck Slocum and Tamara Krinsky), members of the ABC marketing team (Mike Benson and Darren Shillace), and the head of Disney-ABC Digital Media (Albert Cheng). In addition, I examine industry trade materials describing the business and creative roles of the advertising agency Crispin Porter + Bogusky, a Florida-based innovator in the interactive advertising space, which created the popular "Subservient Chicken" ad for Burger King and the visually arresting "Sublymonal.com" campaign for Sprite (one of the four main sponsors of "The Lost Experience"). A textual analysis of the "Sublymonal.com" campaign demonstrates the sponsors' dependence on the television writers to guide their advertising agencies and design firms, which lacked experience crafting these types of cohesive, interactive story worlds. At issue in this essay is whether the resolution reached by the WGA during the strike (which lasted from November 5, 2007, to February 12, 2008) adequately describes and protects the rights of television writers, given their crucial role in crafting these new forms of branded entertainment designed for television in the Internet age.

The concessions made regarding the digital rights of television writers primarily involved mobisodes linked to series. However, in the months and years following the strike, rather than embrace the mass collaborative approach to television explored during the *Lost* moment, the "big three" networks appear to have bolstered their traditional bureaucratic fortresses to maintain singular control over all aspects of the broadcast business. In particular, ABC fired several of the principal executives involved in the *Lost* franchise and reabsorbed many of the functions that the *Lost* writers had performed in collaboration with the network's programming, marketing, licensing, and merchandising divisions. Since 2011, the big three have reverted to more conventional programming choices (reality shows, sit-coms, episodic dramas), reasserted control over their licensed properties (computer games, novels, board games, and the like), and expanded their in-house digital marketing divisions to create and manage digital promotions tied to their series—all, it seems, in an effort to maintain stricter controls over their industry in the post-strike environment. When asked in 2012 if his division at ABC would support "The Lost Experience" today, Albert Cheng, the vice president in charge of the Disney-ABC Digital Media division, replied, "Not likely. It was not cost-effective at the time. It really didn't serve the show that much. It was a lot of effort, requiring management's time, a lot of the producers' and writers' time, but at the end of the day, it wasn't worth it.... Those things tend to be supportive of the super-fans and not serve the main audience. In the future, we have to think critically about whether the effort will truly grow an audience as opposed to meeting the needs of those who are already the converted."[3]

Laboring under False Pretenses

The two *Lost* showrunners took a leadership role in arbitrating how the WGA should negotiate credit and payment for the television writers who create derivative digital content tied to their series. Mounting confusion over whether the networks or the writers should receive credit for creating digital content that doubles as promotions contributed to the breakdown in talks between the AMPTP and the WGA. In the early days of the strike, the WGA West published new media guidelines that spoke directly to this issue: "*Promotional use does not make it free*. Many of these new materials are being justified by the companies by their value in promoting a series. But, unlike the on-air promotion or ad copy of the past, these new forms also cross over the line into content. They are entertainment. This means they must be paid for as WGA-covered writing, including residuals for reuse."[4]

In April 2007, after ABC marketers teamed with the writers to launch "The Lost Experience," the first ARG linked to a major television series, Cuse and Lindelof approached Mark Pedowitz, then head of Disney-ABC Entertainment Group, to help turn the next *Lost* derivative production—"The Lost Diaries" mobisodes (sponsored by Verizon)—into a union production. Acknowledging how difficult it is for the rigidly bureaucratic networks to implement change, Cuse credits Pedowitz for "mov[ing] mountains within the ABC/Disney culture to make it happen."[5] The decision provided a boost for the labor guilds, which had seen their power dissipate in the previous two decades owing to consolidation and deregulation of the big media groups and the rising popularity of a do-it-yourself amateur aesthetic, which prompted big media companies to favor non-union "digital sweatshops" over guild-represented skilled labor.[6] Knowing that they held the advantage because ABC wanted the high-profile *Lost* actors to appear in the webisodes, Lindelof and Cuse insisted that the directors and actors as well as the writers be guaranteed the minimum salaries established by their respective guilds for new media.[7] In a 2010 interview, Cuse explained the significance of the deal ABC-Disney struck with all the guilds (WGA, Directors Guild of America [DGA], and Screen Actors Guild [SAG]): "The LOST deal became an important template for the first new media deals that were negotiated into the guilds' overall agreements with the AMPTP. I was a member of the WGA Negotiating Committee and helped in this process as well. It was of critical importance to establish that writers had the rights to residuals in new media. I think the concessions we won will be a very important source of income for future generations of writers."[8]

While the powerful *Lost* team was busy exerting its considerable influence to assert the rights of above-the-line creative personnel, controversy was surfacing in writers' rooms across Hollywood as all WGA members—not just the television writers—made a series of demands. These included higher DVD residuals, union jurisdiction over animation and reality program writers, and compensation for new media. The latter demand—both payment for original content designed for the web and residuals for reuse of traditional television series content via digital download—became the symbolic center of the strike, capturing the attention of both the blogosphere and the mainstream press. One of the major sticking points was the price to be paid for reuse of traditional television series when they were downloaded on the web. There are two primary categories of digital download: consumer paid versus ad supported. Consumer-paid downloads follow two models: e-rental via a monthly subscription to Netflix or to a cable carrier like Comcast; or electronic sell-through (EST) via a company like iTunes, whereby consumers pay $1.99 per episode and up to $34.99 for an entire season. No advertising appears in these versions. In contrast, ad-supported digital downloads from network websites and Hulu.com (which has controlling interest from NBC-Universal, Fox, and ABC-Disney) require consumers to watch ads before the program starts streaming.

Eager to provide audiences with a legal way to catch up on missed on-air episodes, ABC played a leadership role in both areas of online distribution starting in 2005. By 2008, ABC had sold "nearly 35 million TV show episodes via iTunes," according to Karen Hobson, vice president of corporate communications at Disney-ABC Television Group. At the same time, the network allowed viewers to watch "free" downloads of *Lost, Desperate Housewives,* and

Grey's Anatomy via the ad-supported ABC.com.⁹ Chuck Slocum, a lead negotiator for the WGA, maintains that ABC's refusal in 2005 to contact the guild before implementing these digital distribution platforms represented the first volley leading to the strike two years later.¹⁰ When pressed to compensate the writers, ABC offered the lower home video rates instead of the more appropriate pay television rates requested by WGA. Home video residuals were already a sore topic for WGA members because the studios were still paying writers the rates established in the late 1980s, even though DVDs were much cheaper to manufacture than videocassettes. As a result, the writers had not benefited from the surge in revenues that the studios received for DVD sales/rentals in the 1990s and first half of 2000.¹¹

Cuse and Lindelof were not the only showrunners to use their considerable clout to support the cause of rank-and-file writers. Shonda Rhimes (*Grey's Anatomy, Private Practice*), Steve Levitan (*Back to You*), Shawn Ryan (*The Shield, The Unit*), Greg Daniels (*The Office*), and several other high-profile series writers-producers joined the picket-line despite their status as members of both the WGA and management.¹² Notably, Greg Daniels and several of the actor-writers on *The Office* filmed themselves on the picket line and posted the footage ("The Office Is Closed") on YouTube as a way to educate the public about the paradox of writers earning nothing for writing the online content for their first web series ("The Accountants") while the networks continued to sell ad time even during the strike.¹³ They instructed viewers: "Watch this video. But don't click on the ads."¹⁴ Despite the writers' apparent victory, WGA President Patric Verrone's comments seem overly confident in retrospect: "These advances now give us a foothold in the digital age." He added, "Rather than being shut out of the future of content creation and delivery, writers will lead the way as television migrates to the Internet."¹⁵ In fact, since 2008, more and more studios and networks have created digital marketing divisions and have employed low-cost labor to generate content-promotional hybrids in house.

The *Lost* Experiment in Mass Collaboration

The following analysis of "The Lost Experience" ARG integrates production study and audience study methodologies to underscore the complexity of this highly collaborative, branded entertainment experiment. The four groups examined are: (1) the *Lost* television series writers; (2) the ABC network marketers; (3) the advertisers (focusing on Sprite's advertising firm, Crispin Porter + Bogusky); and, by inference, (4) the ABC network licensors. Special attention will be paid to the writers' interactions with each of the other groups as well as their interactions with fans, who are the target audience for these online promotional experiences.

When the ABC marketing team first approached the *Lost* head writers to collaborate on "The Lost Experience" ARG, the two groups had disparate goals in mind. The writers saw social media (Facebook, Twitter, wikis, and so on) as valuable tools to enhance the traditional television experience; they could plant clues inside digital platforms and invite highly invested fans to use the mind-hive of the web to resolve any of the narrative ambiguities present in the broadcast series without making these digital digressions a requirement for the mass audience.[16] In contrast, the marketers saw "The Lost Experience" and other web-based experiments as a more effective form of grassroots promotion that used social media to appeal to a younger audience where they lived—online.[17] Many of these fan-driven activities attracted the attention of mainstream media outlets (the *New York Times*, *The Jimmy Kimmel Show*), which helped *Lost* become an important water-cooler event for a mass audience—still a necessity in the network television business. In other words, both the writers and the marketers were on the same page when it came to fostering the indirect promotional value of *Lost*'s derivative content, which functioned as what Henry Jenkins calls "spreadable media."[18]

Although Cuse has described "The Lost Experience" as a failed experiment, with the expectations of corporate sponsors not met and a lower number of participants than expected, he concedes that many lessons were learned in the process.[19] By the time work began on the second ARG, "The Dharma Initiative," Cuse's team was much better prepared to accommodate the needs of the sponsors; in fact, Cuse is forthright about his knowing involvement with the network's goal of "harvesting" information about fans as potential consumers of *Lost* DVDs, computer games, novelizations, and other licensed merchandise. Cuse observed: "The second [ARG], 'The Dharma Initiative,' it was like harvesting viewers as people signed up to join.... Now all of a sudden ABC had a database of information from 300–400,000 hardcore fans ... which is an amazing resource.... something [ABC] can monetize like magazines monetize subscribers."[20] This cooperation between television writers and network marketing departments should come as no surprise; after all, the writers' livelihood is tied to the marketing support the networks provide their series. On the other hand, the writers risk fan backlash if they appear too complicit in the strategic designs of their corporate partners.

The Writers

During the 2007–2008 WGA strike, one of the primary issues on the bargaining table was the minimum payment and residuals owed to the writers responsible for creating "derivatives," defined by the guild as the digital content based on a television series. According to Tamara Krinsky, new media program manager for WGA West, "Within the digital world, there are three primary 'buckets of content' that apply to writers. [They are]: original (new programming made for the web), derivative (based on an existing TV show), or traditional reuse (that is, watching an episode of ABC's *Lost* online via Hulu). All three types of new media content are covered by the Guild, with each category possessing specific stipulations under the [Minimum Basic Agreement] pertaining to minimums, residuals, credits, pension and health contributions, separated rights, and more."[21] "The Lost Experience" clearly falls into the category of derivative content and should have produced credit and financial rewards for the showrunners, according to the WGA. The original guidelines posted on the WGA site during the strike stated: "*Showrunner coordination of creative content.* In the same way that showrunners and other guild-represented writer-producers participate in creative decisions relating to the episodes of a series, they should coordinate the decisions relating to these new forms and types of content, which reflect on the series creatively."[22]

Given how busy showrunners are with their primary task of overseeing the writing, production, and promotion of the television series proper, the job of writing derivative content usually goes to lower-level writers. *Lost* showrunners Cuse and Lindelof oversaw "The Lost Experience" storyline, but assigned writers' assistant Jordan Rosenberg to handle the day-to-day writing associated with the ARG and asked *Lost* co-executive producer Javier Grillo-Marxuach to oversee his work. Conversations with Rosenberg and several writers' assistants on other series (*Castle*, *Ugly Betty*, and others) revealed that it is now commonplace for showrunners to ask entry-level writers to contribute to derivative content on social network sites, including character Tweets and Facebook entries. Although these individuals do not receive additional pay for their contributions to digital content, most welcome any opportunity to prove themselves and enhance the likelihood they'll be hired to write for the series proper.

Both Rosenberg and Grillo-Marxuach found themselves doing far more than writing story and dialogue in consultation with the showrunners. Instead, they were fielding calls from representatives of the five main sponsors, sponsors' advertising agencies and design firms, and members of the ABC marketing team. Grillo-Marxuach described a typical workday: "I remember I was driving to my meeting ... and I was on the phone with ... the ad agency for Jeep ... and talking to them about how we were going to do the creative, and they were talking about the demographics for the Jeep Compass and who they wanted to sell to and, I mean, this isn't being a TV writer, this is being an account executive."[23] The once distinct occupation of television writer became intricately intermeshed with the requirements of diverse realms of the corporate landscape involved in the care and management of a vast transmedia franchise.

In his expansive white paper on "The Lost Experience," Ivan Askwith points out that the ARG was not exclusively focused on story and character. Instead, its threefold goals were to serve as: (1) "a promotional campaign to engage fans and generate viral speculation"; (2) "a narrative extension, designed to provide *Lost* fans with additional content . . . to deepen the immersive pleasure of the show"; and (3) "an advertising space, for sponsoring partners to reach the engaged participants of the campaign with brand messages."[24] Mark Warshaw, who has worked on the derivative content for several major television series (*Smallville, Heroes, 90210*), reiterated this claim, arguing that ARGs require "a delicate balance between three things: fans, brands, and content."[25]

Shoring Up the Fourth Wall

Among the storytelling challenges that Rosenberg and Grillo-Marxuach faced while working on "The Lost Experience" was to control the game and integrate the sponsors in a meaningful way without upsetting fans by breaking the fourth wall and exposing their presence as game moderators, also known as "puppet masters." ARG scholar Jane McGonigal defines puppet masters as "the first real-time, digital game designers. An invisible creative team composed of shadowy, often anonymous figures, they work behind the scenes as the writers, programmers, directors and stage managers of live pervasive gameplay."[26] The "shadowy" status of ARG writers underscores the challenges that the WGA faced while negotiating on behalf of television writers assigned to create derivative new media content. Grillo-Marxuach pointed out that the actress playing Rachel Blake in the ARG had to appear in character at Comic-Con to protect the illusion of reality created by this type of immersive game experience. As Grillo-Marxuach explains, "this is where it gets really complicated actually, because you have people playing the game who then assume that the game is actually happening in the real world. . . . That's the assumption among people playing [that] the Hanso Foundation is real. So, to talk about these advertisers as if they're ruining the game—What game?"[27]

Traditionally, the concept of the "fourth wall" refers to the imaginary wall between the theater audience and the fiction unfolding onstage. Today, it has come to mean the implied boundary between any fictional construct and the audience. For Bertolt Brecht, for instance, breaking the fourth wall by having a character speak directly to the audience was a modernist intervention designed to challenge the hegemonic forces at work in realist bourgeois theater. For transmedia production companies, however, many of which are essentially digital marketing companies, the fourth wall means something quite different. A contemporary firm that calls itself Fourth Wall Studios, Inc. explains this new business philosophy: "In the world of marketing, it's the wall of resistance, noise, and clutter that separates your brand story from your customer. Our job is to break through that wall."[28] In other words, from the marketers' perspective, any contest, casual game, or ARG that momentarily distracts a potential consumer is assumed to forge a bond between that grateful individual and the brand. A case in point is the "Subservient Chicken" campaign created by the advertising firm Crispin Porter + Bogusky. Its goal is to establish an alliance with potential Burger King consumers by inviting them to participate in a simple interactive experience—ordering a man dressed in a low-rent chicken suit to obey ridiculous commands, such as "dance like a chicken."[29]

Certainly "The Lost Experience" writers were eager to raise the bar with their ARG, matching the conspiratorial tone and quality of writing inherent in the *Lost* series. From Cuse's perspective, the biggest problem the writers faced was the excessive number of corporate sponsors involved, which resulted in an arcane, byzantine game structure that threatened to discourage all but the most ardent fans.[30] In their efforts, Rosenberg and Grillo-Marxuach had to abide by the two disparate interpretations of the fourth wall, namely, to create a fun, interactive experience between the brand and consumers on behalf of the sponsor *and* to integrate these pleasurable moments into a cohesive storyline that would not call attention to the authors of the game.[31]

Asking a company employee to function both as an internal character and as an external moderator—as employed successfully by the digital media production company Eqal with its Lonelygirl15 web franchise—was not an option. As McGonigal explains, ARG players insist that nothing dispel the myth that "this is not a game."[32] Toward the end of "The Lost Experience," Grillo-Marxuach's solution was to take on the persona of a fictional character named DJ Dan, which allowed him to moderate the game and integrate brands from inside the story. According to Grillo-Marxuach, the simplest way to accomplish the brand integration was to have DJ Dan tell viewers that he drove a Jeep Cherokee to his radio station that morning. However, to create a more immersive experience of the brand—something the sponsors and fans expected—he posted blogs as if he were a Jeep IT employee who had offered to help Rachel Blake expose the malfeasance of the Hanso Corporation by posting her controversial "hacked" videos about the foundation on the Jeep mainframe. Both of the ARG writers explained that this was a typical story strategy designed to show the brand in a positive light *and* to reinforce Rachel's goal "to take down the man." When asked how he and Grillo-Marxuach tried to integrate brands into the ARG's story and characters, Rosenberg explained that stories that convey rebellion or "anything to do with [fighting] corporate America" and "the man" are always popular with fans.[33] Did sponsors express concern about having their brands associated with such negative portraits of big companies? On the contrary, Grillo-Marxuach responded, "they loved it." Rosenberg added that Sprite's Sublymonal campaign was about "brainwashing you so . . . it was easy" to incorporate the Sprite campaign into the storyline of the ARG. However, as will be shown below, that effort was neither obvious nor easy.

The Network Marketing Team

Creation of "The Lost Experience" was proposed to *Lost* showrunners Lindelof and Cuse by ABC marketing heads Mike Benson and Marla Provencio. All the parties recognized that the *Myst*-like game's primary function was promotional, a means of encouraging avid fans to help the network keep the expensive series in the zeitgeist between seasons 2 and 3.[34] However, once Benson realized the huge upfront costs involved in creating the ARG, he resorted to a familiar broadcasting strategy and turned to a handful of product sponsors (Sprite, Jeep, Verizon, and Monster.com) to finance the innovative digital marketing event. In fact, one of the ongoing controversies surrounding ARGs is whether they can exist as an entertainment form outside their function as promotional tools.[35] As Grillo-Marxuach remarked, "Is [the *Lost* ARG] about selling Sprite and Verizon? Yes. Is it about selling *Lost*? Yes. But more than anything else, it's about keeping the word *Lost* in that immeasurable zeitgeist so that when the show isn't on the air, promotionally speaking, somebody is saying something about it somewhere."[36]

Advertisers' interest in expanding their product messages via Web 2.0 social network companies like Facebook and Twitter has grown exponentially in the past decade. Trade journals like *Ad Age* have celebrated "The Lost Experience" and other ARGs as a dynamic new form of advertising, prompting one online journalist to dub this format "advergaming." She writes, "Taking advergaming to a new level, alternate reality gaming (ARG) involves both online and offline participation and can incorporate broader reaching sites such as YouTube, MySpace, and Twitter. Word-of-mouth and viral communication are key factors in the spread of ARG, although, ARG can also be integrated into more traditional media communications. . . . Generally, the games are not overtly commercial; the marketing messages are much more subtle."[37] The reason for the marketing industry's sudden interest in the ARG format is twofold. First, the trend allows advertisers to reach next-generation consumers by casting their product messages across multiple Web 2.0 sites, instead of relying exclusively on broadcasting's passive thirty-second ads. Second, the games create engaged consumers via immersive entertainment worlds that encourage them to unearth hidden-in-plain-sight sponsor messages, scrutinize them for game clues, then digitally download and share them virally with their friends. It's almost as if the fans are doing the marketers' job for them. Media scholars Bernard Cova, Robert V. Kozinets, and Avi Shankar use Michel de Certeau's "construct of hijacking"—a theoretical framework that they ascribe to "consumer tribes as plunderers"—to explain how fans become complicit in the corporate marketing efforts associated with a favorite culture industry product: "[In] order to by-pass a system (an IT system, a political organization, etc.) or to hijack or subvert it, a person must be in love with it. We can see in this statement the same ambivalence, the same paradoxical qualities that inform our conception of the Consumer Tribe as a Double Agent."[38] According to this view, the fan's pleasurable immersion into digital entertainment worlds introduces a greater complacency about who pays for it. Once viewed as cynical and defiant, young millennial fans are now cast as willing participants in a virtual treasure hunt for advertising messages that they are encouraged to "steal" and send to other, equally invested viewers.[39]

Marketers Revisit the *Blair Witch* Hat Trick

Most studio marketing divisions today understand the power of digital campaigns that engage consumers, taking inspiration from the online campaign for *The Blair Witch Project* (1999), in which Artisan Entertainment followed the filmmakers' creative goal to convince viewers that the film's fictional storyline was in fact true. A typical strategy is to aggregate press clippings and other documentary-style elements either online or out in the world. In the case of *The Blair Witch Project*, the filmmakers and the Artisan marketing team, lead by John Hegeman, circulated information online about the filmmakers as if their disappearance were a real-life event. When the stunt was eventually traced back to the filmmakers and to Artisan, many members of the public were impressed by the meta-marketing approach; others were angry about being duped. The job facing the *Lost* ARG writers was somewhat different. Unlike anonymous marketing executives who are forgiven for promotional campaigns gone wrong, Grillo-Marxuach and Rosenberg felt they could not make a "mistake" for fear of alienating the *Lost* franchise's committed fan base or its sponsors. Compounding their problems was the fact that the showrunners and the ABC marketing team kept authorizing additional narrative layers via staged promotional events designed to assert the reality of the ARG's fictional individuals (Rachel Blake), events (*The Bad Twin* novel), and organizations (the Hanso Foundation). For instance, on May 24, 2006, an actor playing Hugh McIntyre, the fictional director of the Hanso Foundation from "The Lost Experience" ARG, was hired to appear via satellite on *The Jimmy Kimmel Show* a few hours after the *Lost* season 2 finale. During the on-air interview, the actor playing McIntyre denounced both the television series (*Lost*) and the novel (*The Bad Twin*) as fictional conceits.[40] Immediately after the broadcast, the fan websites lit up with urgent efforts to shore up the "gaps" prompted by this apparent rupture of the fourth wall of their favorite transmedia franchise. Many answers relied on fans' innate understanding of the rules associated with highly reflexive, postmodernist media practices; others drew on interpretations of literary reader-response theories, including the suspension of disbelief required for readers to incorporate fantastic events. For instance, "fghf" said, "ok . . . basically I am playing along as if Hanso is real. That is what the writers want, to blur the lines of reality and fiction much like the *Blair Witch Project*. In essence, that was not Hugh. He was nervous and gave a B answer when he was asked why the pic didnt match him. He said it was his predecessor, but he is in charge of the website . . . the first thing he would do is change his own pic! I think the entire blurring reality is a stroke of pure genius!"[41]

Another media-savvy fan named "Thro" struggled to reconcile his faith in the fictional reality of "The Lost Experience" game and the competing reality on display in the network-engineered publicity stunt: "I wouldn't have preferred a script for this, otherwise I think it might have come off more fake. The Hugh was a bit unprepared, though. Come on, guys, a little background in improv. . . . But I prefer that they referred to it [*Lost*] as a show, because it sounds like they're actually scared of the show that way. Like it might actually bring them down. Though they should have brought it up in their website or other related things earlier, because they react to Bad Twin like it's the end of the world, but the show that portrays them as crazy and reaches millions of viewers, they can just dismiss that as bad publicity."[42]

Clearly bothered by the apparent rupture in the game's fictional universe, these and other ARG fans spent several days discussing the event online and dissecting the various levels of reality operating in the staged promotional event. Many of the fans wanted to understand whether show producers Cuse and Lindelof (perceived here as the ultimate puppet masters behind the complex, multi-platform franchise) had purposely exposed the inconsistencies in their complex, transmedia storyline by allowing an inept actor to discuss the fictional Hanso Foundation as if it were real. Still, the majority of fan bloggers debating the relevance of the event felt that the complex layering of reality and fiction was satisfying and had further woven them into the ARG's immersive world. Only a handful felt that the event was an overt promotional effort to sell the series and the game to a wider audience via Jimmy Kimmel's show. In other words, while the network marketing team's goal was singular—to keep the show in the popular zeitgeist—it was left to the head writers and the two ARG writers to keep the show's mythology intact and to make sure that the multiple storytelling platforms of the *Lost* world remained a cohesive and hence satisfying narrative experience. As Askwith warns, "While this degree of meta-convolution might be fun for some players, and for the show's creators, it runs the very real risk of confusing, then frustrating, and then alienating potential participants altogether."[43]

Crispin Porter + Bogusky

At around the same time that ABC's Mike Benson was talking to Sprite about becoming a sponsor for the network's ARG, Coca-Cola was signing the cutting-edge advertising agency Crispin Porter + Bogusky to help rebrand the company's aging, 1960s lemon-lime flavored soft drink and make it more desirable to the fourteen-to-twenty-four-year-old demographic. Characterizing their self-reflexive, meta-marketing promotions as part of a total "brand facelift," Crispin Porter + Bogusky jokingly remarked that it would overhaul "everything but the liquid."[44] The firm jumped at the chance to marry the new campaign to "The Lost Experience" ARG in order to access *Lost*'s young, engaged fan base. The campaign consisted of five spots, described as "[m]arrying a cold, Kubrickian visual language with bizarre vignettes . . . [to create] a darker, surrealistic bent."[45] The thirty-second Sublymonal spots first aired during the final, May 24, 2006, broadcast of *Lost*'s second season. Prior to that, on May 10, 2006, a commercial for the fictional Hanso Foundation appeared during a *Lost* episode with the onscreen text, "Paid for by Sprite." Both commercials directed viewers to Sprite's website: http://sublymonal.com/.[46] The Sprite ads were expressly designed to engage consumers enough to get them to pause their DVRs and watch the ads. Regarding the spread of this strategy throughout the industry, TiVo president and CEO Thomas S. Rogers was quoted as saying, "It's all about how you reach those who are fastforwarding, what steps you can take to put creative and different inducements in front of somebody to get them to stop on an ad."[47]

The Sublymonal advertising campaign created an overall impression of modernist rebellion combined with nostalgic references to the past by alluding to the "subliminal advertising" scare of the late 1950s and by using the word "lymon" (a portmanteau of the words "lemon" and "lime") drawn from the 1980s Sprite campaign. Other game clues showcased the word "obey," which was also part of Sprite's slogan, "Obey your thirst."[48] At first glance, these references to "subliminal advertising" and to "obedience" seem designed to engage a teenage audience by inviting the opposite—an attitude of rebellion. However, in spite of alienating images akin to Luis Bunuel's slit eye in *Un Chien Andalou*, the Sprite ads ultimately neutralized any modernist intervention in favor of something closer to postmodernism's pastiche, parody, bricolage, irony, and blank references to a perpetual present. The combination of these two tendencies—modernist and postmodernist—nullified the implied rebellion and replaced it with something akin to John Caldwell's description of *Pee Wee's Playhouse* as a "bourgeois bombshelter," that is, a stylistically excessive text designed to "throw as much radical-looking form at the viewer as possible even as [it] unabashedly promote[s] specific products."[49]

Grillo-Marxuach and Rosenberg were left with the daunting responsibility of containing Crispin Porter + Bogusky's radical-looking content in a narrative of protest against "the man." Put simply, a major part of the writers' job was to help the advertising firms create consumers who function as double agents for brands by incorporating the brands into a storyline of protest. Commenting on an online article about the provocative Sprite spots, twenty-four-year-old

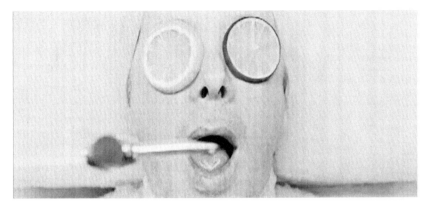

FIG. 5-1 The Sprite Sublymonal advertising campaign's enigmatic "Spa Man" sequence engaged "The *Lost* Experience" ARG players but alienated some consumers. (Frame grab)

"Jinky Williams" displayed a remarkably sophisticated knowledge of how ad-supported digital marketing is used to target his own Gen-Y demographic in contrast to the conventional thirty-second ads aimed at his mother's Gen-X demographic: "The rule used to be never to confuse the viewers with obtuse messages. . . . But yet, more and more people (Generation Y and younger in particular) enjoy 'filling in the blanks,' and take great pride in the feeling that they themselves have discovered something hidden from popular view. . . . Hence, the ongoing [trend] of advertising becoming more and more abstract, until the actual subject of the advertisement becomes an undertone. . . . However, these spots must still be cohesive and maintain a consistant [sic] style so that they can be readily identifiable."[50] Most observers agree that Crispin Porter + Bogusky accomplished the goals of being memorable and resistant to fast-forwarding. However, as Ivan Askwith points out, the campaign gave the ARG fans (and writers) little to work with. When directed to the sublymonal.com website, the ARG fans encountered a static bank of television monitors. The only interactivity provided was the option to choose which of the thirty-second spots to activate.[51]

How do we explain this apparent gap between the ad firm's expectations and the ARG writers' needs? One way is to examine the culture clash revealed in the statements made by the creative personnel from Hollywood and Madison Avenue. The spot directors hired by Crispin, Porter + Bogusky were experienced admen who focused on breaking through the fourth wall (the clutter of ad messages); in contrast, the writers hired by Cuse and Lindelof to oversee the ARG were struggling to keep the narrative's fourth wall intact. The Sublymonal spot director-producer team of Guy Shelmedine and Richard Farmer described their work on the Sprite overhaul as their "most daring": "[We] set up what we called the 'Sprite Lab.' . . . [and] brought in hundreds of props and

FIG. 5-2 "We brought in video cameras and crazy props and TVs and blue screens and whatnot, and it turned into this madhouse" (Richard Farmer, director of the Sublymonal campaign). Photo courtesy of Matthew Santo, cinematographer on the Sublymonal campaign, who identified the crew depicted: Franta Novák (center) was the A-camera first assistant camera operator; Karel Kaliban (left) was the first assistant director and B-camera operator; Chris Jones (right) was the production designer.

lights, gadgets and bits and pieces, and a whole team of people to create extra stock footage so that when we went into the edit, the editor had tons of different things he could throw in there for a brief second or a few frames."[52]

The award-winning Crispin, Porter + Bogusky ad firm is frequently celebrated for its bleeding edge, interactive approach to the thirty-second ad, which consists of two strands: (1) presenting users with minimalist but addictive gameplay (for example, social games like "Subservient Chicken"); and (2) bombarding users with DVR-proof, pop-culture-replete, eye candy. Capitalizing on the growing trend of fans using copyrighted material to create their own mash-ups, Crispin, Porter + Bogusky sought out would-be mad men to create their own Sublymonal messages from the bits and pieces of content created by the advertising directors.[53] The fan who created "the ultimate Sublymonal message" would win a "Sony DVD dream system, including a Bravia LCD HDTV, Playstation 3 with games and noise cancellation headphones."[54] Whereas fan-generated mash-ups are typically motivated by a desire to provide an alternative reading of the narrative (by portraying *Lost*'s Jack and Sawyer as lovers, for instance), most corporately produced contests remain emotionally detached from the subject. By presenting fans with a random compilation of images, the Sublymonal ads were alienating on two fronts: they were unreadable to Gen-Xers as thirty-second ads; and Gen-Yers could not relate them to the *Lost* narrative or to the ARG's puzzle-solving activities.[55]

Collaborating with Network Licensors

Several media scholars, including Derek Johnson and Jonathan Gray in this collection, have chronicled the networks' failure to create well-received licensed computer games tied to their valuable television titles. In his book, *Media Franchising*, Johnson explores the complex institutional and cultural history of the networks' franchise system, tracing it to popular mid-1950s fast-food franchises like McDonald's, which encouraged entrepreneurs to run individual stores but paradoxically asked them to enforce the corporate owner's strict standardization of product (making sure that hamburgers served in Phoenix, Arizona, were identical to the ones served in Downey, California, for instance).[56] Similarly, licensed vendors working for big media companies are contractually obligated to perform creative activities that are aligned with corporate demands. Stripped of any role as autonomous artist, the licensed vendor is more closely aligned with the entrepreneur running a "storefront" within a nationwide chain.

As Michael Clark demonstrates in *Transmedia Television*, studio licensing departments typically require their vendors to meet strict budgetary and schedule constraints, whereas the vendors wish for more creative oversight.[57] In contrast to the licensed vendors, television showrunners hold split roles as creative labor *and* corporate managers, a schism that became evident during the WGA strike, when showrunners were torn between walking the picket line or fulfilling their contractual obligation to complete production on already written episodes for airing by the network. By inserting themselves or other intermediaries from the writers' room into the dialogue between the licensing executives and the licensed vendors, Cuse and Lindelof were adding value to the *Lost* franchise but also threatening the traditional power wielded by the networks. Paradoxically, even though most licensed vendors welcome greater input from creative personnel, the much-admired Montreal-based game designers of Ubisoft appeared flummoxed during their initial meeting with the highly engaged *Lost* showrunners. According to Cuse, instead of pitching their ideas to the two showrunners, "They began the meeting by saying to us, 'Do you have any ideas . . . for what the video games could be?'"[58]

As creative authors who also hold managerial positions, the *Lost* head writers were in an ideal position to apply pressure to network division heads to open up the network's traditional operations while also flexing their entrepreneurial muscles as writers by incorporating interactive social media elements into the conventional television narrative. Cuse and Lindelof found their inspiration to rethink the television industry in part from several Silicon Valley outsiders. In a 2006 interview, Cuse explained his vision of the future of television by referencing the entrepreneurial activities and "rock star" status of YouTube founder Chad Hurley, who, Cuse believed, was reinventing media by introducing a "real paradigm shift." Cuse went on to explain that he and Lindelof were "interested in exploring how to do it, to be part of this wave."[59]

Knowing how overextended the showrunners were, executive producer Grillo-Marxuach proposed to Cuse and Lindelof that he be allowed to work directly with the network licensing department, consulting on ways to incorporate the *Lost* mythology into the network's licensed properties, such as board games, puzzles, books, and other merchandise. He reasoned that "if someone isn't working with licensing on the *Lost* board game, the *Lost* board game isn't going to be very good. They need an inside player who is part of the show's hierarchy."[60] Because he was also assigned to "The Lost Experience," Grillo-Marxuach knew that he could do double duty, enhancing not only the quality of several licensed properties—profit centers for the network—but also the quality of the ARG by placing clues in the board games and puzzles. Along with assigning Rosenberg to oversee the ARG and Grillo-Marxuach to deal with the licensed board games, Cuse and Lindelof hired additional intermediaries to assist with other subsidiary platforms and online promotions; for instance, they asked *Lost* writer-producer Jeff Pinkner to coordinate and manage all the ancillary materials on behalf of the writers, and they hired digital producer Samantha Thomas to oversee the production of "The Lost Diaries" mobisodes and "The Lost Untangled" recaps.[61]

In a 2006 interview, former *Lost* and *Heroes* writer-producer Jesse Alexander expressed a need for what he dubbed "a transmedia czar."[62] Grillo-Marxuach and Rosenberg came to a similar conclusion during a 2008 interview, explaining their need for "a dedicated, core group of producers whose sole job it was to run that game," given the demanding nature of alternate reality games that require a punishing 24/7 schedule to maintain. Grillo-Marxuach said, "Everybody involved was tremendously competent, proficient, and wonderful. Our web design firm was great. Our broadcast partners were great. Our advertising partners were all doing a lot of work on this. But, you didn't have that central nervous system in terms of 'this is where the buck stops.' You know? And, without that, the job became very big, because there was this cat-herding element to it."[63]

With these various creative hires, the *Lost* showrunners acknowledged their need for additional layers of support from inside the writers' ranks rather than from the network's side. Notably, the *Lost* writers arrived at this conclusion several years before the Producers Guild of America (PGA), which in 2010 established a new credit designation called the "transmedia producer"; the title would apply to anyone overseeing "three (or more) narrative storylines existing within the same fictional universe."[64] Transmedia producers like Jeff Gomez from Starlight Runner Entertainment are PGA-represented independent producers and act as unofficial story consultants. Cuse and Lindelof's more radical decision was to enhance the power of WGA-represented television writer-producers (rather than outside consultants), granting guild-represented labor more creative oversight of all story-driven platforms in the franchise.

Recognizing the changes that the *Lost* writers had introduced into the post-strike entertainment industry landscape, the WGA began to offer additional support for writers who wished to be more entrepreneurial by operating outside the studios and networks. The organization's website stated: "In the age of YouTube, Hulu, Crackle, and MyDamnChannel, new media outlets and digital technologies provide writers increased opportunities to become true creative entrepreneurs, armed with the tools and distribution channels necessary to connect directly with audiences—and often without studio/network intervention—like never before."[65] Whereas Cuse and Lindelof proposed making infrastructural changes in the way television writers interfaced with network bureaucracies, the WGA decided not to wage battle with the networks; instead, it advised its constituency to produce low-budget, short-form videos and to promote and distribute these fledgling transmedia franchises outside the studio-network complex by using the internet's DIY Web 2.0 resources (for instance, Kickstarter, Facebook, Twitter, and YouTube). Given the plethora of original short-form content already in circulation on amateur digital streaming sites like YouTube, the WGA's advice was practical but not as daring as what the *Lost* writers proposed. That said, the WGA may have felt unable to offer its constituency more in the post-*Lost* moment; as Cuse observed in 2010, the window of opportunity for writers to flex their entrepreneurial muscles from *inside* the big media corporations appeared to be closed as of May 23, 2010, making him and Lindelof dinosaurs in the history of network television.

The unique achievement of the *Lost* writers was their ability to create a vast, infrastructural sea-change from *inside* the densely bureaucratic network system. However, this hard-won battle appears to have prompted a backlash from the networks once the series was off the air. After Benson was fired, all digital marketing efforts reverted back to Cheng's in-house digital entertainment division, which remains focused on promotions designed for a mass audience. To replicate the *Lost* experiment, the networks would have to agree once again to step up as willing collaborators with the writers; instead, ABC flinched when several new, mixed-genre, hyper-serialized programs (*Flashforward*, *V*, *Alcatraz*) failed to gain traction immediately with a mass audience. By shoring up their traditional analog approach to current programming and by focusing on digital promotions designed for a mass audience, ABC has demonstrated its unwillingness to collaborate with creative partners to the degree seen during the *Lost* moment. Moreover, all of the networks have shown a reluctance to grant additional power to showrunners who wish to engage fans by embracing social media as a viable component of the television experience in the digital age. They have also demonstrated a general unwillingness to partner with outsiders—whether WGA-represented transmedia czars, PGA-represented transmedia producers, super-fans who wish to contribute to the television experience, or the type of creative entrepreneurs that YouTube is hiring for its 100-channel partnerships.[66] Instead, the networks appear to be closing ranks against mass collaboration and, in the process, taking a giant step backward toward their analog past.

Notes

1. "The Hollywood Geek Elite Debates the Future of Entertainment," SCMS workshop, Los Angeles, March 20, 2010.
2. According to the ABC website, Marla Provencio, executive vice president for marketing, "launched the television industry's first multi-platform, brand identity and viewer-navigation system through the 'ABC Start Here' initiative, developed multi-platform marketing strategies, alternative reality games and multi-level engagement activities for the hit series 'Lost,' as well as other programming, and was honored by the Academy of Television Arts & Sciences 2008–2009 Emmy Awards for contribution to the Outstanding Creative Achievement in Interactive Media (fiction) for The Dharma Initiative ('Lost')." See http://www.abcmedianet.com/web/bios/display_bios.aspx?bio_type=executives&bio_id=274.
3. Author interview with Albert Cheng, March 12, 2012.
4. "Guidelines on WGA Coverage for Digitally Distributed Content for TV Writers" (2007). The exact quotation is no longer available on the WGA website (http://www.wga.org/).
5. Gregg Sutter, "Interview with Carlton Cuse," *The Gregg Sutter Website*, November 3, 2010, http://greggsutter.com/wordpress/?p=515.
6. John T. Caldwell, "Worker Blowback: User-Generated, Worker-Generated, and Producer-Generated Content: Within Collapsing Production Workflows," in *Television as Digital Media*, ed. James Bennett and Niki Strange (Durham, NC: Duke University Press, 2011).
7. Edward Wyatt, "Webisodes of 'Lost': Model Deal for Writers?" *New York Times*, November 20, 2007, http://www.nytimes.com/2007/11/20/arts/television/20digi.html?ref=television.
8. Sutter, "Interview with Cuse," 2010.
9. Levi Shapiro, "Hollywood Goes EST," *Video Age International*, February 1, 2008, available at http://www.thefreelibrary.com/Hollywood+goes+EST-a0179160420.
10. Author interview with Tamara Krinsky and Chuck Slocum, May 25, 2011.
11. For a more comprehensive list of the residual rates set after the strike, see entertainment attorney Jonathan Handel's "Residuals Summary Chart-DGA, WGA, SAG, AFTRA and IATSE," http://www.troygould.com/layouts/50/graphics/uploads/jlh_residuals2.pdf.

12 Josef Adalian and Michael Schneider, "TV Shows Quickly Going Dark," *Variety*, November 6, 2007, http://www.variety.com/article/VR1117975483?refCatId=2821.
13 Shawn Ryan, "The Office Is Closed (Shawn Ryan's open letter to showrunners and writers)," *ScribeVibe*, http://weblogs.variety.com/wga_strike_blog/shawn-ryans-letter-open-1.html.
14 "The Office on YouTube," *TV Week Archives*, November 6, 2007, http://www.tvweek.com/blogs/trial-and-error/2007/11/the_office_on_youtube.php.
15 Associated Press, "The Writers Guild Strike is Over," ABC Local News, February 12, 2008.
16 Carlton Cuse observed: "It is exciting for me to talk about how we, as producers are going to be challenged to tell stories in this new environment." Author interview with Carlton Cuse, UCLA, July 18, 2006.
17 While the biggest growth in social media surrounds reality shows like *The Bachelorette* (ABC), *America's Got Talent* (NBC), and *So You Think You Can Dance* (Fox), the trend is evident in lower-rated drama series like ABC's *Rookie Blue* as well. See "The Returning TV Shows That Are Exploding in Social Media," *Advertising Age*, June 6, 2012, http://adage.com/article/trending-topics/returning-tv-shows-exploding-social-media/235210/.
18 Henry Jenkins, Sam Ford, and Joshua Green, *Spreadable Media: Creating Value and Meaning in a Networked Culture* (New York: New York University Press, 2012). Jenkins, Ford, and Green use "spreadable media" in lieu of the more pejorative designation of "viral marketing" to describe the way today's networked media encourage users to circulate ideas and images about a favorite piece of pop culture; in contrast, old media tend to stifle this free-floating exchange of information in their efforts to control distribution.
19 Author interview with Cuse, July 18, 2006.
20 Author interview with Cuse, SCMS workshop, March 20, 2010.
21 Author interview with Krinsky and Slocum, May 25, 2011.
22 "Guidelines on WGA Coverage for Digitally Distributed Content for TV Writers," 2007. Again, the quotation is no longer available in the final version of the WGA "Guide to New Media," http://www.wga.org/content/default.aspx?id=3996.
23 Author interview with Javier Grillo-Marxuach and Jordan Rosenberg, UCLA, January 17, 2008.
24 Ivan Askwith, "Deconstructing the Lost Experience: In-Depth Analysis of an ARG," MIT Convergence Culture Consortium, www.ivanaskwith.com/writing/IvanAskwith_TheLostExperience.pdf.
25 Author interview with Mark Warshaw, SCMS workshop, March 20, 2010.
26 Jane McGonigal, "The Puppet Master Problem: Design for Real-World, Mission-Based Gaming," in *Second Person: Role-Playing and Story in Games and Playable Media*, ed. Pat Harrigan and Noah Wardrip-Fruin (Cambridge, MA: MIT Press, 2007).
27 Author interview with Grillo-Marxuach and Rosenberg, January 17, 2008.
28 Fourth Wall Studios, Inc. website, http://www.fourthwall.net/site/ and http://www.linkedin.com/company/fourth-wall.
29 Burger King's "Subservient Chicken" campaign, http://www.subservientchicken.com/pre_bk_skinned.swf.
30 Author interview with Cuse, July 18, 2006.
31 Askwith observes that the network marketers and head writers announced the ARG, thereby breaking the rules of the format; however, after the launch, Cuse and

Lindelof said they would no longer refer to it publically as a game, and they kept Rosenberg and Grillo-Marxuach's names secret. Askwith, "Deconstructing," 4.

32 Jane McGonigal, "This Might Be a Game: Ubiquitous Play and Performance at the Turn of the Twenty-First Century" (PhD diss., University of California–Berkeley, 2006), http://avantgame.com/dissertation.htm.

33 Author interview with Grillo-Marxuach and Rosenberg, January 17, 2008.

34 *Myst*, a "graphic adventure game" released in 1993, became a surprise hit, praised for its ability to immerse players in its fictional universe. It was the best-selling PC game until *The Sims* exceeded its sales in 2002. For background, see http://en.wikipedia.org/wiki/Myst.

35 Ivan Askwith, "This Is Not (Just) an Advertisement: Understanding Alternate Reality Games," MIT Convergence Culture Consortium, http://www.ivanaskwith.com/writing/IvanAskwith_ThisIsNotJustAnAdvertisement.pdf.

36 Author interview with Grillo-Marxuach and Rosenberg, January 17, 2008.

37 Carrie Urban Kapraun, "Gaming and Alternate Reality Games (ARG)—An Exploration of New Media Sponsorship and Marketing," *IEG Sponsorship.com*, July 24, 2009.

38 Bernard Cova, Robert V. Kozinets, and Avi Shankar, eds., *Consumer Tribes* (Oxford: Butterworth–Heinemann, 2007), 14.

39 Henry Jenkins references a photo caption, "You've got 3 seconds. Impress me," to characterize today's young consumer. Jenkins, *Convergence Culture: Where Old and New Media Collide* (New York: New York University Press, 2006), 64–65.

40 Hyperion Press had published a metafictional book called *The Bad Twin*, written by Laurence Shames, and credited it to the fictional author Gary Troup, who was purportedly a passenger on the doomed Oceanic Airlines, http://www.lostblog.net/lost/tv/show/video-of-hugh-mcintyre-interview.

41 See comments below the blog "The Lost Experience Clue #42: Jimmy Kimmel Interview with Hugh McIntyre," May 26, 2006, http://thelostexperienceclues.blogspot.com/2006/05/clue-42-jimmy-kimmel-interview-with.html.

42 Ibid.

43 Askwith, "Deconstructing," 22.

44 Mark Pytlik, "Gentlemen, Start Your Pause Buttons: Crispin & Happy Put a Spin on Subliminal Advertising for Sprite," *Boards Magazine*, August, 1, 2006, http://www.boardsmag.com/articles/magazine/20060801/spotopsyaug.html.

45 Ibid.

46 The Sublymonal campaign collaborators include Crispin Porter + Bogusky employees Alex Bogusky, Tim Roper, Rob Strasberg, Franklin Tipton, Geordie Stephens, James Dawson-Hollis, and Chris Moore; spot directors Guy Shelmedine and Richard Farmer (also known as Happy); Haines Hall and Damion Clayton, editors at Spotwelder. See Duncan, "Sublymonal Sprite Advertising Campaign," *The Inspiration Room*, June 18, 2006, http://theinspirationroom.com/daily/2006/sublymonal-sprite/.

47 Amy Johannes, "A Cool Front Moves In," *Promo Magazine*, August 1, 2006, http://chiefmarketer.com/mag/marketing_cool_front_moves.

48 Kate Macarthur, "Sprite Steps Away from Hip-Hop, Refocuses on 'Lymon' Roots," *Advertising Age*, May 18, 2006, http://adage.com/article/news/sprite-steps-hip-hop-refocuses-lymon-roots/109277/.

49 John T. Caldwell, *Televisuality: Style, Crisis, and Authority in American Television* (New Brunswick, NJ: Rutgers University Press, 1995), 198.

50 See comments after Duncan, "Sublymonal Sprite Advertising Campaign."
51 Askwith, "Deconstructing," 13.
52 Kevin Richie, "Two Easy Pieces: Smuggler's Happy Injects 70s Attitude into Re-Brands," *Boards Magazine*, May 1, 2007, http://www.softcitizen.com/bin/news/67/Happy%20Boards%20May%202007.pdf. Shelmedine and Farmer's company, Happy World Wide (see http://www.happyworldwide.com/), is managed by Smuggler, a film, television, and commercial production and management company (http://smugglersite.com). The Happy team hired director Jon Watts and cinematographer Matthew Santo.
53 Richie, "Two Easy Pieces."
54 Aditham, "Sprite Offers Mixed Sublymonal Messaging," *Creativity-Online*, July 10, 2007, http://creativity-online.com/news/sprite-offers-mixed-sublymonal-messaging/119131.
55 Crispin Porter + Bogusky later changed the Sublymonal.com campaign because "[w]e found that the target found it difficult to decode." See Eleftheria Parpis, "Sprite Hits Refresh," *Advertising Age*, February 21, 2010, http://www.adweek.com/aw/content_display/creative/features/e3ie935cc06a403502222ac700356c75632.
56 Derek Johnson, *Media Franchising: Creative License and Collaboration in the Culture Industries* (New York: New York University Press, 2013).
57 M. J. Clarke, *Transmedia Television: New Trends in Network Serial Production* (New York: Continuum, 2012).
58 Author interview with Cuse, July 18, 2006.
59 Ibid.
60 Author interview with Grillo-Marxuach and Rosenberg, January 17, 2008.
61 Cuse explained, "The hardest part about doing a series like this is realizing that the mother ship needs constant attention because if the originating show isn't any good, then none of the other [ancillary] stuff is going to be good. Now, we have someone who is our brand manager. Then we have another full-time person, Samantha Thomas. Her job is to coordinate and manage all the ancillary stuff. And then there's another producer on the show, Jeff Pinkner, who is the next in the line of command responsible for overseeing all of this [ancillary] stuff." Author interview with Cuse, July 18, 2006.
62 Author interview with Jesse Alexander, *Heroes* production office, July 9, 2008.
63 Author interview with Grillo-Marxuach and Rosenberg, January 17, 2008.
64 Liz Shannon Miller, "Is the PGA's New 'Transmedia Producer Credit' a Good Thing?" *New Tee Vee*, April 6, 2010, http://newteevee.com/2010/04/06/is-the-pgas-new-transmedia-producer-credit-a-good-thing/. The PGA website explains that a transmedia producer credit pertains to "three (or more) narrative storylines existing within the same fictional universe on any of the following platforms: Film, Television, Short Film, Broadband, Publishing, Comics, Animation, Mobile, Special Venues, DVD/Blu-ray/CD-ROM, Narrative Commercial and Marketing rollouts, and other technologies that may or may not currently exist."
65 See the WGA West New Media website, http://www.wga.org/content/default.aspx?id=3996.
66 Paul Bond and Georg Szalai, "YouTube Announces TV Initiative with 100 Niche Channels," *Hollywood Reporter*, October 28, 2011, http://www.hollywoodreporter.com/news/youtube-tv-channels-kutcher-poehler-254370.

6

Post-Network Reflexivity

•••••••••••••••••••

Viral Marketing and
Labor Management

JOHN T. CALDWELL

Post-network television today is characterized by a set of resilient industrial habits involving collective, critical self-representation. I argue in this chapter that the recent explosive growth and popularity of onscreen self-referencing, self-disclosure, and organizational transparency in the post-network era has been stimulated by at least four general factors: the wide-ranging breakdown of traditional barriers between media professionals and audiences; the new digital technologies that have animated the cross-cultural leaks and blurred borders that once distinguished lay and professional media worlds; the increasingly dense clutter of multimedia markets that require self-referencing metatexts for effective viewer navigation; and the increased competition and task uncertainty that trigger pressures to value craft distinction and innovation symbolically in public ways. Considering but one recent example of excessive self-referencing—the HBO series *Entourage* (2004–2011)—suggests that very close relations now exist between production practices and marketing strategies.

"Don't pitch inside. Audiences don't care about the industry." So one screenwriter cautioned me about proposing a network series. With police procedurals, franchises, and Jerry Bruckheimer dominating primetime, this wisdom made some sense. Yet program schedules are loaded with shows about the entertainment business. This genre includes prestige serials like BBC's *Extras*, HBO's *Curb Your Enthusiasm* and *Entourage*, Aaron Sorkin's hyped dud *Studio 60*, and Steven Spielberg and Mark Burnett's auteurs-in-the-making *On-the-Lot*. Unlike less Emmy-worthy fare (*Next Action Star, Comeback*), *Entourage* rolled into its fourth season as a critical hit, in part because HBO engineered the series to be fan-driven and spun it as "cult TV." But why does *Entourage*—a series that crystallizes many of the themes of this chapter—succeed on HBO when many reality "shows about the biz" have tanked on broadcast or basic cable networks?

The keys to the cult television kingdom and *Entourage* lie in three factors involving self-reference: a strong authoring identity, a complicated narrative form, and lots of opportunities for audience involvement. The last trait is probably the most important. Cult television cannot be made. It must be discovered. Only fans who feel they've discovered an edgy trend can produce the word-of-mouth necessary to build cult status. Hard-sell advertising has the opposite effect (think NBC's *Studio 60*), revealing corporations desperate to hype programming duds. HBO strikes a different pose when it grants fans an interactive online job "interview" with *Entourage*'s manic agent. Ari taunts his website "applicants": "What makes you so special? What can you do for me?" His contempt pulls fans into the fiction's arc, even as the ploy lays bare the show's tech-savvy marketing scheme.

Finding an out-of-the-way hit-in-the-making allows fans to buy into a new phenomenon. Forwarding this find to friends and their friends through a cascade of e-mails and blogs creates buzz. Marketers claim that this "viral" process works only if the "discovery" spotlights the finder's identity more than the show itself. This kind of personal investment, or "under-the-radar" marketing, works especially well in premium cable. There, viewers prepay for production costs regardless of a show's initial popularity or ratings. As a result, HBO can patiently allow costly "quality" series to find their audiences over two to three years, whereas broadcast networks typically hit the kill switch in a matter of weeks or months. *Entourage* fits this niche profile perfectly. The show's writers ask season-long questions about whether the gang and their films will ever make the big screen. These will-they-or-won't-they story arcs exploit the multi-year leash that HBO grants the series to find its audience. *Entourage*'s form also cultivates audiences in ways that go beyond this delaying-the-third-act tactic.

Creators Mark Wahlberg and Doug Ellin pitched the series as a realistic look inside the biz. Yet the show also depicts fans trying to contact and navigate the biz. Beyond each episode's "behind-the-scenes" dish, *Entourage* delivers weekly helpings of fans "approaching the scenes." By collapsing traditional distinctions between audiences and Hollywood, *Entourage* functions as

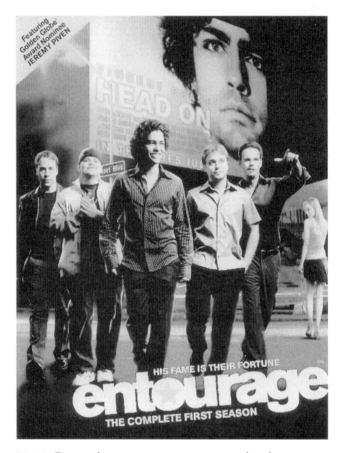

FIG. 6-1 *Entourage*'s onscreen content mirrors its viral marketing campaign. (DVD cover, HBO)

a user's guide to making it in showbiz and a Cliff Notes manual for fan communities. The show underscores the outsider, working-class origins of Turtle, Eric, Vince, and Drama. Over the course of the series they alternately stand gaga-eyed or jaded in the face of celebrities, as fans would. They learn the ropes and traps of public relations from inside and outside. They continually venture into and out of the fame bubble. Each episode shows the posse pressing the flesh with fans, gawkers, the unemployed, and sundry wannabes.

In these ways, *Entourage* dramatizes the porous borderlands of the business, not its inaccessible centers. More than a dramatic premise, teaching fans to travel to and from those centers is very good business. The various multimedia sites where fans interface with *Entourage* show how fan activities drive HBO's primer. As a cultural road map, *Entourage* front-loads useful information for fans and aspirants alike on such topics as managing, agenting, pitching, packaging, casting, marketing, and distribution. It also provides a handy lexicon of employee rhetoric needed to work a room and survive inside *or* outside of the business: acting out, hooking up, networking, negotiating, intimidating, and conspicuous faking-it-until-you-make-it consumption.

A parade of celebrity cameos—James Cameron, Jessica Alba, Val Kilmer, Gary Busey—gives *Entourage* a self-consciousness more like *Entertainment Weekly* and *Access Hollywood* than Bertolt Brecht or Jean-Luc Godard. Mirroring *Entourage*'s "Celebrity-Sighting" fan chat forum on HBO's website, the real work of the underemployed posse making it in Hollywood depends, apparently, on the ability to decipher and leverage celebrity knowledge in public.

The series' big reveal? The posse likes viewers; viewers like the posse; viewers *are* the posse. *Entourage*'s fiction lays out these lessons. HBO's website drives home the same points explicitly. If these road maps are too vague, HBO provides material assistance to help fans *become* the series. HBO's MyEntourage contest on MySpace.com, for example, invited users to round up their own web-based posses and persuade others to link to it. HBO then strip-mined MySpace to select finalists, and MySpace members then voted for the "best" new entourage. By bringing the winning gang to Hollywood for *Entourage*-like "paparazzi treatment," HBO further blurred distinctions between stars and fans in real life just as the series does onscreen. See *Entourage*. Be *Entourage*.

By creating buzz about a series depicting the creation of buzz, HBO's self-portrait shows that Hollywood's real work is not production but rather spin and personal branding. Such concerns now arguably rule the overlapping worlds of both Hollywood and fans. "Pitching inside" may not impress network broadcasters. But it works exceptionally well in HBO's viral, "fan-driven" *Entourage*, where the show's form closely mirrors its marketing. HBO's success at narrativizing and converting viral marketing into A-list programming is merely the tip of a much bigger self-referencing industrial iceberg.

Given the scale and diversity of the film and television industries—a world with at least 250 different official job categories, far more unofficial ones, and a quarter million workers in Southern California—discerning general principles about that world may seem myopic, wishful, or misguided.[1] Yet several trends recur across the many discrete production sectors that I have researched.[2] These persistent tendencies can be mapped on a wide continuum, ranging from institutionalized corporate reflexivity on one end to interpersonalized worker reflexivity on the other. Corporate reflexivity involves top-down self-referencing, values the cultivation of organizational relations, and is closely related to marketing. By contrast, worker reflexivity can be understood via more local forms of craft self-referencing, socio-professional relations, and individual expression. Corporate reflexivity encompasses a set of institutional strategies: branding, marketing, making-ofs, metatexts, franchising, DVD extras, electronic press kits (EPKs), and conglomeration. Worker reflexivity, on the other hand, includes interpersonal activities, mentoring, how-to panels, trade stories, technical retreats, and craft meritocracies. While it may be tempting to force this corporate-versus-worker continuum into the longstanding methodological mold that opposes macro political economies and micro sociologies of work, I argue that both extremes involve a rich array of cultural self-representation and visual expression. Cultural texts, that is, are fundamental parts of both corporate practices and labor activities. Any attempt to understand the industry's economy without understanding these cultural textual practices will provide only a partial picture of film/television work worlds. Likewise, attempting to understand the production culture's reflexive texts onscreen without also examining the economy that animates them is shortsighted. It is difficult, furthermore, to mark precisely where human agency ends on the continuum and corporate control begins.

In the model used here, top-down self-referencing can be understood as an outgrowth of two contemporary corporate goals: first, to "level industrial distinctions" in the production/labor chain; and second, to "level hierarchical distinctions" in the market/distribution chain. While the immediate, abstract goals of the first strategy are to lower corporate costs and eliminate costly labor entitlements, something more profound and unsettling unfolds alongside these benefits. As I have argued elsewhere, and will suggest in the pages that follow, the leveling of labor distinctions in the production workforce—between colorists and timers, directors of photography and directors, producers and executives, editors and sound designers, story editors and screenwriters, reality production assistants and union editors, production designers and visual effects supervisors—keeps much of the workforce off-balance and stirred up with intercraft contention. The manic pace of production today—in union and non-union work, on location and in post-production work—reflects this objective. Desperate filmmaking increases productivity and, some now argue, creativity. The erosion of job classifications also helps drive down production costs and scale, as production companies unintentionally emulate the reform ethos of the game community's *Scratchware Manifesto* in order to create innovative titles with small teams for affordable purchase.[3] Such tactics—downsizing, outsourcing, "team building"—reinforce what I now consider to be the new uber-fantasy and goal of Hollywood: to acquire content for little or no cost and to get everyone to work for free.

The second strategy suggested above—leveling hierarchies in the market/distribution chain—aims to cultivate direct and efficient economic relations with media consumers by cutting out middlemen. But it too has a more lasting cultural impact. Endless publications hail the new consumer power that now ostensibly drives film and television in the United States due to digital interactivity. Fewer accounts acknowledge that Internet-driven media, blogging, and uploading also provide ideal conditions within which the media conglomerates can succeed in utterly traditional business activities like direct-to-consumer marketing. The collapse of the barriers between media producers and consumers, then, is less about democracy than it is about facilitating two specific tactical film/television goals: first, the creation of information cascades on multiple platforms (a publicity-driven viral process needed to maximize fan buzz around exceptional, blockbuster franchises); and second, the cross-promotion of less exceptional conglomerate properties (advertising-driven promotions needed to raise mundane, syndicated content above the media market clutter). The shrinkage of the theatrical distribution window from six months to four months before DVD release and, in some cases, the release of content on all media platforms simultaneously have made finding the audience a chaotic industrial free-for-all. The new direct-to-consumer imperative requires self-referencing and hyper-marketing as necessary tools in the corporate skill set. The deer-in-the-headlights assault of YouTube by Sumner Redstone in March 2007 illustrates the business logic of unsettling the distribution chain and the strategic importance of fan activities. By suing YouTube for $1 billion at the same time that its corporate partner CBS struck a major deal with YouTube, Viacom's schizophrenic relations placed chains on the uploaders even as the company opened its own distribution floodgates. As the majors wring their hands about no longer owning distribution, they textually carpet-bomb video-sharing sites, thereby making every uploader and downloader a potential distributor.

One contradiction of this dynamic is that the industrial leveling of distinction in both the labor chain and the market chain occurs even as the same conglomerates generate excessive degrees of distinction in the form of abundant and ever-narrower consumer niches. Industrial job categories blur. Consumer taste mutates endlessly. Corporate reflexivity provides one effective way to proliferate the sense of distinctiveness among consumers. To do this, *distinction-invoking* texts dominate *external* corporate and consumer discourses. At the same time, however, managers drag out an array of *internal distinction-reducing* rituals to enable workers to multitask, distribute cognition, disperse authority, share competencies, inspire the workplace community, and convert labor itself into artistic expression. Management tomes praise the benefits of nonhierarchical workplace sharing, and managers attempt to precipitate it by creating agitated, adrenalin-driven hot spots within companies. Good for management, perhaps, but alarming for most workers.[4] Others conflate this unstable leveling and sharing with a corporation's moral purpose and altruism.[5] These sorts of two-faced organizational contradictions help conglomerates move light and fast on their feet. The major conglomerates remain profitable, despite their institutional and fiscal inertia, by pursuing the two fundamental requirements of post-Fordism business: to externalize risk (through co-productions, presales, crowdsourcing, merchandizing, and ancillary markets); and to cultivate flexibility (through outsourcing, contract labor, short-term labor commitments, and rapid project-based incorporation cycles). Wooing *consumers* in this kind of cluttered market requires sophisticated metatextual abilities and corporate transparency. Because NBC, CBS, and Fox can no longer expect young viewers to find them in the clutter, they now send some screen content to as many media-sharing sites as possible. At the same time, the attempt to solicit *workers* reeling from industrial flux demands as much in the way of innovative self-referencing and self-disclosure.

The collision of reflexive corporate strategies and reflexive worker countermeasures that I am comparing here can be gauged in three arenas in which the conflict unfolds, each of which is governed by informal principles of self-referencing. Specifically, corporate activities stimulate *unruly work worlds, unruly technologies*, and *unruly audiences*. As corporations make each of these worlds more volatile, they use reflexivity to manage and exploit the same three areas. For example, cultivating unruly work worlds, the first industrial arena, precipitates labor competition, which in turn creates an "oversupply" of content (and workers) at the industry's "input boundaries." An anxious work world of underemployment results, but this anxiety, far from derailing productivity, actually spurs an increase in the production of screen content. Given the excessive failure rate of pilots and films, film and television need a vast oversupply of content, and therefore personnel, at their input boundaries. Stirring up confusion about job assignments makes craft workers especially vigilant and invested in developing projects and turf. For example, movie work shifts from film stock to "digital magazines," and new jobs like "data capture" collapse big parts of at least three traditional job classifications (video assist operator, video playback, and assistant camera). One data capture worker exemplifies the seat-of-the-pants theorizing (and self-justification) that results: "[David] Fincher started doing tests.... He called and said 'this is how I want to shoot, so go learn how to do it.' There was no class." In dialogue with the manufacturer, the self-taught worker established a conceptual approach that director Fincher could then import from show to show. Theorizing how a new tool like this should be used also justifies one kind of labor over others. The rationale provides an upside for data capture, but a downside for video assist. *From a corporate perspective, the reflexive principle of unruly work suggests that as production costs decrease and task uncertainty increases, critical theoretical self-justifications increase.*[6] Every individual is a salesperson, every crew member is a brand.

Second, corporations typically promote new, "unruly technologies" as "user-friendly" devices capable of achieving efficiencies and economies by "collapsing workflows." One company, for example, disingenuously brags, "We offer an end to workflow," and shows cinematographers shooting with MacBook Pros running Final Cut Pro (FCP) bolted to their tripods. This is a suspect fantasy that ostensibly allows for simultaneous production and post-production.[7] Of course, the recent reduction of electronic news gathering from one-person crews and on-camera reporters to video jockeys who shoot and edit their own stories is greeted warmly by aspirants and the unemployed but not by "real professionals." Corporations also actively discipline unruly new technologies by reflexively theorizing them within traditional aesthetic frameworks and standard film/television business and product categories. Tens of thousands of amateurs now produce shows on sites like Live365.com, for example, cheered on by the free-range opportunities of the Internet and pod-casting. No longer demonized as pirates, they are referenced in the industry's new rhetoric (backed by court orders in 2007) as clients, forced to pay licensing and royalty fees.[8] Once media brigands, they are now simply media distributors who are "harvested" by the legal department. *From a corporate perspective, the reflexive principle of unruly technologies suggests that the greater any new technology's disruptiveness, the more extreme the corporate theorizing needed to justify and discipline it.*

Third, film and television have for some time now gladly entered the vast gold mine of "unruly audiences." While the thousands of "cease-and-desist" orders from NBC and News Corp. against YouTube and MySpace uploaders get the lion's share of attention in the trades, the same production companies that whine about piracy are fully involved as unannounced corporate lurkers, outed manipulators, or explicit partners of the same online sites. In 2007, Endemol, a television production company heavyweight, began using the fan uploading site Break.com as a cheap proving ground for risky, experimental program concepts that could later be migrated to primetime. In this sphere, such corporations brand themselves by circulating ancillary content and engaging in viral marketing in order to build more intimate psychological relationships with fans. *From a corporate perspective, the reflexive principle of unruly audiences presupposes that fan loyalty increases as the degree of corporate disclosure and transparency increases.*

Of course the picture of corporate-versus-worker warfare that I've sketched thus far may look too causal, top-down, and deterministic. In practice, any fieldwork in the world of production shows that there is as much ground-up worker agency and rhetorical resistance as there is top-down corporate control and acquiescence. Reconsidering the three unruly industrial spaces just described—but now from below—proves this point. First, in the off-balance, increasingly division-less, unruly work world of crews, worker reflexivity constantly negotiates and resuscitates technical and craft identities for vocational survival. Much reality programming is non-union, and all reality shows are scripted, the majority surreptitiously by story editors and members of producers' staffs. Fearing the networks' plans to use reality shows as warehoused "filler" to circumvent the unions if labor strikes ensued, the Writers Guild of

America (WGA) and the International Alliance of Theatrical Stage Employees (IATSE) went after CW's top show, *Top Model*, by asking the National Labor Relations Board (NLRB) to intervene in a 2006 representation vote. The NLRB sided with the unions, prompting IATSE president Thomas Short to draw out the moral: "This election points up the importance of bottoms-up organizing and grass-roots representation . . . these [types of] employees have always belonged in the IA, and we are pleased to bargain on their behalf."[9] This victory was a warning shot across the bow to networks everywhere against collapsing job descriptions. The IATSE succeeded by critically challenging the industrial "genre theory" that had made spurious aesthetic/labor distinctions between fiction and reality screen content. *From labor's perspective, the reflexive principle of unruly work worlds means that the histories, hierarchies, and cultural rhetoric justifying and explaining crafts increase in prominence and intensity as the oversupply of labor increases.*

Second, in response to the disruptiveness of new and unruly digital technologies, worker reflexivity is deployed to legitimize one technical or craft group over another, usually by establishing superior competence and thus exclusivity. ABC, Fox, and NBC opened up the new digital technology gates in the fall of 2006 to allow viewers to download expensive primetime programs like *Heroes*, *Prison Break*, and *Ugly Betty* for free a day after airing. One problem. In the networks' giddy embrace of video sharing and video iPods, the half-century-old precedents for paying screenwriters and actors syndication royalties or residuals on the shows they had made were thrown out the window. "We've learned from history that when these new technologies emerge we can be left behind," said the president of the Screen Actors Guild.[10] Again, only critical legal arguments from the unions—these downloads were syndicated end uses, not just marketing—forced the networks to back away from the digital free-for-all. Once again, labor had to make convincing assertions (now about technologies) that the new portable and mobile media fit the *old* definition of syndication and distribution windows. Industry's calculated, well-managed confusion between its self-referential marketing and its content was at the root of the conflict. *From labor's perspective, the reflexive principle behind these unruly technologies is that craft and worker theorizing, self-referencing, and collective cultural activities increase as the pace of technical obsolescence accelerates.* Distinctions between workers do in fact matter.

Third, worker reflexivity also churns in response to the unruly audiences that now threaten the lucrative job guarantees once securely held by organized production labor. Users, fans, and digital uploaders increasingly share production and aesthetic competencies with film/television workers. The capabilities that make production workers distinctive and economically valuable, therefore, now matter a great deal. As one wearied, middle-aged videographer complained, "The sad part is that . . . a lot of good, talented people will suffer. . . . It seems that [our profession] is constantly being undermined by wannabes."[11] *From labor's perspective, the reflexive principle of unruly audiences means that worker claims of exclusivity and "professionalism" increase as the popularity and circulation of user-generated content increases.* Dismissing amateurs, independents, and outsiders is a time-honored cultural habit in Hollywood, one that goes hand-in-hand with high production values and the cult of technical superiority. Behind the trade harangues against amateur uploaders and one-person crews, however, many film/television workers in the lesser ranks are quietly migrating to the lower-stakes world of the Internet. Break.com now pays pros to produce uploads, YouTube compensates the best of its uploaders from ad schemes, and Jack Black and his Hollywood "entourage" pose as outsiders who—with authentic outsiders from Channel101.com—co-produce "TV pilots" for the VH1 series *Acceptable TV* (part of the giant Viacom conglomerate).

This worker flip side of corporate reflexivity shows that there is not one type of reflexivity but many, as Georgina Born discovered at the BBC.[12] In Los Angeles, some forms circulate around the interpersonal ground zero of disenchanted employees who "de-fame" their producer/executive bosses at DeFamer.com, who take down former on-set celebrities at TMZ.com, or who bitch and moan anonymously online about horrible conditions caused by specific producers on the set. Remember that for every fan spoiler who ruins future plot episodes of *Lost* through cultural espionage, there are crew member spoilers who sabotage a series story-arc secrets and skewer showrunner hubris by leaking episode information before broadcast. But it gets even more complicated than that. Fans, lay critics, and production employees are not the only ones launching cynical criticisms and unauthorized snarking against shows and producers on Televisionwithoutpity.com (TWOP). Producers and executives themselves anonymously wade through this site, facing critical deconstructions of their shows in order to monitor reception and, if possible, influence it positively. Bravo, a network in the Universal-NBC conglomerate, purchased TWOP in 2007 precisely to harness the churning, agitated buzz of worker and corporate reflexivity as part of the industrially incestuous programming flow that has come to define the Bravo brand.

On one level, the distinction between corporate and worker reflexivity can be expressed as follows. Corporate initiatives circulate metatexts to manage instabilities and unruliness conceptually once labor and consumption distinctions are leveled. In effect, corporate reflexivity builds on the blurring of consumer identities *and* job descriptions in order to mine the economic confusion that follows on both ends of the spectrum. By contrast, worker reflexivity tends to resuscitate many of the distinctions that corporations have tried to level or erase in the production/labor and market/distribution chains. In the world of industrially blurred distinctions, jobs, craft legitimacy, and careers are clearly always at stake. Workers know this and seldom limit themselves to the physical job at hand. Along with online and on-set griping, self-defining statements and meta-commentaries continuously issue from the labor unions, guilds, and professional gatherings in an attempt to manage the volatility from the ground up.

Yet the politics of worker-versus-corporate reflexivity are not as clear and unproblematic as the top-down versus ground-up model may imply. At least in union production, worker reflexivity emerges from professional communities that make their craft, association, or guild self-perpetuating through a quasi-medieval system requiring protracted mentoring. In Los Angeles, such groups still codify "scientific management" efficiencies in order to maximize the degree to which production tasks and subroutines are divided and distributed across department areas and crew. Given the nomadic system of rapid start-up/shut-down production incorporation, workers need to network in order to survive the unending cycles of unemployment. These instabilities provide the groundwork from which many of production's cultural activities, as described in this chapter, are launched. As a social and economic problem-solving operation, production culture now persistently cultivates ideals of unified industry in collusion with management in order to protect incomes after contracts are signed; converts work (paid or unpaid) into cultural capital via socio-professional rituals, demonstrations of craft ancestry, and meritocracy; and buffers underemployment by showing off and leveraging cultural capital via credits and demo reels. From this perspective, the industry uses aesthetic and cultural capital to limit workers. But, given the yearly incomes of many in the industry, including below-the-line workers, it is difficult to explain this shortchanging as a form of victimization. Yes, organized labor is under attack, and jobs are threatened. But even as the old labor system slips, slides, and regroups, many of its practices remain exclusionary. Labor's old guard—still predominantly white, male, and upper middle class—seldom gets much sympathy from the tens of thousands of non-union workers, industry aspirants, women, and people of color in Los Angeles. The resilience of the old system results in part because production labor maintains high costs of entry and exclusivity. Thus, production's cultural rhetoric preaches collectivity even as it bars aspirants and outsiders from entry. Yes, production is anxious. But it is anxious about the many invisible aspirants and underemployed individuals off the set and outside the studio as well.

The growing confusion and fascination about what the industry now is, and is not—dramatized by the proliferations of "shows about the biz" in prime-time—begs larger social questions. Five industrial tendencies can be understood as defensive responses by post-network labor to this rising confusion and anxiety over identity and limits. First, though, two general relationship-building tendencies and collective rituals tendencies lie at the heart of these responses: regeneration, keyed to the logic of the group; and legitimation, keyed to the logic of career and craft. These tendencies tend to *segregate* professional practices from audience activities, spotlighting their differences, by continuously redefining and re-valuing the otherwise uncertain futures of creative communities through expressions of professionalism.

From the perspective of *regeneration*, many cooperative activities of film/video workers can be understood as socio-professional forms of consensus building or dissensus making. These unifying and segregating tendencies tend to emerge in response to broader institutional and industrial changes.[13] Effectiveness in controlling these processes proves crucial in the formation, survival, and re-creation of production groups, firms, and associations. Both consensus and dissensus activities enable the diverse and heterogeneous coalition of craft communities to redefine and regenerate themselves constantly—through cultural expressions of willed affinity—as a temporarily unified industry. Such identity boundaries hold at least until competition or economic and technological changes threaten consensual relationships and temporary, tactical affinities. Practitioner networks also participate in broader social and professional processes through rituals and events that define the industry in a symbolic and public relations sense. Members of unions and guilds, that is, do not just participate interactively in the technical tasks that define production in the workplace. They also represent different groups that constantly cultivate either a sense of professional autonomy (when needed) or common cause and industry-wide consensus (when coalition is needed to cross craft and company lines). Rapid technological change—and the threat of obsolescence—has destabilized the traditional ways that tasks are distributed during a production. In response to these instabilities, companies regularly stage handholding events. One studio hosts holiday "tree-lighting ceremonies" to forge common "studio family" identities. Yet many of the attendees are transient producers and contract employees soon to be replaced by the next group of migratory tenants on the lot.[14] Without the real long-term employee loyalty that defined and enriched the studios in the classical era, contemporary companies work overtime to concoct imaginary families for their brands in ways that cover over the depersonalizing churn of tenants and contract employees.

Industrial changes also animate the social rituals that are used to cultivate a craft or trade group's legitimacy. These *legitimizing activities* frequently seek to underscore the ostensibly necessary role the given craft or trade group plays within a common industry (even though this consensus and commonality can be largely symbolic). The production stories that practitioners tell are not just narratives about "what happened"; they also provide legitimacy for careers and crafts. Allegories are also arguments, and sometimes parables, that legitimize one or more perspectives even as they discredit others. I include in this proposition not simply the stories and anecdotes one finds in trade publications or the career "war stories" that professionals tell in public appearances, but also the icons, images, demo tapes, and self-representations that practitioners make across a wide range of formats. Invariably, by telling stories or making demos, practitioners individualize industrial phenomena for career reasons. This tendency to personalize is common in almost any field of work, since stories are among the most efficient ways to anchor more complicated ideas about any industry.[15] I have been interested in both the career and the institutional logics of trade narratives, iconographies, rituals, and spaces. Such forms and practices serve as self-reflections or group self-portraits, thus providing scholars with evidence of workplace analysis and self-interrogation that is frequently as provocative as the films or series that the group produces for the public at large. Trade stories are told to value and resuscitate careers, but they are also emblematic expressions of trade-group cultural preoccupations.

Three other tendencies—distributed cognition, producers-as-audiences, consumerism-as-production (or producer-generated users)—are antithetical to regeneration and legitimation because they have the opposite effect. They *blur* lines between producer and consumer. These trends operate systematically at a broader cultural level even though they are based on very local work practices. They are fueled by the need to *desegregate* professional and audience activities. This separation makes these last three tendencies particularly effective in popularizing the kinds of self-referencing that typify both industry and consumers at large. Professional knowledge about film/television production now asserts a widespread cultural competence *and* consumer activity. These final three tendencies directly inform reflexive primetime shows like *Entourage* and *On the Lot*.

First, while the organization and working methods of film/video production crews can be understood as examples of *distributed cognition*, so too can the activities of consumer *users* of production software and studio/network websites, whom the industry now values as "networked externalities."[16] Once derisively dismissed as a mob, users today are welcomed by film/television corporations in their efforts to harvest productive work from audiences. Rather than closely guarding intellectual property, the smart guys in film/television today actively attempt to harvest the power of the online audience or "hive" through strategies of "crowdsourcing" or "hive sourcing." The collapse of traditional distinctions between entertainment content and marketing, a fundamental concern of this chapter, now spurs corporations to go with the flow of the audience rather than to fight it, to tap the hive as a source for production, not just consumption. Production groups conceptualize a film or scene interactively through worker networks that pursue stylistic or narrative effects based on pre-established schemas and problem-solving roles. The circulation of computer software and video games—along with a corresponding collapse of distinctions between work, leisure, and creativity—has pushed these forms of distributed cognition far beyond the bounded sets of old media or Hollywood. Web and PC software applications mimic aesthetic modes of interaction via interfaces derived from theatrical, gaming, televisual, and cinematic icons and traditions. Apple sells tens of thousands of adolescent YouTube-fixated, iLife-cocooned software buyers on the complex "meta-editing" aesthetics (the sizzle) behind its FCP and iDVD (the steak), thus hawking film theory as a home electronics user's guide. Even as they invite users into culturally coded, collective forms of thinking and creating, some products and services sell the utterly practical posture of the critical analysis they provide. As one product announced: "No Abstract Theory—Just 200 Proven Techniques."[17] Yet the same product provides a dizzying list of theoretical prompts for the frustrated, aspiring screenwriter who has lost "the big picture."[18] Another software's demo traces the academic origins of its patented "story engine" to theories of artificial intelligence, sexuality, physiology and biochemistry, and psychiatry.[19] The company's website flaunts its theoretical utility to imagine cinematic worlds in explicitly scholarly terms: "If Collaborator is the conscientious grad student Teaching Assistant, then Dramatica Pro is the worldly writing professor with a philosophical bent. Dramatica takes theory to a new level Collaborator never dares approach."[20] Theory is the ghost-in-the-machine of production and screenwriting software. Sometimes it churns along quietly. Sometimes it guides the conjuring writer Socratically. At other times film theory mentors the mash-up-producing YouTube rookie posed as a Hollywood insider. Off-the-shelf software embeds these theoretical discussions in its design, and each new software user becomes (in computer science terminology) an externality that adds value to the ever-growing, interconnected, mutating network that ensues.

Second, we seldom acknowledge the instrumental role that *producers as audience members* play or the role that the industry as cultural interpreter plays. Film/video makers are also audiences, and film/video encoders are also decoders. Media scholarship tends to disregard the inevitability of maker-viewer multitasking and the industry's competence as an interpretive "audience."[21] Many favored binaries fall by the wayside when one recognizes the diverse ways in which those who design sets, write scripts, direct scenes, shoot images, and edit pictures also fully participate in the economy, political landscape, and educational systems of the culture and society as a whole. Above-the-line producers, directors, and executives are especially good at intentionally confusing the audience/producer split. Executives frequently invoke hard numbers from research departments when useful—but ignore them when the data contradict their personal hunches or intuitions. To break this research/intuition quandary (and the managerial conflict that necessarily follows from it), executives employ a favored tactic. They master the pose of "speaking for the audience" in order to prevail in contentious production, development, and programming meetings. Arguments that the "audience wants this" or "that" trump all others—at least if the person saying so has enough institutional power to ignore evidence to the contrary.[22] Production personnel also make decisions within the parameters and constraints of the consumer electronic and home viewing environments that they personally know. Media conglomerates, in turn, have shifted to direct audience merchandising and the DVD as the key to delivering features, given the inability of ratings and advertising systems to track the multichannel flow into the home accurately.[23] Directors/editors use split screens and frenetic editing to keep apace of viewers' sensory acceleration, while producers design online components to engage the multitasking activities of increasingly distracted viewers.[24] To exploit these new digital options, television creators develop mobile content and "snack TV" for hand-held "third screens"[25] and insert Internet referrals for the audience within screenplay dialogue and primetime scenes.[26] At the same time, online blogs by series characters help viewers solve fictional crimes before sending them back to view the next episode.[27] Finally, production personnel circulate publicly in consumer culture. Many professional associations and guilds have speakers bureaus, educational divisions, and publicized internship opportunities. Other companies cultivate and interact with the public through timely topical media events or by co-sponsoring local quasi-Sundance film festivals. Colleges host alumni meetings in Los Angeles for industry networking (aka fund-raising) and produce "Alumni in Hollywood" features for their alumni magazines. Far from Los Angeles and New York, film/television professionals circulate as short-term artists-in-residence, while film/video equipment companies, star directors of photography, editors, and directors travel widely to participate in regional production workshops and technical demonstrations.[28] All of these producer-as-audience initiatives work to merge audience identification with industrial identity.

Third, although much has been made recently about *user-generated content* (UGC), far less attention has been focused on two other trends: what I term UGC's evil twin, *producer-generated users* (PGU); and the many ways that production analysis serves as a form of hyper-consumerism. Even as production theory and media aesthetics now circulate widely in and as consumer discourses, audiences themselves frequently serve as self-conscious media producers and critics. More than simply a technology-driven phenomenon, many popular press outlets and websites now promulgate film/television theory as fan discourses. As discussed earlier, Metacritic.com and Televisionwithoutpity.com compile second-order reflections on critical trends and biting deconstructions of film/television style and content.[29] *Entertainment Weekly* and newspapers formulate film/television canons through annotated lists of the "most important DVDs, films, TV shows you should own."[30] Making-ofs and behind-the-scenes documentaries promote "production thinking" as staples

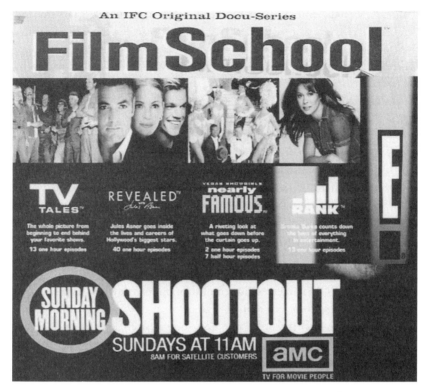

FIG. 6-2 Endless onscreen deconstructions and paratexts drive *Film School*. (Publicity material, IFC)

on many channels and networks (AMC, the Sci-Fi Channel, HBO, Bravo, Discovery, TNT, IFC, and so on). Plus, through reality television, "making-ofs" and "makeovers" have become entertainment programs. We don't just get *Extreme Makeover: Home Edition*, we get *Extreme Makeover Home Editions: How'd They Do That*. We don't just get MTV's *Making of the Band* (about music video production), we get ABC's *Next Action Hero* (blockbuster casting and acting), Jon Favreau's *Dinner for Five* (critical debates about cinema), AMC's *Shoot-out* (industry trends and film development), and HBO's *Project Greenlight*. (producing, directing, and managing a feature film). Given this televised cineastic milieu, the series *Film School* (2004) and *Film Festival* (2005) on IFC and *After Film School* (2006) on Fox Movie Classics are actually fairly unremarkable exercises, since production pedagogy is *constantly* churning on many other channels as well. The theorizing bent also spills over into blockbuster films, where some journalists pull out their Philosophy 101 Cliff Notes to score smug big-screen references to "metaphysics," "ontological" inquiry, the "psychoanalytic id," "ancient philosophical conundrums about the nature of free will," and celebrity "intellectuals" from Columbia, Harvard, and Princeton.[31] Others paint a far darker picture of onscreen theoretical exhibitionism: "This summer, millions of teenagers have been invited to experience *the tedium and pedantry of graduate school in Dolby-surround*, accompanied by the latest special effects."

The industry also solicits and "welcomes" contact with viewers. Test screenings in "Glendale" or "Peoria" have been a part of the Hollywood mystique for decades. But current economic conditions have made the solicitation of viewers for service as online focus groups or test screening participants even more intense.[32] These public solicitations posture audience research less as a "deal with the devil" than as a unique opportunity to change culture and "serve the entire viewing public" by building personal relationships with producers who are (apparently) standing by, waiting for each viewer's every decision. Critical acumen about production, therefore, doesn't just travel from Los Angeles to the heartland. It supposedly surges back to Hollywood in an ecstasy of shared, staged critical analysis from the provinces as well.[33] Viral multimedia marketing and PGU strategies also employ numerous new media formats to perpetuate production's conceptual frameworks *as* viewing frameworks.[34] The widespread network/studio practices discussed earlier of planting faux personal videos by fans on MySpace.com and YouTube.com in order to market forthcoming features virally, or of harvesting antagonistic personal video mashes on the same sites as part of anti-marketing campaigns, provide merely the latest evidence that lines between producers and consumers have been irrevocably breached.[35] Producers generate faux-amateur content, buy and distribute amateur content professionally, provide online learning in film/video aesthetics, spin blogs and online discussions, spoil ostensible secrets as stealth marketing, snark and defame competitors, pose as fans, award fans, and *are* fans. Welcome to the brave new world of PGU.

The classic binaries separating culture industry and citizenry, producer and consumer, and ideological perpetrator and victim are not satisfactory explanations for these cultural-industrial tendencies. This insufficiency is especially evident when one talks to screenwriters, producers, and directors about how personal, private, and familial concerns they deem important inform, inflect, or percolate up through the films and series they make. With rare exception, such creators have little problem discussing why and where the themes they deal with come from. Certainly this heightened form of analysis and interpretation by film/video professionals—together with the audience's continuous awareness and reading of production nuance in shows like *Entourage, On the Lot*, and *After Film School*—provides a form of critical interrogation every bit as complex as those of professional critics. Yet we seldom grant industry this critical capability—or recognize that industry's obsession with self-reference and self-disclosure has proven corporate, not just aesthetic, benefits.

Notes

1. This figure of 250 "craft classifications" is for film, video, and CGI only, not for television, broadcasting, or the music industry, and is from the definitive text by William E. Hines, *Job Descriptions for Film, Video, and CGI* (Los Angeles: Ed-Venture Books, 1999).
2. This essay and my book *Production Culture: Industry Reflexivity and Critical Practice in Film and Television* (Durham, NC: Duke University Press, 2008) resulted from film and television industries fieldwork done in Los Angeles between 1995 and 2006, primarily among below-the-line workers. Methodologically, I used an integrated cultural-industrial approach that combined ethnography, textual analysis, interviews, archival research, and economic analysis.
3. "The phrase *scratchware game* essentially means a computer game, created by a microteam, with pro quality art, game design, programming and sound to be sold at paperback book store prices." *Scratchware Manifesto*, http://www.homeofthe underdogs.net/scratch.php.
4. See Lynda Gratton, *Hot Spots: Why Some Teams, Workplaces, and Organizations Buzz with Engery—And Others Don't* (London: Berrett-Koehler, 2007).
5. See Nikos Mourkogiannis, *Purpose: The Starting Point of Great Companies* (London: Palgrave Macmillan, 2007).
6. This statement and the five italicized propositions that immediately follow in this section (about what spurs competing forms of reflexivity in both corporate and labor contexts) have been excerpted from the "Conclusion" of *Production Culture* and significantly revised for inclusion in this chapter. I thank Duke University Press for allowing me to adapt and summarize this material for a different context and argument.
7. See the ad for Tekserve, *TV Technology*, January 10, 2007, 30.
8. See Alana Semuels, "In Cyberspace, Nobody Can Hear You Stream," *Los Angeles Times*, March 20, 2007, http://articles.latimes.com/2007/mar/20/business/fi-webdjs20.
9. This quote and account is from Carl DiOrio, "'Model' a Step Closer to Unionization," www.HollywoodReporter.com/hr/content_display/news, posted December 5, 2006.
10. This quote is from Richard Verrier, "Residuals Debate: Old Script on New Set," *Los Angeles Times*, October 19, 2006, http://articles.latimes.com/2006/oct/19/business/fi-newmedia19.
11. Letter to the editor of *TV Technology*, April 2, 2007, 4.
12. See Georgina Born, "Reflexivity and Ambivalence: Culture, Creativity, and Government in the BBC," *Cultural Values* 6, 1–2 (January 1, 2002): 65–90.
13. Chapters 2, 5, and 6, in particular, of *Production Culture* demonstrate this point.
14. At the December 5, 2005, "holiday tree lighting" at Paramount, for example, studio executives celebrated and thanked the "Paramount family," even though many of those gathered were transient producers and independent company employees who would be replaced by other tenants on the lot in a matter of weeks or months. This account is from a television producer on the lot, describing the odd ethnic/religious contradictions of such an event.
15. In fact Angela McRobbie makes a key distinction between careers in the old industries, which were based on managing career "narratives," and the new creative industries, which are driven by what she terms "portfolio careers." See McRobbie,

"Clubs to Companies: Notes on the Decline of Political Culture in Speeded Up Creative Worlds," *Cultural Studies* 16, 4 (2002): 517–531. Hollywood workers show that trade storytelling is almost as important as the demos they make and use to professionalize and manage their careers.

16 The networked externalities principle states that each new user added to an interactive or communication network "adds value" to the network as a whole. One of the themes I examined in chapter 5 of *Production Culture* and elsewhere involved the symbolic struggle over authorship that pervades the film and television industries. One form of authorship confrontation occurs when numerous executives, agents, and rights holders demand and receive producers' credits in a film, despite not having been involved substantively in the actual production process. This unruly struggle can involve scores of producers and peripheral individuals, and is only the most obvious form of distributed cognition I am describing.

17 This statement/title and the quotations that follow are from David S. Freeman, "Beyond Structure: L.A. and New York's Most Popular Screenwriting and Fiction Workshop!" a direct mail flyer, summer 2004.

18 The workshop offers to teach "11 ways of modifying any idea to develop radically alternative story-lines," "5 ways to express a character's dimensionality," "4 ways to inject depth or subtext," and so on.

19 Produced to help writers craft better screenplays, Dramatica Pro includes demo tapes that explain the complicated theoretical and scientific origins of the manufacturer's sophisticated "story engine." The demo articulates how the software is: a form of self-analysis; an AI consistency engine based on a simulated user-in-the-text; a form of visual modeling that anticipates imagined cinematic worlds before they are produced; and a programmed cognitive map of all possible logical arguments and choices available to users of the software.

20 The conflation of software with one's personal "professor with a philosophical bent" is from a review by *Back Stage West* (n.d.), posted at http://www.dramatica.com/review/quotes/index.html.

21 Film and television have never been produced in a vacuum that completely walls off "the industry" from "the audience." The general journalistic and academic tendency to segregate industry and culture methodologically, therefore, is shortsighted and problematic at the least.

22 No matter how disingenuous this rhetoric can be at times, this kind of assertion is commonplace in everything from production meetings to "Q&As" with producers after public screenings. The "producer as audience" trope also percolates through various behind-the-scenes genres, such as ABC's mid-1970s hour-long primetime news special entitled *ABC News Close-up: Primetime TV, The Decision Makers*. See PVA#2340T at the UCLA Film and Television Archives.

23 Sixty percent of revenues for films come from electronic media, and DVD merchandise is now by far the most lucrative of these home delivery formats. As a result, DVD opportunities are exploited during all major film productions in Los Angeles, both in terms of what the film/video looks like, but also in terms of all of the metatexts and making-of segments that will be produced and packaged with the DVD. Perhaps the best example of this domestication of film studio thinking is the first release of the boxed set of *King Kong* in December 2005, which consisted almost entirely of making-of stories and was released simultaneously with the feature film premiere.

24 A consortium of major studios established the Digital Cinema Initiatives (DCI; the

"Digital Cinema Laboratory") to conduct various comparative tests of new digital projection systems with test audiences. DCI underscored to the public the fact that "engineering knowledge is NOT required," even as it stroked the egos of viewers who had the "opportunity to participate in an historic test that will influence the way motion pictures will look on the screen for the next 50–100 years!" (This statement and solicitation are from an e-mail blast sent by Charles Schwartz, CEO of the Entertainment Technology Center, April 26, 2004.) Far from dismissing "amateur" technical knowledge, DCI enjoined the public to participate critically in the development of media consumption technologies. Finally, companies like GoldPocket Interactive provide television producers and network programmers with "turnkey" interactive systems to augment programming. In providing an easy way "to connect internet audiences to your TV broadcast" via "competitions, chat sessions, [and] streaming audio/video," the one-stop service allows programmers to "reach millions of people," in order to "get the audience to reach back to your programming—before they go somewhere else." From GoldPocket Interactive, direct marketing brochure, Los Angeles, 2003.

25 With television viewing and box office numbers declining, many entertainment providers began developing content that could be efficiently dispersed to mobile phone users. The use of cellphones for film/television consumption follows concerted industrial efforts to find and harness a "third screen" in the viewer's hands, sometimes referred to as "snack TV." The television and the PC/web are deemed the other two screens. Fox Television, for example, developed *24: Conspiracy* to launch the new season of primetime's *24*. Verizon cellphone users in the United Kingdom and United States received twenty-four one-minute "mobisodes" that corresponded to the primetime series, albeit with a different cast. A mobile division of NBC produces up to twenty news stories per day for mob-casting. Fans can also watch updates from CNN, regular sports reports, or outtakes from *The Simple Life*—all custom-produced to work effectively on the two-inch screens of cellphones. One marketing analyst described the efficient financial logic of mobile phone usage: "You're not necessarily looking to channel surf ... you've got five to 10 minutes to kill." This quote is from Linda Barrabee, senior analyst at the Yankee Group, a market research firm, as quoted in Matea Gold, "I Can't Talk, I'm Watching My Cellphone," *Los Angeles Times*, May 7, 2005, http://articles.latimes.com/2005/may/07/entertainment/et-mobiletv7.

26 This characterization of NBC's *Crossing Jordan* is from the "TV Reaches Millions of People" advertisement in the *Hollywood Reporter*, February 6–12, 2001, 19.

27 This "circular" strategy is discussed further in Chris Gaither, "The Plot Thickens Online," *Los Angeles Times*, February 25, 2005, http://articles.latimes.com/2005/feb/25/business/fi-webtv25. Television creators now design stories that directly target the audience's "second shift" life outside of primetime, through multitasking technologies that mirror increasingly accelerated lifestyles. Contemporary on-the-run consumers are deemed too unpredictable and migratory for traditional television programmers and advertisers. Largely circumventing the problem of content clutter and grazing, pod- and mob-casting promise cost-effective ways to find and grab transient viewers with bursts of customized niche programming.

28 Examples include HDCamp and DVExpo/Chicago/Dallas. Film/television associations also regularly host production award shows that function as primary forms of onscreen consumer entertainment.

29 See http://www.televisionwithoutpity.com/ and http://www.metacritic.com.

30 See, for example, the "Special Edition: Overwhelmed by Extras and Specials and Extra, Extra Specials at Your Local Video Store? We Make It Easy with This Guide to 50 New and Revisited Titles," *Entertainment Weekly*, April 15, 2005, 37–50.

31 The 2003 arrival of the features *The Hulk* and *Matrix Reloaded* and their references to film theorist James Schamus and philosopher Cornel West generated endless "think pieces," along with concept- and name-dropping throughout the popular press. These phrases and the quotation that follow are from A. O. Scott, "The Pretentious Summer Superhero," *New York Times*, July 13, 2003, http://www.nytimes.com/2003/07/13/movies/13SCOT.html (italics added).

32 The following promise strokes the viewers' awareness of their role in production: "Your participation in a Television Preview screening will serve the interests of the entire viewing public, as you will be providing direct feedback to producers, directors, sponsors, and other people behind television, who in turn will be better able to understand and respond to your viewing preferences." This statement begins with, "You have been selected . . . ," and is included in printed correspondence (direct mail) to the author from G. B. Edwards, Director, Audience Selection Staff, Television Preview, Inc., Hollywood California, Spring 2004.

33 Admittedly, film/video production cultures do not always approximate or mirror the cultures of consumption. Yet there are many significant overlaps between the two cultural sets, as these eight industrial examples indicate. When one finds marked differences between cultures of film/video production and cultures of consumption, important and/or problematic issues can be found and critical questions raised. The politics of race, for example, offer a classic, recurring example in this regard. Racial representations produced by the progressive, white, liberal Hollywood establishment seldom mirror the actual degree of racial diversity in American society at large or align easily with public pressures to achieve ethnic and racial diversity in the industry. As a result, public activists emerge every few years in repeated attempts to pressure Hollywood to change and diversify. Many of these confrontations end in head-scratching or disbelief, mostly because industry employment still does not reflect the racial diversity of society at large. Even so, the question of industrial status and autonomy in this racial "mismatch" is not an either/or question.

34 Bonus tracks and making-ofs function like film appreciation or intro to production courses. Hidden DVD "Easter eggs" facilitate film/television production pedagogy among consumers. The ability to choose among multiple shot angles or endings to films on DVD cultivates production competencies among users. Directors' tracks encourage viewers to see production as a working process from its makers' point of view.

35 In August 2006, I was asked by a marketing staffer at Columbia Tri-Star studios if I could provide film students who would be willing to produce "their own" unpaid videos for YouTube.com that would have the effect of marketing a forthcoming feature film.

7

Fan Creep

• • • • • • • • • • • • • • • • • • • •

Why Brands Suddenly
Need "Fans"

ROBERT V. KOZINETS

In the world of new product development and innovation, the term "feature creep" is given to the tendency of designers and engineers to keep adding features to a product. For example, a cellphone manufacturer might first add a high-definition video camera to the phone, then a digital voice recorder, then a remote car ignition apparatus, a bottle opener on the side, and finally, a small flamethrower for emergencies. These extra features are infamous in the worlds of high technology and computer software development. Websites are particularly prone to this enthusiasm for add-ons. Feature creep is commonly blamed for cost overruns and missed schedules. It overcomplicates the elegance of basic designs and leads to trade-offs in efficiency. It also complicates the world of marketers, which thrives on simplicity. A simple and clear message, based on a clear feature or feature set, can cut through the clutter of thousands of competing claims and is usually the best one.

When used in the world of fandom, the word "*creep*" often possesses a different meaning. A creep is an unpleasant, undesirable, or obnoxious person, usually a man. As the adjective "undesirable" would suggest, the term has negative sexual connotations as well: creeps can be not only lonely but also predatory; they tend to try too hard, but they are stigmatized; they find acceptance difficult. In the Radiohead song of the same name, the über-creep speaks for all outsiders and marginalized Others: "I wish I was special / But I'm a creep / I'm a weirdo / What the hell am I doing here / I don't belong here."

As the "get-a-life" responses of Henry Jenkins indicate, and as practically all fan scholars point out, various media fan communities have for many decades suffered from this stereotype of being composed of creepy, mainly male, and sexually unattractive losers.[1] Whether the negative image is accurate is not the issue here, although fan scholars tend to argue convincingly against it (see the extensive treatment in Jenkins's *Textual Poachers*, for example). As recent movies like Kyle Newman's *Fanboys* suggest, the negative image of the fan persists in popular culture and in popular thought. The advent of information technology and the rise of video gaming and anime seem only to have spawned new forms of bias, new alleged proofs of the inherent nerdiness, uncoolness, and desperation of the fan.

Given the strong notion of fan-as-creep, why have we seen the very obvious embrace of fan creep in marketing and business? If feature creep is the tendency for new product engineers to add more and more features to a product, I propose in this chapter to use the term "*fan creep*" to reflect the idea that business managers, strategists, and particularly marketers have hit upon the notion that consumers should be courted as more than mere consumers; they need to become "fans." Accompanying the cultural shift that has elevated the fan to a far less stigmatic status than ever before has been the colonization of this highly committed communal-emotional-commercial identity space. In this chapter, I explore how the popular, culturally embedded notion of fan identity has been increasingly analyzed, synthesized, and regenerated as a strategic objective of contemporary marketing managers. Partially due to the idea of the fan, and partially reflected within it, consumption in contemporary postmodern society has been reconceptualized as an affective, engagement-seeking commitment that has somewhat suddenly become a desirable, if not a mandatory, goal of effective marketing practice.

From a cultural perspective, I find the notion of fan creep fascinating. With some minor exceptions, media and cultural scholars rarely venture into the considerable and, I believe, significant business school literature on consumer behavior, marketing, and "Consumer Culture Theory."[2] Yet marketing and consumption scholars have been examining, and contributing to, managerial thinking on the image-laden, meaning-drenched, and symbolic nature of consumer responses to product for over half a century. Sidney J. Levy, the spiritual godfather of consumer culture theory, asserted in the 1950s that marketing is mainly about the selling of symbols. He advised managers that "if the manufacturer understands that he is selling symbols as well as goods, he can view his product more completely. He can understand not only how the object he sells satisfies certain practical needs but also how it fits meaningfully into today's culture. Both he and the consumer stand to profit."[3]

Thus there is considerable impetus to track, in however rudimentary a fashion, the outlines of the history of the development of the consumer-as-fan idea in consumption and marketing. This idea stands at the junction of cultural and media studies and of consumer culture and marketing scholarship. These notions of affective attachment and "brand fandom" thus can serve to expose important and dimly understood tendencies underlying the changing consumption commitments requisite in the nature of both consumer and media society. Why would marketers want consumers to be fans? What does it mean for a consumer to be a fan? Why would consumers want to be fans? Are there risks to marketers? Are there risks to consumers? What does the notion of consumer-as-fan indicate about contemporary consumer culture? How might it change our understanding and our theorizing about media culture, popular culture, and consumer culture? In this chapter, I will touch upon answers to these questions first by analyzing several key managerial texts pointing to the growth and diffusion of the consumer-as-fan idea and identity-goal. Next, I will offer an analysis of fan identity that seeks to synthesize notions from fan studies with those from consumer research. Finally, the chapter will present some provisional conclusions and suggestions that may lead to further understanding of this phenomenon.

Stark Raving Fans

The completely unrevolutionary 1993 book *Raving Fans: A Revolutionary Approach to Customer Service*, by popular "One-Minute Manager" author and consultant Ken Blanchard, is based on the simple idea that satisfying customers is no longer sufficient for contemporary businesses.[4] The goal has changed to creating "raving fans." Although Blanchard is vague about the specific characteristics of the customer fan, he uses the metaphor constantly and consistently throughout the book to illustrate how particular companies have created advocates, or evangelical supporters, out of consumer interactions. The idea of evangelism, as well as links to sports fans' devotional fervor, also play very important roles in the book. Blanchard discusses the notion of "owning" consumers, by which he means not just gaining their purchasing power for their entire lifetimes, but also their hearts, minds, and, yes, even spiritual devotion. His key points pertain to customer service. By exceeding expectations, shifting to what he calls "Legendary Service" (thus introducing, albeit merely by implication, the important mythic component), companies can own consumers completely and turn mere fans into raving fans.

The adjective *"raving"* is particularly important and interesting as used here.

In common discourse it often precedes the term "maniac" to add force to a description of someone who acts and talks irrationally and incessantly. The term has delirious overtones of the feral, and sometimes of admiration, as in "a raving beauty." Raving, we can conclude, is about extremely strong emotions that cause a powerful reaction: awe and the need to talk about it. Blanchard's raving fans are in love with a company, with a brand, so much so that they cannot stop thinking and talking about it. It is this addictive idea of loss of control, perhaps more than anything, that makes the concept attractive to business. If a manager can hit upon the correct customer service formula, Blanchard seems to suggest, consumers will go out of their minds with mad, crazy love for a brand.

Of course, as has been overanalyzed in the fan studies literature, the term "fan" originates from the same root as "fanatic," the Latin *fanaticus*, the one who belongs, heart and soul, to a particular temple. We see the same emphasis on affective commitment in popular definitions of fandom and fans, such as in Cornel Sandvoss's specification of "fandom as the regular, emotionally involved consumption of a given popular narrative or text."[5] Sandvoss, however, later pointed out that his definition may have been deceptively narrow: "We do not describe popular icons such as musicians, actors, or athlete, or other fan objects such as sports teams, as deliberately authored texts."[6] Yet fans will follow particular icons or objects across multiple media, reading about a sports team on the Internet, subscribing to magazines and newspapers, buying books, traveling to stadiums, and watching games on television. Thus deliberate authorship is not necessary for the text, and the reader her or himself seems to be an operator in this transmedia space, drawing together iconic symbols and images across a range of platforms. As Sandvoss suggests, "The fan text is thus constituted through a multiplicity of textual elements; it is by definition intertextual and formed between and across texts."[7] The guiding principle, however, is one of emotional involvement, the all-important affective commitment. And it is this highly committed and evangelical element of fandom, of the fan experience, that has drawn marketers, business consultants, and business managers to embrace wholeheartedly the identity of consumer-as-fan.

Another important illumination comes from the influential book *The Entertainment Economy: How Mega-Media Forces Are Transforming Our Lives* by Booz-Allen & Hamilton consultant Michael J. Wolf.[8] Wolf offers a single, simple, big idea: media and entertainment have become the driving forces of the global economy. Therefore, all businesses need to act as if they were in show business. At the early morning of the Internet age, everyone, he claims, needed to be in the content business. Although, amazingly, he never mentions or analyzes fans and fandom in his book, Wolf cites the Harley-Davidson "pilgrimage" to Sturgis, South Dakota, to illustrate his idea that "entertainment entrepreneurs are exploiting the communitarian impulse that was formerly the exclusive province of religion."[9] Fanaticus devotees appear yet again.[10]

This explicit relation of brands to religious devotion is the thesis of *The Culting of Brands: When Customers Become True Believers*, by Douglas Atkin (another advertising manager).[11] Atkin's premise is intriguing, if a little bit frightening. Because cults inspire such loyalty and devotion, perhaps we can learn what they do and replicate their practices in business to build deep, lifelong, irrational, and emotional commitments. He studied a range of cults, such as the Unification Church, in order to derive basic principles to guide brand marketing, then showed how these principles of "cult branding" have been followed by companies such as Harley-Davidson and Apple. Atkin is noteworthy not only because he does not shy away from the unification of church, cult, and commerce, but also because he studies fan clubs and repeatedly holds fans to be a sort of ideal consumer type, one filled with longing and devotional desire for a commercial offering.

Loving Big Brand-ther

Another key text in this movement toward the acceptability, and near inevitability, of cultivating deeply desirous consumers and delirious customers as the fans of brands is *Lovemarks* by Kevin Roberts, CEO of the global advertising firm Saatchi & Saatchi.[12] Employing the common problem-solution format reminiscent of award-winning advertising, Roberts begins by pointing out how brands are in trouble around the world: shares are declining, loyalty is buckling, and so-called "private label brands" (the ones that retailers sell with their own store brands) are on the rise. Why is this so? What can be done? The problem, Roberts claims, is that brands have been managed in an overly rational manner. But consumers are not rational, and they do not seek a rational relationship with brands. Rather, they yearn for a deep, passionate, devoted, and loyal relationship with brands. Consumers want to love brands.

The remainder of the book charts Roberts's journey to discover the ingredients behind "lovemarks," his new term for brands. No longer shall a brand merely be a designation of origin, of quality, or of attributes. It will be a mark of deep emotional commitment. Unlike mere products, brands, or fads, lovemarks must command both respect and love, a power they can achieve through a magical trinity of mystery, sensuality, and intimacy.

The details of Roberts's formula are not as important as his identification of the absolutely clear and highly influential shift away from the desire for purchase to the unmitigated desire for desire. The difference between a brand, which is respected but not loved, and a lovemark, which is both loved and respected, is, of course, love. And what could exemplify this love more in the

Church of Mass Culture than the relation of the fan to the object of unquenchable desire? Economically, socially, culturally—in terms of devotion, there is no one quite like the fan. Evocative of Blanchard's hyperbolic "raving maniac" and also of the single-minded *fanaticus*, Roberts frequently describes the consumer's relation to lovemarks as "loyalty beyond reason."[13] Mystery? Sensuality? Intimacy? In a brand? In the world of contemporary marketers, the desire for consumer irrationality reigns supreme. Indeed, it is more than just a desire. As any episode of *Mad Men* (or *Bewitched*, for that matter) clearly illustrates, marketers study consumers and deliberately plot and plan how to induce irrationality. In a world where private label brands often contain nearly if not exactly the same components as name brands at significantly lower prices, the alleged irrationality of consumer response to the brand is absolutely key.

The lovemarks approach is certainly not without its critics. To activist Anne Elizabeth Moore, "it becomes clear what the strategy of lovemarks really is. It's a forged connection to authenticity, a borrowed veil of integrity, a disingenuous stab at honesty."[14] If Roberts describes the addition of love as the introduction of heat, or emotional temperature,[15] Moore sees inauthenticity: "Because, in the end, 'lovemarks' describes not an emotional connection but a financial transaction."[16] That critique is mistaken, however. As any comic book fan who has just purchased a precious title can tell you, or a sports fan who has paid big bucks for a hot ticket, financial transactions and emotional commitments are intertwined in the consumer culture of today. To assume, as Moore does throughout her book, that marketplace transactions are somehow separable from other human relationships is to commit a logical error of the first order. They are not.[17] And the fan, ensconced and enraptured by the consumer culture marketplace that surrounds her, is the iconic representation of that (I believe, probably along with Roberts and Wolf) glorious postmodern reality.

Grabbing the Engagement Ring

When considering the increasing inextricability of marketplace cultures, popular cultures, and the daily culture of lives lived, we cannot ignore the influence of the Internet and, in particular, social media. The Internet plays an important part, certainly, in upgrading and diversifying the media consumption of fan communities, the so-called long-tail effect included. Beyond that, the Internet and its connectivity are absolutely complicit in the social phenomena that turn consumers into fans. I believe that the way in which social media have been conceptualized has, in fact, helped to create and propagate the phenomenon of brand fandom, brand communities, virtual communities of consumption, consumer tribes, and the range of such phenomena.[18] There is little doubt that the intertwining of material and popular cultures has important macro and social implications for our world, trends that ramify increasingly on a global scale as the spreading multiple springs of information and communications technologies widen and magnify.

With the rise of social media has come the rise of social media marketing. And the so-called, much-hyped social media marketing revolution has spawned hundreds of self-proclaimed social media gurus who sell, for the most part, remarkably similar and similarly unremarkable tidbits of advice based upon anecdotal evidence. One of the more influential of these self-proclaimed gurus is yet another advertising manager, Brian Solis. His 2010 book *Engage! The Complete Guide for Brands and Businesses to Build, Cultivate, and Measure Success in the New Web* contains a typical packaging of the sort already described and thus can serve as a convenient exemplar of the flavor of this offering.[19] The salient notion to derive from the genre is the emphasis on the idea and principle of "engagement." Solis hammers it home in his title, and it assists him as a consultant interested not only in explaining but also in claiming particular intellectual terrain. The term "engagement" is notoriously, and typically, ill defined. Solis, in fact, uses the term repeatedly and in various ways without ever offering a definition. However, his guiding argument runs something like this: the social media revolution empowers consumers to act on their own, writing, publishing, rating, reviewing, recommending, critiquing, organizing, and so on. Consumers are creating communities through social media that have definite business implications and impacts on the way businesses and brands are perceived. Therefore, marketers must relearn their dark arts. They must move from being broadcasters to being engaged in a conversation. In the cutthroat era of what Solis terms "digital Darwinism" managers and marketers must work to gain the support of online champions.

The term "employee engagement" is widely used to describe the enthusiasm, dedication, and willingness to perform and exceed expectations on the job. It is a love of the company, similar to the love that sadly eluded Winston Smith in George Orwell's classic *1984*. The goal of consumer engagement is parallel. Consumer engagement takes place in and through an active, positive online community whose members are deeply emotionally committed to a company, a product, or a brand, who are dedicated and devoted to it, who discuss it widely, evangelically share information about it, and may even contribute to it creatively and productively, as in the burgeoning literature on consumer innovation.[20] The goal of engagement strategies and tactics is to help foster and create a platform where consumers will respond collectively with a deep emotional response that is indistinguishable in effect from fandom in its characteristics of loyalty, commitment, dedication, devotion, and productivity. Although "engagement" is frequently referenced in relation to "metrics" or the measurement of various online activities, such as views, comments, or Facebook "likes," the term is clearly about internalization and affect.

The fan moves throughout the management of social media literature at first like a wisp, but takes form in the phenomenon of the Facebook fan page. In their heyday, probably 2009, almost every major company had a presence on Facebook's "fan" pages (which have subsequently been changed simply to pages). Fans could like—but never dislike—brands, products, events, or just about anything. Major media channels and specialty sites like Mashable.com commonly spoke about the fan pages of brands and companies and offered advice on how best to utilize them. With the Facebook fan page, the notions of the fan, the social media community, and the practice of branding became entangled. The transformation of consumers into fans was practically complete.

The Anthropology of Ideology Symbology

All of this management bafflegab about fans leaves unanswered the important question about what, exactly, constitutes a fan. I will broaden that question to include the nature of fannish behavior in contemporary consumer culture. I have written elsewhere that we can conceptualize consumer culture as an interconnected system of "commercially produced images, texts, and objects" that various types of groups use to make collective sense of their environments and to orient their members' experiences and lives.[21] This sense-making is accomplished through the uniquely cultural construction of overlapping and even conflicting practices, identities, and meanings. A certain historical and social configuration anchors consumer culture, and increasingly, this configuration is linked to the offerings of sports, news, entertainment, and "media" industries and companies.

Because consumer culture is related to the consumption of market-made commodities and desire-inducing marketing symbols, its relationship to popular culture is obvious. As with sports, television, movies, and actors, consumers-as-fans are able to locate themselves inside the local universe orbiting mediated consumption objects and to find identity positions within that system that they perceive as authentic. Contained within the notion of consumer culture is the notion of ideological symbologies—not just symbols, but interlocking and systematic systems and sets of symbols. These symbols are both flexible and global. Their constant cultural mimesis and adaptation allow them to take on the appearance of the local, even as they retain the ideological power of the global. As Taçlı Yazıcıoğlu demonstrates on the national and ethnic level with the adaptation of English-language rock in the Turkish context, these pop cultural meanings have considerable fluidity and are prone to local adaptation and co-optation, partly because of their partiality;[22] they are ever fragmented and always frustratingly incomplete.[23] Authenticity and self-expression blend with localization, and music, like many forms of popular culture around the globe, "smoothly incorporates itself into and adds imagination to the daily lives of people."[24]

In this world of mutating commercial meme-scapes driven by desire for desire's sake, the fan identity fits like a glove. For if fan culture is linked to ideological symbologies, these symbologies are themselves articulated with interconnected systems of commercially produced images, texts, and objects. In addition, and especially in a social media age, they are linked to the various ancillary materials produced by multifarious corporate and other social actors as well as by consumers themselves relating to, expounding upon, and expanding that system of meaning, sometimes very significantly.[25] Using, as one example, how Islamic Iranian women defy religious law to sneak into the soccer stadium and root for their home team, Franklin Foer powerfully illustrates how the fan identity can be used to challenge heavily entrenched and enforced cultural identities.[26]

Being a fan means responding to the invitation to improvisation, participating in the opportunities offered by a particular popular product. Consumer culture is as much seduction as persuasion.[27] It seeks to impel action, rather than compel it, and it is thus no surprise that marketers since the propaganda-informed birth of the post–World War II advertising age have turned increasingly to the inner fantasy lives of consumers to sell their wares. So if the word "fan" derives from a religious meaning, but now is colloquial for a person who is enthusiastic about a specific sport, hobby, leisure activity, or performer, it indicates the fluidity not only of language but also of the play of playful and meaningful meanings in our ever-evolving culture. Companies, brand managers, and the ubiquitous book-writing advertising managers—all are following the contagion of commitment and the amazing activation of attention from one cultural power center to the next.

Pioneering fan studies scholars C. Lee Harrington and Denise D. Bielby found a range of characteristics that distinguished fans from consumers, primarily "their degree of emotional psychological and/or behavioral *investment* in media texts . . . and/or their active *engagement* with media texts . . . [which they relate to] issues of community, sociality, self-identification, and regularity of consumption."[28] Engagement appears to be a sort of link that raises the stakes for being a fan and, for a company, industry, artist, or brand, the potential rewards from having fans. For if, as John Fiske noted two decades ago, "being a fan involves active, enthusiastic, partisan, participatory engagement,"[29] then fan engagement seems to lead to production, a production that may have a very long history, beginning "as a medium of political and social protest in the seventeenth century."[30] Where engagement ends in the age of social media, however, is in ever deeper and wider networks of social sharing, word of mouth, communal commitment, and co-creation. The consumer-as-fan becomes advertiser, entrepreneur, marketer, and producer. The consumer, intrinsically motivated and loyal to the brand for life, entrenched in networks bound to the brand, becomes even more committed to the brand than any merely career-driven marketer or executive ever could.

Conclusion: Four Factors of Brand Fandom

In the scientific labs of global advertisers and market researchers, the grand experiment to turn consumers into fans continues. Building on the original insights of P. T. Barnum and other marketing/entertainment pioneers, brand scientists are exploring the ways in which mythmaking and storytelling are key to consumer engagement and involvement. Meanwhile, the world of academia struggles to keep up, to inform, and to understand the implications of this rapidly changing world. It is in the spirit of such an enterprise that this chapter was written. The areas of fan studies, consumer culture research, and marketing research and practice have barely begun an interchange. But I believe such an exchange will be extremely fruitful to each of these fields. Drawing on the work summarized within this chapter, I can identify four key factors that lead brand managers to seek brand fandom rather than mere brand purchase or consumption. These factors are usefully informed by the fan literature, but the fan literature and scholarship also might benefit from consumer and marketing research in related areas.

The first area is the internalization of affect. Purchase of one brand rather than another might reflect a supposedly rational decision. Preference might be the result of a superior set of product or service attributes. But being a fan of a brand, a member of a brand community or brand tribe, is a manifestation of an identification, a "consciousness of kind," an acceptance of a tribal identity, the adoption and spread of various interrelated myths, an adherence to rituals, a moral stance, and a sense that the brand matters.[31] Above all, brand fandom signals an affective commitment, an internalization of emotion, the love that Roberts talks about and espouses worldwide.[32]

The second factor in brand fandom is the identification of consumers not as isolated individuals interacting with a product or service but as part of a community. Through the brand, fans participate with various communities or groups. Like the *Star Trek* fans who formed their own communities, building on the extant infrastructure of science fiction conventions and fan clubs in the 1950s and 1960s, Apple computer users, beer home brewers, Harley riders, and other types of fans followed the grassroots path. However, marketing in the past decade has forever altered the nature of spontaneous organization. Seeking to stoke community online and off (but especially through social media), marketers and managers have created various forms of top-down community that attempt, with varying levels of success, to create a community superstructure that consumers will see as not only appealing but also authentic.

Third is a vastly enhanced appreciation for the role of content in the development of brand fandom. Drawing on the work of cultural theorist Walter Benjamin, I and my co-authors Stephen Brown and John F. Sherry Jr. argue that powerful brands possess a detailed story structure: "Aura (brand essence), Allegory (brand stories), and Arcadia (idealized community) are the character, plot, and the setting, respectively, of brand meaning."[33] This line of branding research builds upon the idea that brand allegories are stories, narratives, or extended metaphors in symbolic form. Successful branding is successful world building, and that world can be a mythical and imaginative play land, a window into the brand's own (often rosy-colored or stereotyped) past, or a delight-filled imagined future. In addition, a strong brand will manifest "antimony" by combining opposing elements of a deeply meaningful product.[34] The brand that can tap into some of the insoluble puzzles of life or offer fragments of clues to moral and cultural conundrums alongside consumption delights will achieve lasting engagement and meaningful connection that can persist through decades. Brand texts, like all texts, are animated by mystery. The difference between an entertainment text and a brand story in these cases is that the brand texts are stories sutured onto existing products and services; brand texts are stories manufactured for consumption objects.

This chapter suggests that these changes are already well under way. When a brand becomes strong or iconic, as in the case of the understandably oft-cited Harley-Davidson example, it is well on its way to being open source and the center of a brand community or brand tribe.[35] What is the difference between a brand community member or a brand tribe member and a fan? When viewed from the rarefied altitudes of fandom, those differences seem moot. All of which raises questions about the implications of this development for various stakeholders, such as television producers, brand managers, fans, and consumers. For television and other content producers, the acknowledgment that the consumer can be upshifted to the status of fan seems to be spotlighted in the emergent and dynamic field of transmedia and transmedia storytelling.[36] For example, consultants like Jeff Gomez help to spice up and build the imaginary worlds around brands such as Coca-Cola by creating richly storied places and characters, such as those that populate the "Happiness Factory." If consumers are to be turned into fans, they need richer brand stories. As brand management increasingly and deliberately morphs into a sophisticated form of storytelling management, a modern mythic and literally epic endeavor, professional imagineers and flexible producers of television and other content will find themselves in ever higher demand.

The final factor in the inculcation of a fannish consciousness concerning products and brands is to encourage production. One of the hallmarks of fannish engagement is the fans' collective production of various additional texts and objects that simultaneously customize and extend the original sets of texts and objects. In *S/Z*, Roland Barthes provides a useful distinction between a readerly text, where there is little room for the reader to enter the text, and a writerly text, one that is open to the reader's input.[37] That writerly text is a participative document that lives in the present; and as a reflection of our own writing, it never stops or completely closes but remains in play and in motion. Building on Barthes, Kristina Busse and Karen Hellekson speculate that fans may be open text readers and that the "serial production" of Internet fan activity is in this sense "the ultimate writerly text."[38] Stepping from the fan realm to the realm of marketing, these notions are reminiscent of the current social media–driven fascination with what marketing scholars often term "consumer co-creation," prosumption, and consumer innovation.

Marketing professor Leyland Pitt and his colleagues looked at the Linux and Red Hat cases to explore the idea of the "open source brand" using remarkably similar concepts.[39] Focusing on the difference between the closed, readerly "corporate brand" and the open, writerly "OS brand," the authors speculate that the basis of open source's openness can be physical, textual, experiential, or meaning related. They examined mass customization, interactivity, the staging of experiences, and brand communities as progressive steps that can lead companies from having closed brands to creating open brands. However, their work misses huge swaths of consumer culture research that could inform and nuance their perspectives. In particular, they fail to comprehend, acknowledge, or theorize the internal shifts required in the consumer's collective and individual psyche, and ramifying through consumer culture itself, for this shift in brand experience to occur. Their analysis squarely misses the importance of the concept, phenomenon, and corporate colonization of the fan identity.

Ruminating and expounding on related topics, Jenkins asks: "Who isn't a fan? What doesn't constitute fan culture? Where does grassroots culture end and commercial culture begin? Where does niche media start to blend over into the mainstream.... As fandom becomes part of the normal way that the creative industries operate, then fandom may cease to function as a meaningful category of cultural analysis."[40] Perhaps Jenkins might have gone even further. If, fed by scholarly work such as that cited in this chapter, business managers and marketing executives tap into fan studies and fan literature in an attempt to crack the cultural code leading to the elusive goal of engagement, then the fan identity, distinct as it may be, will become the prevailing standard by which marketers measure consumers' collective and individual responses. A brand will not be successful unless it has one million likes on Facebook. An advertisement will not be a success unless it goes viral. A commercial will be authentic only when it has been produced by an actual fannish "brand community."[41] Where, then, does this leave us as scholars interested in the interrelated rise of media and consumer culture?

Where indeed? These factors deserve further discussion, debate, and careful study. The end result for consumers, for audiences, and for fans may be an increasingly sophisticated world of offerings, not just of products and services, but also of the richly storied tales that wag (behind) them and the communities and identities that share and value them. Can anyone simply use a laundry detergent any more? Wear a pair of jeans? What about a coffee? A cola? After the full ripening of the postmodern Age of the Fan-Consumer—the raving, maniacal, unquenchably and deeply loving mark/fan—contemporary branding, contemporary marketing, and the simple everyday act of consumption may never again be straightforward.

Notes

1. Henry Jenkins, *Textual Poachers: Television Fans and Participatory Culture* (New York: Routledge, 1992).
2. Eric J. Arnould and Craig J. Thompson, "Consumer Culture Theory (CCT): Twenty Years of Research," *Journal of Consumer Research* 31 (March 2005): 868–882.
3. Sidney J. Levy, "Symbols for Sale," *Harvard Business Review* 37 (July–August 1959): 124.
4. Ken Blanchard, *Raving Fans: A Revolutionary Approach to Customer Service* (New York: William Morrow, 1993).
5. Cornel Sandvoss, *Fans: The Mirror of Consumption* (Malden, MA: Polity, 2005); Lawrence Grossberg, "'Is There a Fan in the House?': The Affective Sensibility of Fandom," in *The Adoring Audience: Fan Culture and Popular Media*, ed. A. Lisa Lewis (London: Routledge, 1992), 50–65.
6. Cornel Sandvoss, "The Death of the Reader? Literary Theory and the Study of Texts in Popular Culture," in *Fandom: Identities and Communities in a Mediated World*, ed. Jonathan Gray, Cornell Sandvoss, and C. Lee Harrington (New York: New York University Press, 2007), 22–23.
7. Ibid., 23.
8. Michael J. Wolf, *The Entertainment Economy: How Mega-Media Forces Are Transforming Our Lives* (New York: Times Books, 1999).
9. Ibid., 282.
10. Also see B. Joseph Pine and Kames H. Gilmore, *The Experience Economy: Work Is Theatre & Every Business a Stage* (Boston: Harvard Business School Press, 1999).
11. Douglas Atkin, *The Culting of Brands: When Customers Become True Believers* (New York: Portfolio, 2004).
12. Kevin Roberts, *Lovemarks: The Future Beyond Brands* (New York: Power House, 2004).
13. Blanchard, *Raving Fans*; Roberts, *Lovemarks*.
14. Anne Elizabeth Moore, *Unmarketable: Brandalism, Copyfighting, Mocketing, and the Erosion of Integrity* (New York: New Press, 2007), 29.
15. Roberts, *Lovemarks*.
16. Moore, *Unmarketable*, 29.
17. For details, see Robert V. Kozinets, "Can Consumers Escape the Market? Emancipatory Illuminations from Burning Man," *Journal of Consumer Research* 29 (June 2002): 20–38.

18 See Bernard Cova, Robert V. Kozinets, and Avi Shankar, eds., *Consumer Tribes* (Oxford and Burlington, MA: Butterworth-Heinemann, 2007); Albert M. Muñiz Jr. and Thomas C. O'Guinn, "Brand Community," *Journal of Consumer Research* 27 (March 2001): 412–432; Robert V. Kozinets, "E-Tribalized Marketing? The Strategic Implications of Virtual Communities of Consumption," *European Management Journal* 17, 3 (1999): 252–264.
19 Brian Solis, *Engage! The Complete Guide for Brands and Businesses to Build, Cultivate, and Measure Success in the New Web* (Hoboken, NJ: Wiley, 2010).
20 See Johann Füller, Gregor Jawecki, and Hans Mühlbacher, "Innovation Creation by Online Basketball Communities," *Journal of Business Research* 60, 1 (2006): 60–71.
21 Robert V. Kozinets, "Utopian Enterprise: Articulating the Meanings of *Star Trek*'s Culture of Consumption," *Journal of Consumer Research* 28 (June 2001): 67–88.
22 E. Taçli Yazicioğlu, "Contesting the Global Consumption Ethos: Reterritorialization of Rock in Turkey," *Journal of Macromarketing* 30, 3 (2010): 238–253.
23 See Jenkins, *Textual Poachers*.
24 Yazicioğlu, "Contesting the Global Consumption Ethos," 249.
25 See Fuller et al., "Innovation Creation"; Robert V. Kozinets, Andrea Hemetsberger, and Hope Schau, "The Wisdom of Consumer Crowds: Collective Innovation in the Age of Networked Marketing," *Journal of Macromarketing* 28 (December 2008): 339–354; and Leyland F. Pitt, Richard T. Watson, Pierre Berthon, Donald Wynn, and George Zinkhan, "The Penguin's Window: Corporate Brands from an Open-Source Perspective," *Journal of the Academy of Marketing Science* 34, 2 (2006): 115–127.
26 Franklin Foer, *How Soccer Explains the World: An [Unlikely] Theory of Globalization* (New York: HarperCollins, 2004).
27 John Deighton and Kent Grayson, "Marketing and Seduction: Building Exchange Relationships by Managing Social Consensus," *Journal of Consumer Research* 21, 4 (1995): 660–676.
28 C. Lee Harrington and Denise D. Bielby, "Global Fandom/Global Fan Studies," in Gray et al., eds., *Fandom*, 186 (emphasis in original).
29 John Fiske, *Understanding Popular Culture* (London: Routledge, 1991), 139.
30 Abigail Derecho, "Archontic Literature: A Definition, a History, and Several Theories of Fan Fiction," in *Fan Fiction and Fan Communities in the Age of the Internet*, ed. Karen Hellekson and Kristina Busse (Jefferson, NC: McFarland, 2006), 67.
31 Cova, Kozinets, and Shankar, *Consumer Tribes*; Grossberg, "Is There a Fan," 50–65; J. H. McAlexander, J. W. Schouten, and H. F. Koening, "Building Brand Community," *Journal of Marketing* 66 (January 2002): 38–54; Muñiz and O'Guinn, "Brand Community," 412–432.
32 Roberts, *Lovemarks*.
33 Stephen Brown, Robert V. Kozinets, and John F. Sherry Jr., "Teaching Old Brands New Tricks: Retro Branding and the Revival of Brand Meaning," *Journal of Marketing* 67 (July 2003): 30.
34 Ibid.
35 See James H. McAlexander, John W. Schouten, and Harold F. Koening, "Building Brand Community," *Journal of Marketing* 66 (January 2002): 38–54.
36 Behice Ece Ilhan, "Transmedia Consumption Experiences: Consuming and Co-Creating Interrelated Stories Across Media" (PhD diss., University of Illinois–Urbana Champaign, 2011).
37 Roland Bathes, *S/Z*, trans. Richard Miller (New York: Noonday Press, 1974).

38 Kristina Busse and Karen Hellekson, "Introduction: Work in Progress," in Hellekson and Busse, eds., *Fan Fiction and Fan Communities*, 6–7.
39 Pitt et al., "The Penguin's Window," 115–127.
40 Henry Jenkins, "Afterword: The Future of Fandom," in Gray et al., eds., *Fandom*, 364.
41 Muñiz and O'Guinn, "Brand Community," 412–432.

8

Outsourcing *The Office*

• •

M. J. CLARKE

In a 2002 episode of *The Simpsons*, the animated family is forced to flee their termite-infested home and find refuge by being cast in a new reality television program, "The 1895 Experiment."[1] The show within the show, which challenges contestants to live as if it were 1895, is the brainchild of an executive of "The Reality Channel" who calls himself the program's creator but admits in an aside, "by creator I mean I saw it on Dutch television and tweaked the title." The program is an instant hit, but only until the Simpsons become too comfortable with their old-fashioned lifestyle, resulting in boring television and viewer drop-off. Desperate to retool the series, the executive gathers his inner cabal of corporate thinkers and urges them, "Fixing the show is going to take some original thinking . . . everyone pull out your TV and start flipping around."

As a work of parody, this episode addresses several developments in the production of network television content. For example, the lampooning of the executives' creative process is a dig at television's stereotypical unoriginality and unchecked imitation. But more pointedly, this depiction is also an allusion to the increasingly global trade in television formats, by which television producers and distributors respectively import and export, not programs themselves, but the rights to license and reuse the idea of a preexisting program in a new territory. The fact that "The 1895 Experiment" began life as a Dutch series is a coy nod to the Dutch television company Endemol, originator of *Big Brother* and other reality and talent game shows; by 2006, it netted

€1.1 billion in sales, of which only 5 percent was made in the company's native Netherlands.² The fictional executive's global poaching reflects a contemporaneous trend in the business of television. The episode also comically underscores the executive's frantic and immediate need to retool and reconfigure the program to make it "work." This impulse culminates with the show producers dropping the Simpsons' temporary home into a river to enhance the program's drama. While there is nothing particularly new about television producers' willingness to rework material, there is something contextually specific about the executive's obsession with speed and instant feedback. In an era when the growth rates of advertising revenue have leveled or, in some cases, drastically fallen, networks have increasingly committed themselves to the never-ending search for the oft-invoked new business models for television. Whether this means making labor leaner and meaner, as in NBC's "NBC 2.0" agenda of layoffs, or rethinking production, as in the reduction of preproduction rituals and overspending through the format trade, or expanding distribution, as in the instantaneous arrival of digital downloads or online streaming, the underlying theme is a preoccupation with quickness itself.

The *Simpsons* episode concludes with a violent rebellion against the producers of "The 1895 Experiment," in which the family is aided by a group of reality television losers who have been voted off previous programs and literally cast out into the wilderness. Here, the rapidity of production and change is haunted by the unceremonious disposal of seasons past. Such a dilemma clearly resonates with the contemporary condition of "liquid modernity," described by sociologist Zygmunt Bauman, which points to a key shift in the theorization of power, no longer linked to solidity and immutability, but rather to speed and fluidity. The dark underside of this shift is the problem of waste: "waste disposal [is] one of the two major challenges liquid life has to confront and tackle . . . the other major challenge is the threat of being consigned to waste."³ Network television confronts precisely these two problems in its modifications of the production process and in its reappraisal of the significance and purpose of its own mission. Simply put, this small segment of *The Simpsons* includes discussions of increasing global trade, current network mandates of speed and corporate agility, and a preoccupation with the problem of waste. However, this reflexivity is not benign; the episode quite emphatically ridicules and critiques these new trends in the business of television.

This brief, introductory example demonstrates (1) how certain television programming clearly, though comically, depicts several key transformations taking place within the field of network television production and (2) how attitudes toward these changes have become intertwined with the manifest content of the programs themselves. Armed with this dual-purposed model, this chapter presents a deep investigation of NBC-Reveille's *The Office* (2005– 2013), a program whose production is at the bleeding edge of the new global market for network television and the just-in-time strategies of format sales. As the television networks upgrade themselves from terrestrial broadcasters to international entertainment nexuses, it falls on researchers in television studies to elucidate how these differences alter the practices of production and the division of labor. In the first half of this chapter I will elucidate these changes with respect to the creation of *The Office*; the second half will bring this context into a consideration of the style and themes of the program texts themselves.

The exact relationship between the practices of cultural industries and their output is a highly contested issue. The debate goes back at least to Max Horkheimer and Theodor Adorno's *Dialectic of Enlightenment*, which argued that the rationalization of cultural production led to crippling sameness in both the style of the works produced and the mentalities of those watching. However, studies in this vein, despite the novelty of the results, contain the fatal flaw of reducing cultural expression to the means of production: in the final instance, the superstructure is never more than the icing on the cake of the infrastructure. One can detect the folly in such a distanced gaze by observing the Weberian conundrum that change within network television seems to have both an economic (in the changing architecture of business) and a cultural/ideational (in the preoccupation with speed as a virtue and waste as a vice) component.

Recently, authors working in the broad field of television studies have made several advances beyond the Frankfurt School model by suggesting a more nuanced understanding of the dynamic between the work of cultural industries and their output. Such analytic innovations go back at least as far as CBS's oft-discussed 1971 purge of successful hayseed comedies in favor of more sophisticated fare believed to attract a so-called quality demographic. If nothing else, this industrial truism suggests that the link between producers and their product is a fruitful site of analysis and not as simple and predictable as Adorno and Horkheimer would have us believe. However, it is one thing to claim that television content simply reflects a rational, measured business strategy; it is another thing altogether to interpret a specific television program based on observations made about the professional culture responsible for creating it. This second method is approached in the work of anthropologist Barry Dornfield, who has examined how layers of self-interpretation and self-understanding among media producers influenced the production of a single PBS docu-series.[4] Indeed, the field of cultural anthropology, given its preoccupation with the relationship between native cultures and their unique cultural expressions, has much to offer television scholars seeking to understand media professionals and their complex relationship to their productions.

More specifically, this essay suggests that the work of anthropologist Clifford Geertz offers much to the study of network television production in general and to the case of *The Office* in particular. An adaptation of Geertz's approach treats networks and creative talent not simply as rationally operated businesses/individuals with purely economic needs and desires, but as a distinct culture with shared (and contested) values, meanings, and attitudes that are manifested in particular instances of symbolic action (whether in trade interviews or in the content of television programs themselves). In his discussion of ritualized actions, Geertz claimed that these outward expressions are always examples of "culture of" and "culture for"; that is, they function both as a representative transcription of a worldview and its ideal reconstruction.[5] Similarly, *The Office*, as a cultural expression, serves both as a representation of the current state of business as well as an argument for certain attitudes toward business in general. Additionally, one can apply Geertz's examination of actual cultural expressions, such as his work on the complex Balinese calendar system, which, he argued, points to a sense of empty, uncumulative time.[6] From that famous analysis we can draw several corollaries. First, Geertz's essay demonstrated how close readings of texts and their makers, when combined, produce novel and insightful analyses of these cultures. Second, Geertz insisted that the interlocking calendars prove nothing in themselves. They are not symptoms, psycho-cultural peculiarities, or reflections of a material base or national culture; rather, they are observable only in their actual use. In other words, the consistency of culture is always a matter of situated instances of use and practice. This principle becomes all the more clear in the case of network television, which, with increasingly frequency, posits its programming as solutions to the query: "What is the function of television in the so-called digital age?"

The following sections assemble a narrative of *The Office*'s backstage by way of a thorough investigation of the context of British and U.S. television as seen in industry trade and governmental reports. Admittedly, the scope and scale of this project is well beyond what would typically be classified as Geertzian. However, the multi-sited, networked organization that is typical of television production necessitates such a broad approach in order to render sufficient depth in observation. Using the Geertzian model allows us to cast *The Office* as a cultural expression, one that casts light on certain native attitudes. More specifically, *The Office*, with its preoccupations with business failure and waste (through the tropes of failed authority, surveillance, and technology), can be understood as both a reflection and a construction of the global, speed-addicted business culture of television that hatched it.

Big Ben and the Global Market for Television

The Office began as a thirteen-episode series for BBC2. To understand how this program jumped the ocean to be featured on NBC, one must account for the changing attitudes toward television in both the United States and the United Kingdom. Facing the pitfalls outlined in earlier chapters in this volume,

FIG. 8-1 *The Office* (NBC) is preoccupied with business failure and waste through the tropes of failed authority, surveillance, and technology. (Frame grab)

television workers in the United States have become dissatisfied with the ubiquitous old models of television production and financing. As a consequence, American producers and distributors have increasingly looked to developing foreign markets (spurred by the spread of satellite, cable, and commercial television in general) to sell their wares; high-end dramas have reportedly begun to fetch as much as $1.5 million per episode in Western Europe.[7] This export bonanza is mirrored in an equally lucrative imported format market. Since the 1980s, the U.K. television industry has taken steps to change the very way it does business by moving away from a public model of broadcast and toward a privatized, commercial system bolstered by a focus on the international trade of content. Simply put, U.K. and U.S. television have become more interlocked in a global partnership, which the production of *The Office* clearly reflects. However, the strength of format sales also points to a specific balance of power and division of labor between players in the respective nations.

In their work on global culture, sociologists Scott Lash and Celia Lury claim that one of the more distinctive features of the current cultural moment is the tendency of media to become "thingified," meaning that culture, once subsumed by the technical rationality of industrialism, is now becoming the determining logic guiding the economic base.[8] As a result, such ephemeral goods as concepts and ideas take on an increased materiality. This theme is mirrored in the work of Manuel Castells, who has argued that our "informational economy" is one in which "what has changed is not the kind of activities humankind is engaged in, but its technological ability to use as a direct productive force what distinguishes our species as a biological oddity: its superior capacity to process symbols."[9] The global trade in television formats, which is essentially a trade in ideas thingified, is indicative of these larger theoretical trends. Literally, a format is the blueprint for a television program and includes all the elements of preproduction (the creative products of development) as well as elements of production itself (going so far as to encompass production schedules, budgets, and lighting designs). Equally important is the fact that formats function as licensing agreements: a format sale is the purchase of the right to produce and exploit an identical or similar program in a different region of the world or for a different linguistic group. This is precisely the sort of deal that brought *The Office* to the United States.

More specifically, this format deal was brokered by Ben Silverman, executive producer of *The Office* and former CEO of Reveille, a company whose mission is "to develop shows with multiple sources of financing including foreign production companies and advertisers to avoid the usual television industry structure."[10] Driven by this desire to alter the division of labor in television, Reveille has significantly reconceptualized the traditional role of advertisers, not only by utilizing product integration, but also by enlisting advertisers in preproduction and by using ad revenue from integration as the profit center for content.[11] Production costs for *The Office*, for example, are defrayed by the investment of the program's integration partner, Staples, Inc. Also, since its inception in 2002, Reveille has profited greatly from the global export market, becoming a key entity responsible for the distribution of many independently produced reality programs; in the brief period that it managed the NBC-Universal catalogue, the company made fifty sales in thirty countries over the course of a single year.[12]

As the former head of international packaging for William Morris in London, Silverman had a hand in the format sales to the United States of *Who Wants to Be a Millionaire*, *The Weakest Link*, *Dog Eat Dog*, *Big Brother*, and *Queer as Folk*. The phenomenal success of several of these programs prompted Silverman to form Reveille in order to escape the prohibition denying agencies and agents profit participation. He has described this business model succinctly: "I scour the world for the best ideas and the game-changing hit shows."[13] Notably, Reveille set up shop under the umbrella of Universal Television and under the supervision of USA Networks head Michael Jackson, the former head of Channel 4 in the U.K. Despite the company's focus on reality television, Silverman took a chance on scripted television by producing a remake of the British hit *Coupling* in the fall of 2003 for NBC, a network desperate for a *Friends* replacement. Even with the program's very clear and very public failure, Reveille continued its foray into fictional fare with the mid-season premiere of *The Office* in March 2004. Reportedly, Silverman's informal ties with the British media industry gave him access to the rights to the program. Silverman himself claims that he tracked down Ricky Gervais and Stephen Merchant, the creators and international rights holders of the original BBC2 series, at a London Starbucks before marching them to their agent's office to broker a deal.[14] This informal network led to a more concrete business relationship when in July 2003 the BBC signed a two-year first-look deal with Reveille to market its scripted fare in the United States.[15]

Enamored with Silverman and his mastery of "new models" and "alternative paradigms," NBC-Universal president Jeff Zucker handed him the reins of both the NBC network and its affiliated production arm in May 2007. Giving control of the network to a largely untested, independent producer signals a clear shift in the business culture of the U.S. television industry. The new mandate is simply to find a way to make more money, faster, in a climate of uncertainty, regardless of whether revenues accrue through overseas exports, exported formats, early syndication deals to cable, product integration, more aggressive licensing, digital distribution, or online advertising. Notably, Silverman's expertise in imported formats facilitated the industry's new focus on speed and provided a mechanism for reducing many of the costs and risks associated with preproduction by outsourcing a number of shows. Not surprisingly, Silverman's ascension to network head was quickly followed by the announcement that NBC would increase its development of imported formats, including *Phenomenon* from Israel, *Kath and Kim* from Australia, and *Sin Tetas* and *No Hay Paraiso* from Colombia, all of which were able to bypass the slow-moving, ritualized machinery of American television's pilot production.[16] Similarly, Silverman preordered thirteen-episode seasons of *Fear Itself* and *Robinson Crusoe* without a frame of footage being shot, again maneuvering around the institutional architecture of television's old models of doing business.[17]

This intent to change the traditional means of doing business extended to *The Office*, as evidenced in the altered responsibilities and concerns of the show's creator and principal executive producer. In 2007 showrunner Greg Daniels called *The Office* the "show of the future," referring to its high-profile presence on the Internet and its early adoption of the iTunes-based sales model.[18] However, the production of such off-air add-ons was partially halted when it became clear that NBC had no plans to reimburse the writers for their additional work. The controversy over Internet-based webisodes erupted into legal action in February 2007 and subsequently became a significant point of contention in the November 2007 Writers Guild of America (WGA) strike.[19] It is not surprising that Daniels, whose program has aggressively sought out alternative methods to find viewers and make money, was a particularly vocal supporter of the 2007 strike, given the guild's efforts to extend new media residuals to writers. Mocking what he considered the feigned uncertainty of network and studio heads, Daniels stated, "They've already figured out the business model, but they haven't told us. . . . [Essentially] it's the Internet."[20] This aggressive attitude toward change adopted by the networks and the writers had an appreciable effect on both the business architecture of the television industry and the creative process itself. In contrast to *The Simpsons* example, which faulted reality television for its lack of originality, Daniels described the work conducted by himself and his writing staff as follows: "[We] identify things seen on sitcoms and eliminate them."[21] In one sense, this description is representative of the incremental innovation of television production; yet the writers' almost irrational fear of being too "sitcomy" also points to a general reluctance to follow older models, in this case aesthetic ones, and older ways of making television in order to avoid being consigned to television's waste pile.

In its industrial rhetoric, Reveille casts itself as a leader in the so-called alternative paradigms. The company's very name is meant to signify a literal wake-up call to the television industry. This search for a new, workable division of labor was also very much on the minds of network executives when NBC purchased Universal Studios from French media conglomerate Vivendi in April 2004. While the motivations and justifications behind such an important corporate maneuver are beyond the scope of this chapter, the acquisition was indicative of the generalized search for viable new business models in an expanded, conglomerated, and global economy. In 2004, 90 percent of NBC's revenues came from advertising; in the same year, Universal made 90 percent of its revenue in box office and subscription services. Thus, their merger gave the new entity an even 50–50 split between these revenue streams, demonstrating each firm's desire to diversify its media holdings and business practices.[22] The merger also signaled another shift in policy, summed up succinctly by former NBC head Bob Wright: "We elected to be a company that's content based, not distribution based" (or, in Lash and Lury's construction, media is thingified).[23] It is only with a greater stake in the ownership of content that television networks put themselves in a position to participate in global (and digital) markets.

When evaluating the logic of format sales to the United States, it is reasonable to ask why American networks would be interested in outside sources when there is already an absurd glut of domestic product? Some 500 pitches a year are filtered down to a handful of aired pilots for prospective series, of which very few remain on the air for successive seasons.[24] A few possible answers can be posited. First, an overall concern with speed and breaking away from business as usual, as sketched above, currently drives the American television industry. Second, anything that can reduce the risk associated with U.S. television production, such as proven success in a foreign market via specific (demographic) research data, will logically (or at least psychologically) facilitate production. Third, there is a tremendous human and financial cost associated with television development, with little opportunity for reimbursement. Format sales in television (like acquisitions in features) bypass this dilemma because development costs can be defrayed over the entire length of a series through individual episode fees, as opposed to the enormous initial layout needed for a traditionally developed series. Fourth, U.S. television workers are likely drawn to the unequal trading positions between U.K. and U.S. executives/producers. When Parliament enacted the Communications Act 2003, there were fears among British industry pundits that native media producers would be swallowed up by international conglomerates. So far, these anxieties have proved to be unfounded; the British market has worked more as an incorporated labor source and aftermarket for the global network. And while the relative market positions of U.K. and U.S. companies are mutually lucrative, they are, as we shall see, unbalanced. As a result, the sort of deals established between British producers (who are desperate to sell) and American broadcasters (who are overburdened by content in the form of "broken pilots") cannot help but be reflected in the terms of sale.

Faced with a leveling of advertising revenue and the threat of digital obsolescence, U.S. television producers and networks have increased the global exploitation of their product significantly. This general move toward global synchronization of media is identified by the authors of *Global Hollywood 2*, who note a similarity between the New International Division of Labor in traditional work, which exploits markets and labor pools across borders, and what they call the New International Division of Cultural Labor (NICL) within media, which works equally to integrate foreign sources of funding and cheap labor.[25] Simply put, this theory argues that freeing production from the confines of space and time allows capital to circumnavigate the globe in pursuit of maximum profit. *Global Hollywood* concentrates on global capital's use and perpetuation of colonialism, or more clearly, the power imbalance of nations through the micro exploitation of preexisting economic disparity in sweatshop and runaway labor as well as through the macro coercive forces of intellectual property accords and liberalized trade procedures. However, this more conspiratorial, political-economic model does not account for the complexity of globalization on a number of counts. More specifically, this model explains the relationship of powerful nations to ex-colonies and impoverished nations but is less applicable in describing harmonization in agreements between metropoles, as in the case of the United Kingdom's relationship to the United States. It becomes analytically skewed to cast the United Kingdom, a successful media center in its own right, as simply a victim of U.S. domination. Instead, one can use the theories of social thinker Max Weber—who claimed that domination must be understood as the situation where the commands of the powerful are carried out by actors who participate according to their own investment in this authority—to see that policy changes and international business dealings are often better understood as actions undertaken to participate in global media flows and not necessarily as instances of subservience to American powerbrokers.[26] This view is echoed in the earlier quotation from Castells, who suggests that a position in global markets ("social morphology") and not necessarily its domination becomes the paramount concern for competitive nations/corporations/television power players.

As a result of distorted statistics, the actual import/export relationship between the United Kingdom and the United States in regard to television is a slightly contentious matter. What does seem clear is that even though the United States is the world leader in fictional television exports, accounting for 72 percent of the total drama market, these exports account for only 9 percent of the top-rated primetime programs in the European Union.[27] Further, it has been reported that in 2006 U.S. interests accounted for $6 billion of the $10 billion that make up the global television market and that in the same year British exports netted a total of £593 million, or over $1.1 billion.[28] Likewise, even though the United Kingdom made between 36 and 43 percent of its export sales to the United States in the years between 2003 and 2006, these purchases represented a small percentage of U.S. transmission. Reading too much into these statistics is misleading, given that each survey is slanted to exclude certain considerations. For instance, how are countries of origin determined? Does one count factual content (a lucrative market for the United Kingdom)? Does one focus on total hours transmitted, or are schedule times also taken into consideration? How are intangible results, like influence, measured? Further, the numbers collected by a partnership between the U.K. government's Department of Culture, Media and Sport (DCMS) and the British trade association Producers Alliance for Cinema and Television (PACT) are reported in such way that each previous year's net revenue is reduced to ensure annual positive gain (on paper if not in fact). For example, overseas monies gained in 2005 added up to £632 million in a 2006 annual report, but the following year's statistics dropped this figure to £494 million, thereby guaranteeing percentage gains for the industry.[29] In other words, numbers lie.

Ultimately, one could argue that the cultural efficacy of these statistics—specifically, their use by a business culture that is desperate to make its significance in the global market patently explicit—is more important than their nominal accuracy. This being said, a few general statements about the U.S.–U.K. media relationship can be posited. First, although the relationship between the two is clearly lopsided, it is also mutually beneficial. Securing American export monies allows the United Kingdom to outpace Australia and France as the number two media exporting nation in the world. As noted in 2006 by Tessa Jowell, the former secretary of state for culture, media and sport, British media industries accounted for 4.1 percent of the country's gross domestic product (GDP), growing by a yearly average of 6 percent between

1997 and 2003 and outdistancing overall economic growth by 300 percent.[30] Here, as media scholar Silivio Waisbord contends, the impulse to liberalize and harmonize the market is viewed as a shared economic interest supported by "massive changes in the structure of television systems [that] in the 1980s and 1990s . . . connected television systems that, until then, functioned in relative isolation."[31] Additionally, it can be stated that the United States is an extremely lucrative market for the United Kingdom, accounting for 43 percent of its export sales in 2003; yet, as stated above, this volume of trade has very little impact on the overall picture of U.S. television.[32] In other words, there is money to be made in the international television market, yet the broadcast opportunities in the United States are limited. The result is an increase in co-productions and, as described below, format sales.

These two massive changes in the U.K. television system were facilitated by several key policy initiatives, beginning with the European Union's 1989 document "Television Without Frontiers" (TVWF). This directive was an essential step in changing European television from a dominantly public service model to a commercial broadcasting model. Drafted during the expansion of cable and satellite services within Western Europe, TVWF was, in part, a response to an increased volume of imported American product. The document's preamble indicates its interest in closing proto-E.U. ranks: "All restrictions on freedom to provide broadcasting services within the [European] Community must be abolished."[33] A harmonized European market is imagined as a counterweight to anxieties over Americanization. Yet, ultimately, the directive is less about defending European cultural exception and more about achieving global harmonization—or "intensified interconnectivity," in the words of Waisbord—through its stress on independent production, loose definitions of ownership and country of origin, and focus on cross-border intellectual property protection. In the first case, Chapter IV of TVWF declares that at least 10 percent of broadcast time in a member state should be devoted to independent production, a figure that is currently set at a much higher rate of 25 percent in the United Kingdom.[34] This move toward privatization and commercialization increased the opportunities for cross-border investment, creating alternative modes of production and financing outside of the old models of the BBC and Italy's RAI. Waisbord goes so far as to claim that the increased global reach of television markets is, in part, "the unintended by product of the existence of protectionist laws."[35]

Despite the expressed intention of TVWF, the U.K. television market responded to the rise of cable, satellite, and round-the-clock programming with, in most cases, an even larger diet of U.S. imports. In *Selling Television*, Jeanette Steemers does an excellent job of describing the two factors that led to the increased import of American content to the United Kingdom in this period: first, these new services were subscription based and relied on high-gloss American fiction as well as residual marketing power to draw in subscribers; second, British independent production remained extremely underdeveloped and simply could not keep pace with demand.[36] At approximately the same time, British exporters found what little space they occupied on American television diminishing, given the decreased participation of PBS, their historic co-producer, and the shift in American cable television away from acquisition and toward original production. Further, a revision of financial syndication regulations in the United States encouraged television networks to create their own content, leaving even less space for British content. As an alternative solution to the global market, many producers turned to the international sale of formats.

The economic logic behind British format sales makes immediate sense, given the opportunities for global expansion described above. The little space that remained for British content in American transmissions has become even smaller. This decline has led to an increase in gains by U.K. content producers who exploit other revenue streams outside of traditional program sales (formats, licensing, DVD sales, and so on). A 2007 report by British media analysts Oliver & Ohlbaum Associates pointed out that the trade in formats is a "cost effective answer to the traditional dilemma of all European broadcasters," namely, that "most European markets are not large enough to sustain a large volume of expensive new programming."[37] In format sales, these costs are bet against the ability to sell the concept internationally. This same report claimed that the European market for formats had grown 18 percent a year since 2002 and that the United Kingdom, in 2006, netted 43 percent of the world's market, making the country, by far, the global leader.[38] Similarly, it has been reported that the BBC increased product sales to the United States by seven times between 1998 and 2003, spurred largely by format and DVD sales.[39] The reimbursement affords British distributors a licensing fee anywhere from 5 to 15 percent per episode created by the format rights holder.[40] The sale of formats from the United Kingdom to the United States also carries the chance, however slight, that British producers could become major cogs in American television. The BBC's production of ABC's *Dancing with the Stars* is the one positive example in this regard. What is more likely, however, is the scenario suggested by the NICL and Castells's network society, which has the United Kingdom constructing itself as an offshore site of creative outsourcing and a media center of manufactured savvy—or, as Ben Silverman calls it, an invaluable "40,000 person graduate school."[41]

The United Kingdom's specific place within the global division of labor is maintained by a series of strategies. First, one must observe the highly social structure of media transactions and deal-making. Without access to network executives and other American cultural gatekeepers, British producers must do their business through powerful American intermediaries such as Silverman.

Waisbord describes this collaboration as a globalized homogenization of television professionals. However, such a broad view ignores the micro politics and the disparate mobility (Bauman's maker of power in liquid modernity) among global television workers. The term "cultural discount" is often deployed by players like Silverman to discredit the playability of U.K. content in foreign markets. This xenophobic designation demands American control of formats under the pretext that foreign producers do not understand the American market (a questionable claim in the face of British moves toward increased global awareness) and that popular American audiences will never embrace culturally specific programming (a claim that is empirically unproven). This ideational leverage, coupled with increased economic and productive capacities, spells out a power imbalance on the American side of the content trade.

The United Kingdom's willingness to participate in these deals can be read as evidence of a massive cultural change within the country's insecure television production culture. On the one hand, it signals a return to what Leo Ching referred to as "mercantile capitalism," wherein spatial difference becomes the essential key to value creation.[42] This claim counters the ideas of thinkers such as Waisbord, who bemoan the troublesome global homogenization of television, a condition combated by slight national articulations in format adaptations. However, arguments of a creeping global cultural sameness, such as Waisbord's "McTV," are observations from a decidedly macro perspective that fails to appreciate the cultivation of national difference referenced in Ching's notions of regionalism. A more micro view can posit that exogenous programming provides a veneer of difference to an American system that thrives on shocks of change compounded with swatches of sameness in terms of both content and business practice. This attitude can be seen in the cultivation of "Britishness" in varied forms, such as period pieces of the Merchant-Ivory mold and the tradition of bureaucratic critique (remember *Monty Python*) in *The Office*. Equally likely is that format sales characterize an industry desperate to be a participant in the global network (in the Castellian sense) by any means necessary, even if that requires selling ideas and not programs. Indeed, the principle behind the Oliver & Ohlbaum report is that participation in overseas markets is the key to the development of U.K. broadcasting, which will (in their analysis) not only eclipse the old pubcaster way of doing business, but also make London, along with Mumbai and Tokyo, the new world centers for media production.

Celebrating Obsolescence

The preceding examination of the synergies taking place between the television industries of the United Kingdom and the United States—their mutual efforts to achieve global synchronization and shake off the ubiquitous old models—provides a glimpse of the inner workings of the business context surrounding the production of *The Office*. Within the American production culture, we have witnessed a general distrust toward business as usual and a search for a quick solution for what industry journals commonly depict as a desperate situation. Trade in global formats is posited as one possible remedy. In the case of U.K. television, we have traced a similar theme in various institutional efforts to cast off an outdated television system, in this case public broadcasting, and to support global trade as the key to change and advancement. In key policy and business manifestos, we have seen how formats have become fetishized in the United Kingdom, becoming the embodiment of the country's efforts to secure a position in the global media network. The emergence of *The Office* on American television represents the convergence of the hopes and aspirations of these two interlocking production cultures.

Yet there is an immediate irony between this description of each native production culture and the content of the American version of *The Office* as its cultural expression. Set within the regional office of a paper middleman, Dunder Mifflin, *The Office* depicts everything that is wrong with business as usual and the disastrous consequences that befall those who fail to embrace the new flexible, liquid way of doing business. The series' first season was haunted by the threat of "downsizing," a threat that was made good in the subsequent firing of employees and dissolution of a neighboring branch. Other episodes tackle contemporary workplace issues: the corporate office's cost-cutting measures lead to reduced employee benefits;[43] union busting is broached when warehouse workers seeking unionization prompt the corporate office to threaten closure of the branch office;[44] and two employees try to make ends meet by taking on second jobs, including a shift at a fly-by-night telemarketing office.[45]

Yet *The Office* is not solely a depiction of the New Economy's losers. The main characters are complicit in their own failure as a result of their ineptitude (to be expected in a comedy) and, even more so, because of an untenable business model—after all, Dunder Mifflin is a paper company in a digital age. I would argue that this negative representation of business is a function of the specific cultural-industrial circumstances of its production. Sociologist Sharon Kinsella, in her study of the Japanese publishing industry, traced a relationship between the changing cultural reputation of manga/comic books, the rise of the editor as the chief artistic agent in the medium, and the preeminence of so-called "adult manga," which frequently portrays the adventures of corporate CEOs and economists in its storylines.[46] Similarly, a nexus exists among business culture, the division of labor, and story content in the case of *The Office*. The series, as a cultural expression, is both an example of a "culture of" (it is a program that has facilitated new ways of doing business) and a "culture for" (its content emphatically demonstrates the obsolescence of certain business models). In the process, the show's creators can be seen tacitly endorsing the sort of globalized, speed-addicted maneuvers that gave rise to the series in the first place.

This argument can be traced first in the program's representation of bureaucratic authority and individual investment in work. To observe these features one must return to Max Weber to establish some terminology. Weberian social science insists that all social actions be rationalized; that is, they must be made in accord with a legitimate form of authority. For Weber, the three forms of authority are: the charismatic (having to do with the messianic call to duty toward those of a seeming "superhuman" quality); the traditional (having to do with the hand-to-mouth existence imposed by the rule of elders); and the legal (having to do with the obedience to impersonal, rule-bound order).[47] The structure of bureaucracy becomes, for Weber, the ideal-typical form of legal authority that he characterizes according to the following logic: regular activity in the form of duties; a clear hierarchy; a management system based on written documents; an organization of workers according to their area of specialization; and an authority based on rules that can be learned.[48]

Notably, *The Office*'s frequent deviations from Weber's definition of an ideal bureaucracy constitute the show's choicest moments of comic irony. For instance, regular activity in the form of duties is studiously avoided by the show's principals, as suggested by the titles of various episodes, which include "Basketball," "The Fight," "Diversity Day," "Christmas Party," "Beach Games," and "Launch Party." These titles suggest a seemingly inexhaustible supply of interruptions to the work of *The Office* denizens. Whereas so many fictionalized workplace dramas invoke an existential bureaucratic critique concerned with eternal sameness, here the workers' frustration is reversed in a Sisyphean twist as work itself is always just out of reach. Branch manager Michael Scott further sabotages bureaucratic order by consistently contending that business is personal and by posing as the branch's benevolent father, such as when he insists on giving away employee Phyllis at her wedding.[49] Here, legitimate authority is tempered by traditionalism in the form of the family, Weber's ideal-typical traditional social formation. Likewise, the distribution of hierarchy is one of the show's running gags, as office twit Dwight wrestles over the difference between the titles "assistant manager" and "assistant to the manager." Yet, in *The Office*, Weber's ideal-type lives on through its own self-conflicted institutional inertia: "where administration has been completely bureaucratized, the resulting system of domination is practically indestructible."[50]

The Office, in this regard, belongs to a tradition of bureaucratic critique manifested in the social science of postwar America, most prominently in the work of David Reisman (*The Lonely Crowd*) and William Whyte (*The Organization Man*). Whyte, in particular, proposed to update Weber's *Protestant Ethic and the Spirit of Capitalism*. He argued that the "Social Ethic," which finds social utility as the principle motivation for all action, had entirely eclipsed the Protestant Ethic, which stressed the importance of individual calling and vocation.[51] This shift included an organizational dismantling of thrift and the fear of sloth in the name of consumer spending and leisure, respectively, as well as the philosophical dismantling of self-reliance in the work of William James and John Dewey, who each called for the organization of action around the strict dictates of social utility. Whyte, surprisingly, parroted the concerns of Marxist commentators, who insisted that modern work was responsible for dehumanization; however, Whyte argued that the cause was stunted individualism and not capitalist greed. This revised critique of the workplace followed from several management studies that observed that productivity was not ultimately related to working conditions or pay, but rather to the social organization of work. All problems need only be talked out rather than solved, as all disagreements are seen only as misunderstandings.[52] In other words, *The Office* references Whyte's criticism of the Social Ethic, albeit in a parodic mode.

Stylistically, the program, which is hypothetically shot by a documentary crew, operates on several different valences. Characters explicitly and implicitly perform for the camera by carrying out their lives seemingly unaware of its presence (while calling attention to it on occasion); as a result, there is always the possibility that they will shift between these modalities. The latter occurs frequently in the small moments of direct address wherein characters clandestinely mug for the camera (which may or may not be noticed by other characters within the diegesis) in response to particularly absurd occurrences or ridiculous requests, made most often by their boss. On a very crude level, this technique suggests that the "problem" posed by the narrative cannot be solved within the social organization depicted in the program's workplace. *The Office*, however, diverges from Whyte, suggesting that even individuality, Whyte's antidote for forestalled social utility, does not offer an escape. Dwight, the only character who displays the hope of upward mobility, is repeatedly ridiculed and even compared to a fascist in the character's memorable reenactment of a speech by Benito Mussolini.[53] The only hope for the characters of *The Office* is to look wistfully outside the confines of their moribund bureaucracy toward the "documentary cameraperson," who presumably functions as a symbol of the possibilities of creative labor and a prototype for success and upward mobility in the New Economy.

This backward depiction of the fictional workplace contrasts sharply with the ostensibly forward-thinking organization of the production cultures of the U.K./U.S. television industries. The incorporation of U.K. formats into the global market in general and into the U.S. television industry in particular pushes the architecture of the customary television business away from clear, unilateral hierarchies that previously existed toward a "network," in the sociological sense, which is composed of a loose formation of actors with strong, yet informal and mutable ties with one another. While the advantages of such a formation are varied, one can point to the ability of television production communities to excise and outsource unnecessary institutional baggage and wasted human capital (with creative decisions made on the fly); furthermore, the workplace cultures associated with the New Economy are uniquely constructed to encourage innovation through the promotion of a rapid-fire transfer of information that can take on novel forms through any number of unique combinations. These are precisely the two competitive advantages brought to bear in *The Office*, advantages that are notably absent in the depiction of business in the program proper. *The Office* elaborates on this contrast by rendering a clearly negative representation of Fordist business practices. The critique of linear hierarchy and long-term career security is underscored in an episode depicting the termination of the office pest, Dwight. The character responds to his dismissal by lamenting the fact that his dream of dying at his desk at work will never be fulfilled.[54]

It is not only authority and bureaucratic organization that are being attacked in the program's narratives, but also the exercise of surveillance—bureaucracy's primary technology of social control. In Marxist theory, wage labor began as an exchange of time for money. It is only with the invention of Taylorist ("scientific") management that laborers became accountable for performance, and it was through the constant surveillance of work that managers (the intermediaries between capitalists and laborers) were able to quantify this performance. In this schematic, the modern institution of management can be said to begin with surveillance. Similarly, Michel Foucault describes social observation and the deployment of surveillance as a total apparatus whose "disciplinary power became an 'integrating' system, linked from the inside to the economy and linked to the aims of the mechanism in which it was practiced."[55] Watching is not simply a monitoring, but an invaluable machinery that recapitulates power clandestinely on the underside of legitimate law. Yet the perfect functioning of the panopticon is more a Weberian ideal-type than a historical or social reality. For as much as Foucauldian surveillance may have been a model for social control, we can now look to Zygmunt Bauman for a contemporary evaluation that lumps together Weberian bureaucracy, totalitarianism, and panoptic domination as fictions of modernity that have little bearing on the current condition of so-called liquid modernity.[56] As a product of this contemporary mode of existence, *The Office* takes great pains to discredit the disciplinary function of surveillance. In the episode entitled "Health Care," branch manager Michael Scott (Steve Carell) hides from his irate employees in his office by claiming that he is swamped with work. However, the errant gaze of the in-program documentary camera clandestinely captures Scott pretending to be "busy" while he is in fact playing with a toy truck. Subsequently, Scott calls his assistant, Pam, to avoid having to exit his office and encounter his angry employees. Twice, Scott tries to preempt a dialogue with his employees by claiming he is receiving another call; however, Pam calls his bluff by announcing that no such call is registered on her phone bank. In this example, surveillance—both via the commentary by the unseen documentary camera operator (through camera movement and framing) and Pam's narrative actions—is depicted as a reversible force that allows employees to undermine the potency of their employer. Notably, this type of resistance to institutional control is allowed for in Foucault's theory, as the practice of surveillance knows no master. The poles of watcher and watched can be reversed. However, what is more interesting in this example is the way in which surveillance is cast as powerless; when the unseen cameraperson and the assistant Pam expose Michael, there is no consequence. In fact, there are multiple examples in this and in subsequent episodes of surveillance as a meaningless mechanism of control and a technology ill equipped to ensure productivity. In the larger context of the television production culture, the surveillance model would be equally ill suited to monitor the looser, networked relations among international co-workers in the New Economy. Equally relevant is the U.S. television industry's oft-stated skepticism regarding hopelessly outdated mechanisms for measuring audience share in the face of TIVO, digital downloads, and other alternatives viewing practices. Ben Silverman has been especially vocal about the failure of this measurement system, advocating revenue earnings over the metric of Nielsen ratings to gauge network performance. Finally, the executives responsible for transforming the American television industry via the importation of outsourced formats are able to maintain a position of moral superiority (and visual deniability) by ignoring the ugly realities of economic disparity inherent in globalized labor.[57]

While discussing the nature of the current information economy and its relation to new technologies, Castells has stated that "what characterizes the current technological revolution is not the centrality of knowledge and information, but the application of such knowledge and information to knowledge generation and information processing/communication devices, in a cumulative feedback loop between innovation and uses of innovation."[58] *The Office* (as a format) can be understood as just such a piece of information generation facilitated by the growing global television market and made possible through rapid advances in information technology. As a form of cultural expression, the series parodies a Luddite sentiment and depicts its characters as lost amidst even modest improvements in information technology. The series' 2007 fall season began with a lampooned launch of Dunder Mifflin's new website, which was constructed by a corporate up-and-comer, as a "floor to ceiling streamlining of our business model" and as an attempt to make the company "younger, sleeker, more agile."[59] In response, Dwight and several other employees resisted the implementation of these new technological forms by waging a sales contest against the website. Although the "human" (Dwight) wins, he does so only by exhausting all of his sales leads, making a protracted battle impossible and giving the episode the tone of a pyrrhic victory. Other episodes depict boss Michael Scott as unable to master even the simplest functions of e-mail ("Back from Vacation") or Microsoft Powerpoint ("Money").

The series' depiction of workers who are unable to adapt to technology—much like the failure of managerial authority and surveillance in the earlier example—point back to the larger production culture responsible for creating *The Office*. While Silverman, Reveille, and the American networks in general are singularly preoccupied with the reformatting of entertainment business models through global markets and a focus on the virtue of speed itself, a Geertzian reading of *The Office* finds these preoccupations being scrutinized in the text via a cast of characters who find themselves unable to keep pace with the frenzied rate of change taking place in the larger context of the global marketplace and the New Economy.

In conclusion, the losers of *The Office* are in a position analogous to that of the feral castaways of "The 1895 Experiment." Both sets of fictional characters are depictions of the improper disposal of antiquated lifestyles. Both are embodiments of the leftovers/left-behinds of a system that has been drastically updated. *The Office*, in this sense, is a clear example of a rather commonplace set of practices or behavior—that of a ritualized disavowal of waste. These fictional types are the cultural scapegoats offered up by a U.S. television network as self-assurance that it shares nothing in common with the characters depicted. The more *The Office* is ridiculed, and thereby distanced from its makers, the easier those working in television can breathe. In this regard, the program functions as a cultural expression that is no different from Geertz's cockfighters, who practice brutality to prove their humanity. In *The Office*, television, as a distinct production culture, demonstrates obsolescence to prove its vitality. Responding to a rapidly changing marketplace that seems poised to throw network television into the cultural waste bin as a result of an encroaching digital revolution, *The Office* is part of a chorus of emphatic naysayers whose message is always the same: television is not dying.

Notes

1 "Helter Shelter," December 1, 2002.
2 Oliver & Ohlbaum Associates, *Prospects for the European TV Content Sector to 2012* (London: October Oliver & Ohlbaum, 2007), 21.
3 Zygmunt Bauman, *Liquid Life* (London: Polity, 2005), 9.
4 Barry Dornfield, *Producing Public Television, Producing Public Culture* (Princeton, NJ: Princeton University Press, 1998).
5 Clifford Geertz, "Religion as a Cultural System," in Geertz, *The Interpretation of Cultures: Selected Essays* (New York: Basic Books, 1973), 93–94.

6 Clifford Geertz, "Person, Time and Conduct in Bali," in Geertz, *Interpretation of Cultures*, 391–398.
7 Elizabeth Guider, "H'wood Dramas Draw Big Bucks in Cannes," *Daily Variety*, October 12, 2006, 8.
8 Scott Lash and Celia Lury, *Global Culture Industry: The Mediation of Things* (London: Polity, 2007), 7–9.
9 Manuel Castells, *The Rise of the Network Society* (Cambridge, MA: Blackwell, 1996), 92.
10 Bill Carter, "Agent Offers One-Stop TV Production," *New York Times*, March 11, 2002, 14.
11 Todd Wasserman, "[Hollywood Meets Madison Avenue]," *Hollywood Reporter*, September 27, 2005.
12 Josef Adalain, "Reality Chief Turning Indie," *Daily Variety*, February 25, 2005, 1.
13 Quoted in " 'Paradise' Found by NBC Honcho," *Daily Variety*, June 15, 2007, 1.
14 Bill Carter, " 'The Office' Transfers to a New Cubicle," *New York Times*, March 20, 2005, 12.
15 Mimi Turner, "BBC, Reveille in Concert," *Hollywood Reporter*, July 8, 2003.
16 Josef Adalian and Michael Schneider, "Ben's Primetime Debut," *Daily Variety*, July 17, 2007, 1.
17 Nellie Andreeva and Kimberly Nordyke, "NBC Sets Sail with 13-Episode 'Crusoe,' " *Hollywood Reporter*, October 1, 2007.
18 Greg Daniels, presentation at UCLA, November 8, 2007.
19 WGAW press release, "NRLB Finds Favor with Writers Guild, Dismisses NBC/Universal Webisodes Complaint," February 21, 2007.
20 Daniels, presentation at UCLA.
21 Ibid.
22 Tim Burt, "Marriage Guidance from GE Merging NBC and Universal," *Financial Times*, May 10, 2005, 13.
23 Quoted in George Szalai and Shiraz Sidhva, "NBC Universal Born as Vivendi, GE Make Definite Deal," *Hollywoodreporter.com*, October 9, 2003.
24 Pamela Douglas, *Writing the TV Drama Series: How to Succeed as a Professional Writer in TV* (Studio City, CA: Michael Wiese Productions, 2005), 44.
25 Toby Miller, Nitin Govil, John McMurria, Richard Maxwell, and Tina Wang, *Global Hollywood 2* (London: BFI Publishing, 2005).
26 Max Weber, *Economy and Society: An Outline of Interpretive Sociology*, ed. Guenther Roth and Claus Wittich, trans. Ephram Fischoff et al., 2 vols. (Berkeley: University of California Press, 1978), 2:946.
27 Jeanette Steemers, *Selling Television: British Television in the Global Marketplace* (London: BFI Publishing, 2004), 32.
28 Elizabeth Guider, "Banks, Bombs and 'Betty,' " *Variety*, October 2, 2006, B1; Benedetta Maggi, "PACT UK Television Exports Survey 2006," last modified April 11, 2007, http://www.culture.gov.uk/NR/rdonlyres/AFFC3C9B-8BFD-4AEB-B027-981F05C0AF91/0/TVExports2006StatisticalRelease.pdf.
29 DCMS and PACT, "Revenues from the Export of UK TV Programmes Jump 21%," last modified April 18, 2006, http://www.pact.co.uk/uploads/file_bank/2005exportstats.pdf.
30 Tessa Jowell, "Speech to the Oxford Media Convention," last modified January 19, 2006, http://www.ippr.org/uploadedFiles/events/Tessa%20Jowell%20Speech.doc.

31 Silvio Waisbord, "McTV: Understanding the Global Popularity of Television Formats," in *Television: The Critical View*, ed. Horace Newcomb, 7th ed. (New York: Oxford University Press, 2007), 376.
32 Steemers, *Selling Television*, 43.
33 Office for Official Publications of the European Communities, "Television Without Frontiers," http://europa.eu.int/comm/avpolicy/regul/regul_en.htm.
34 Oliver & Ohlbaum Associates, *Prospects*, 14.
35 Waisbord, "McTV," 378.
36 Steemers, *Selling Television*, 24–25.
37 Oliver & Ohlbaum Associates, *Prospects*, 23.
38 Ibid., 3, 19.
39 Turner, "BBC, Reveille in Concert."
40 Steemers, *Selling Television*, 40.
41 Ben Silverman, "Broadcasting: Pitching Across the Pond," *The Independent*, November 14, 2005, 13.
42 Leo Ching, "Globalizing the Regional, Regionalizing the Global: Mass Culture and Asianism in the Age of Late Capital," in *Globalization*, ed. Arjun Appadurai (Durham, NC: Duke University Press, 2001), 287–288.
43 "Health Care," April 5, 2005.
44 "Boys and Girls," February 2, 2006.
45 "Money," October 18, 2007.
46 Sharon Kinsella, *Adult Manga: Culture & Power in Contemporary Japanese Society* (Honolulu: University of Hawai'i Press, 2000).
47 Weber, *Economy and Society*, vol. 1.
48 Weber, *Economy and Society*, 2:956–958.
49 "Phyllis' Wedding," February 8, 2007.
50 Weber, *Economy and Society*, 1:487.
51 William Whyte, *The Organization Man* (Garden City, NY: Doubleday Anchor Press, 1956), 7.
52 Ibid., 38.
53 "Dwight's Speech," March 2, 2006.
54 "Traveling Salesman," January 11 2007.
55 Michel Foucault, "The Means of Correct Training," in *The Foucault Reader*, ed. Paul Rabinow (New York: Pantheon Books, 1984), 192.
56 Zymunt Bauman and Keith Tester, *Conversations with Zygmunt Bauman* (Cambridge: Polity Press, 2001), 74.
57 Brian Lowry, "GE and NBC: 'We're Number oh, Never Mind,'" *Daily Variety*, July 23, 2007, 16.
58 Castells, *Rise of the Network Society*, 32.
59 "Dunder Mifflin Infinity," October 4, 2007. Even corporate techno-speak is played for laughs, most likely pointing to the fact that Dunder Mifflin's attempts to upgrade are woefully late.

9

Convergent Ethnicity and the Neo-Platoon Show

••••••••••••••••••••

Recombining Difference
in the Post-Network Era

VINCENT BROOK

Convergence in mass media has typically been analyzed in economic and technological terms: tightly diversified conglomeration and the rise of Big Media on the one hand, merging of media platforms on the other. The combined effects of these industrial forces on programming forms and audience response have been extensively theorized through notions of televisuality and narrowcasting, commercial intertext, textual poaching/semiotic democracy, postmodernism/globalization.[1] While race/ethnicity has not been neglected in these and other theoretical interventions, it has not been extensively examined in terms of media convergence.[2]

This essay explores how ethno-racial diversity in American network television in the multichannel age has responded to convergence through a type of multicultural, ensemble-cast show I term the "neo-platoon" show. Formed around a tight-knit or fatefully intertwined cohort of ethno-racially diverse characters, with a complex, soap-like narrative structure, significant interracial romance, and a sophisticated televisual aesthetic, a flurry of neo-platoon dramas has emerged in the first decade of the new millennium, including some of the period's biggest hits: *Boston Public* (2000–2004), *Kevin Hill*

(2004–2005), *Lost* (2004–2010), *Grey's Anatomy* (2005–), *Heroes* (2006–2010), and *Ugly Betty* (2006–2010).

Both reflecting and bolstering the dual moves of the post-network/postmodern era toward fragmentation and recombination, these shows, although generically and aesthetically distinct, have clear antecedents in the classical network period. The "platoon" sitcoms of the 1970s—for example, *Barney Miller* (1975–1982), *Welcome Back, Kotter* (1975–1979), *Taxi* (1978–1983)—themselves small-screen variants of the platoon combat films of World War II, parlayed the "'relevant' TV" turn of the early 1970s into a trend toward ethnoracial diversity in casting for continuing roles. *Hill Street Blues* (1981–1989), in its dramatic generic mode and self-conscious televisual style, and *The Cosby Show* (1984–1989), in the African American television trend it generated during a period of institutional change, serve as significant bridges between the platoon sitcoms of the 1970s and the neo-platoon dramas of the 2000s.

Miami Vice (1984–1989), *NYPD Blue* (1993–2005), *Homicide: Life on the Street* (1993–1999), *New York Undercover* (1994–1998), *ER* (1994–2009), *City of Angels* (2000), and *Gideon's Crossing* (2000–2001) can be taken as *proto* neo-platoon shows, lacking the soap structure (except for *ER* and *City of Angels*) and interracial romance (except for *ER* in its later incarnation) that characterize the full-fledged neo-platoon show. The *Law and Order* (1990–) and *CSI* (2000–) franchises and their crime drama cousins (*Cold Case* [2002–2010], *NCIS* [2003–], *Without a Trace* [2003–2009], *Bones* [2005–], *Medium* [2005–2011], *Criminal Minds* [2005–], *Shark* [2006–2008], and *The Unit* [2006–2009]), though seemingly made to order for neo-platoonism, spend too much time solving crimes and missing persons cases to allow for much romance among the regular cast, much less the interracial variety.

Several cable shows have also developed along neo-platoon lines: *Oz* (1997–2003), *Strong Medicine* (2000–2006), *The Division* (2001–2004), *The Wire* (2002–2008), *The Shield* (2002–2008), *Battlestar Galactica* (2003–2008), *The L Word* (2004–2009), *Weeds* (2005–2012), and *Dexter* (2006–). However, while a comparative analysis of the cable companies' and the broadcast networks' investment in the neo-platoon formula would prove useful, this study focuses on network shows for two primary reasons: first, they attract the most controversy and discourse in general around race and ethnicity in American television; and second, the networks maintain a ratings hegemony, versus cable's more concerted niche programming strategies, and the federal government's regulatory oversight is mainly focused on networks. Media monitoring groups, for example, survey representations of their constituencies only on network shows because such groups' ability to affect licensing decisions by the Federal Communications Commission (FCC) is limited to the networks and their affiliates.[3]

The network/cable divide was highlighted in the most immediate determinant of the neo-platoon show trend: the so-called Lily White controversy of 1999, which arose from media monitoring groups' aggressive attempt to reverse a perceived backslide in minority representation on network shows for the coming season. Pressure put on the networks by the monitoring groups, working together rather than separately for the first time, forced the networks not only to tilt their upcoming lineups toward greater diversity, but also to commit to long-term upgrades, both onscreen and behind the scenes.

Further socio-political and economic support for the neo-platoon trend comes from the increased pervasiveness and acceptance of multiculturalism in U.S. society, especially among the younger, consumerist audience that advertisers, and thus programmers, most covet. Significantly also, this audience is no longer judged to be exclusively white: people of color (especially Hispanics and Asian Americans) have increased their presence in the population since the 1970s not only quantitatively but also qualitatively as a consumer demographic prized for its growing market share. Finally, recombining difference within a single dramatic show, rather than doling out difference (a "black" show, a "Hispanic" show, an "Asian" show), typically on sitcoms, benefits television programmers: first, they can showcase diversity more cost-effectively; second, they can avoid the genre "ghettoization" stigma; and third, they can "load up" audience appeal, both domestically and internationally. The inclusion of non-American characters in shows such as *Lost* and *Heroes*, for example, may not be based exclusively on a global distribution agenda, but political economic concerns clearly encourage an ecumenical approach.

The line between commerce and creativity may remain fuzzy when *Heroes* creator Tom Kring claims, "I was trying to address something that would really connect in a kind of international way.... I knew I wanted the show to bridge cultures and borders."[4] But the comment of former NBC Entertainment co-chairman Ben Silverman reveals that business considerations were decisive: "I wanted to bring an entrepreneurial energy to our broadcast channel and work with foreign partners because the foreign market place is incredibly rich right now, and if we can come up with ideas that sell globally from the beginning, like *Heroes*, it benefits how we finance them."[5]

The very real, if still insufficient, progress in representational diversity is problematized not only by the commercial motivation that accompanies it but also by the contradictory pressures experienced by marginalized groups positioned between multiculturalism and assimilation, that is, between the desire to assert difference yet also to be accepted into the mainstream. A similar tension is prevalent in the notion of "color-blind" casting, which, as the description suggests, tends to erase the very difference it purports to assert. Through historical and socio-cultural analysis, information gleaned from industry

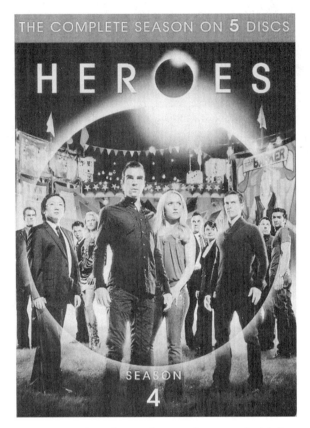

FIG. 9-1 *Heroes* (NBC): Interethnic couplings ... with global scope. (DVD cover)

insiders and media monitors, and glosses of key neo-platoon shows, this essay examines both the upside and the downside of convergent ethnicity, intending thereby to shed heat as well as light on one of the more striking aspects of the new media landscape.

From Platoon Films to Platoon Television

Wake Island (1942) is generally considered the first true World War II combat film, according to Jeanine Basinger. In focusing not only on military battle but also on the combat unit—"that unique group of mixed individuals, so carefully organized to represent typical Americans"—this film also launched a prime subgenre of the combat film, the platoon film. Subsequent combat films, notably *Air Force* and *Bataan* (both 1943), further established the ethnoracially diverse, microcosmic military unit as a fixture of the platoon film.[6] For

Basinger, the latter is "clearly the seminal film," because "the nature and composition of the combat unit in *Bataan* became a veritable paradigm for subsequent films, along various social and cultural lines—the ethnic, racial, and religious background of unit members; their ideological, economic, and cultural status; their geographical and regional origins; and their military rank, experience, and professionalism." *Bataan* set other standards for the platoon film regarding the hierarchical order and "structure of authority" of the unit, which would reemerge in the "platoon" and "neo-platoon" television shows from the 1970s on.[7] Most significantly for our purposes, *Bataan* turned the integrated platoon film into a tool for propaganda, what Lewis Jacobs has termed "a national collective hero," and what I, following Michael Schudson and Thomas H. Zynda, call "capitalist realism."[8]

Both Schudson, who applies the concept to American advertising, and Zynda, who applies it to *The Mary Tyler Moore Show* (1970–1977), base their notions of capitalist realism on the precepts of the Soviet Union's First Socialist Writers' Conference of 1934. Socialist realist art, according to the conference's guidelines, was to represent contemporary reality: (1) not as it is so much as it should become; (2) in a simplified and pleasing manner that is accessible to the masses; (3) with an air of optimism and as something worth emulating; and (4) in a way that reveals and endorses new features of society and thus aids the masses in assimilating them.[9] "Highly idealistic in theme and technique, rather than literally faithful"—and thus a misnomer, as J. R. Bowlt points out—socialist realism was a "visual rhetoric" that "strove to transmit the idea of the imminent fulfillment of a utopian dream through lyrical distortion of reality."[10]

Of course, one can debate whether the differences between Stalinist fiat and Hollywood commercial constraints are ones of kind rather than degree, and whether all classical Hollywood cinema—with its marketing of the American Dream, its idealized representation of the capitalist system, and its pleasurably accessible aesthetic—meets capitalist realist criteria to some extent. Inarguably, however, the platoon film, made at a particularly fraught historical moment and under the added pressure of Office of War Information directives, took the ideological and representational precepts of capitalist realism to new levels. First, in the ability of the disparate platoon members "to work in harmony for the survival of all," as Patricia Erens explains, "we are given a lesson in democracy at work. In truth, however, the ethnic differences in the Platoon Film are rather superficial. The men differ little from one another in a significant way. The key word is unity; seldom does an issue develop that causes major dissension. In essence, then, Platoon Films are another manifestation of the melting pot."[11] Second, in seeming contradiction, yet fundamental to the all-American ethos, the platoon film's "democratic mix" includes a strong, entrepreneurial leader, invariably a WASP, "who is part of the group, but is forced to separate himself from the group because of the demands of leadership."[12] Finally, and most tellingly, the fully integrated platoon was, both during and in the first years after the war, not merely a fantasy but truly impossible. U.S. military units of World War II might conceivably have spanned the spectrum of ethnic hyphenates (Italian-Americans, Greek-Americans, Jewish-Americans, even Puerto Rican-Americans), but the armed forces were strictly segregated along color lines until President Harry Truman issued Executive Order No. 9981 in 1948.

During a time of domestic and international crisis, with the United States emerging from a prolonged economic depression and the "free world" fighting for its very existence, projecting an image of interracial harmony and cross-cultural solidarity clearly served to fortify patriotism and boost morale both within the military and among the populace at large. Cold War rivalry between the United States and the Soviet Union, which expanded the need to present an image of ethno-racial tolerance from the national to the international sphere, propelled a brief resurgence of the World War II platoon film in the late 1940s (for example, *Battleground*, *Twelve O'Clock High*, *Sands of Iwo Jima* [all 1949]). By the 1950s, however, another hot war in Korea and the rising fear of global nuclear annihilation led to a more mythic, less threatening dispenser of pro-American/anticommunist propaganda in the decade's dominant movie genre: the western. Then, following the shattering of the ideological underpinnings of both the western and the war film in the 1960s (*Dr. Strangelove* [1964], *The Wild Bunch* [1969]), the platoon film resurfaced in the late 1960s and early 1970s, albeit in a new generic form and in a new medium.

The groundbreaking sci-fi series *Star Trek* (1966–1969), with its military command structure and diverse ethno-racial and gendered cast—albeit transposed from an idealized present or past to an imagined future—gave the first indication of the small screen's receptivity to the platoon film format. But it would take a confluence of industrial and societal factors in the early 1970s to usher in a full-fledged platoon show trend. The 1970s saw a revolution in minority representation on American television, with more major ethnic characters appearing on primetime than ever before. Media historian Howard Suber, writing in 1975 for the Jewish journal *Davka*, listed a spate of series featuring African Americans (eleven shows), Italians (three), Asians (two), Chicanos, Irish, Greek, Poles, Swedes, and Eskimos (one each). Suber bemoaned the dearth of prominent Jewish characters in the ethnic surge, although two popular shows with Jewish protagonists, *Bridget Loves Bernie* (1972–1973) and *Rhoda* (1974–1979), had already aired.[13]

However, in another article written in the same year, Suber contradicted, or at least qualified, his previous critique. Titled "Television's Interchangeable Ethnics," this piece decried the homogenizing of *all* ethnic television depictions, not merely Jewish ones. Likening television's "obsession with minorities" to the platoon films of World War II, Suber found that "it didn't really matter which ethnic groups were represented.... Characters 'happened' to be Jewish, or 'happened' to be Polish, or 'happened' to be black . . . as if by accident."[14] *Barney Miller* was the epitome of the platoon approach, with the police precinct standing in for a military unit and exhibiting the same multiethnic configurations: one black, one Asian, one Puerto Rican, one Pole, one (or maybe two, if Barney himself is included) Jew.[15] As in the platoon movies, the ethnic characters on *Barney Miller*, and on television generally, seldom appeared in a manner that had "anything to do with their number, their historical importance, or their relation to the society itself." Television's "accidental minorities" remained "like colorful locations or weapons . . . interchangeable."[16] Other prominent 1970s platoon shows included *Welcome Back, Kotter*, with high school special-ed instructor Gabe Kotter in charge of a band of interethnic "sweat hogs," and *Taxi*, with Alex Rieger most prominent among a ragtag contingent of rainbow-colored cabbies.

The platoon shows, like other ethno-racially driven shows of the decade (*Sanford and Son* [1972–1977], *Good Times* [1974–1979], *The Jeffersons* [1975–1985], *What's Happening!!* [1976–1979], *Chico and the Man* [1974–1978]), while certainly a response to the identity politics zeitgeist, also stemmed from structural changes in the television industry, most notably: the increased clout of media monitoring groups resulting from the WLBT case of 1969, which forced the FCC to allow for public input on licensing decisions; FCC-instigated industrial reforms (financial interest and syndication restrictions, the primetime access rule) to encourage independent production; and a shift in audience targeting toward a younger, more sophisticated, "quality" (in taste, culture, and buying power) demographic.[17]

The foregrounding and, as Suber noted, the eliding of difference in these shows can also be read as another form of capitalist realism. Zynda, as previously mentioned, chose *The Mary Tyler Moore Show* as the prime exemplar of capitalist realist television. Although it features a cadre of workplace cohorts, *Mary Tyler Moore* is not a platoon show in the classic sense, since all the main characters are white. However, the capitalist realist element that Zynda highlights in the show—its idealized "vision of the workplace as a primary setting [and] . . . as the main activity of building . . . a new, postindustrial society of fulfilling, nonsexist, and socially constructive work"—can be extended usefully to the platoon shows as well.[18] The emphasis in the platoon shows is on race/ethnicity rather than gender, and the "society" depicted—police station, inner-city public school, cab company—is more evocative of the industrial than the post-industrial sphere. However, the communal yet regimented workplace setting and the cultural pluralist, "we can all get along" ethos (still headed, significantly, by white males) bespeak just as clearly the "visual rhetoric" and "utopian dreams" of the time.

Hill Street Blues and the "*Cosby* Moment"

Although the upbeat Reaganist 1980s would lead, in general, to a turn away from the "relevance" of the 1970s, the decade began with a show—*Hill Street Blues*—that can be seen as both an extension of the previous decade's platoon shows and as a precursor of the neo-platoon shows of the 1990s and 2000s. Where *Hill Street Blues* diverges most from its predecessors and converges most with its successors is in its genre and style. As a crime drama rather than a sitcom, as were all the previous successful platoon shows, *Hill Street Blues* marked a major breakthrough for ethno-racial diversity on American television, shifting issues of difference, significantly for the first time, from the comfortable confines of comedy to a more grittily realistic milieu. The stylistic distinction was even more extreme. Contrasting sharply with the flat "zero-degree" style of the stage-set, live-video sitcom or the low-budget, B-movie style of the typical television *policier*, *Hill Street Blues* adopted a hand-held, direct-cinema format with overlapped dialogue that derived from and resonated with documentary film and the avant-garde. This more cinematically sophisticated, "televisual" approach would proceed apace in subsequent 1980s "quality" shows (*Miami Vice* [1984–1989], *Moonlighting* [1985–1989], *thirtysomething* [1987–1991]).[19] But *Hill Street Blues*' most influential formal innovation for the neo-platoon shows of the 1990s and 2000s derived not from the lofty heights of art cinema but from the lowly daytime soap.

Along with the unabashed primetime soaps *Dallas* (1978–1991) and *Dynasty* (1981–1989), yet in a more earnest, less campy vein, *Hill Street Blues* pioneered the multiple-protagonists/interlocking-narratives/serial-storyline formula that has since come to dominate American television drama. This convergence of high and low cultural elements and of generic forms, which now stand as key stylistic markers of postmodernism, has become a prime enabler—from a narrative and therefore also from a casting perspective—of convergent ethnicity. *Hill Street Blues* may still have a white male "on top," in Captain Frank Furillo, but prominent off-white, non-Christian, and female underlings—African American Officer Bobby Hill and Detective Neal Washington; Latino Lieutenant Ray Calletano; Jewish Officer Andy Renko and Detective Mick Belker; Sergeant Lucille Bates and Attorney Joyce Davenport—take up significant chunks of diegetic space and time.[20]

What *Hill Street Blues* had done for the cop show, *St. Elsewhere* (1982–1988) did for the medical drama. Rather than a single (or partnered) lead, as had been standard for such shows—*Dragnet*'s Sergeant Joe Friday, the eponymous Kojak, Starsky and Hutch, Doctors Kildare and Ben Casey—*St. Elsewhere* featured an ensemble. The ethno-racial mix may not have been as great as on *Hill Street Blues*, but *St. Elsewhere* did include the first regular African

American doctor on primetime (played by a young Denzel Washington), three Jewish doctors, and a bevy of women doctors. The greatest spur to convergent ethnicity in the 1980s came not from a drama, however, but from a comedy, one that both marked the return of the ethnic sitcom and became the decade's greatest hit: *The Cosby Show* (1984–1992).

The spate of successful African American comedy shows that followed— for example, *227* (1985–1990), *Amen* (1986–1991), *Frank's Place* (1987–1988), *A Different World* (1987–1993), *In Living Color* (1990–1994), *The Fresh Prince of Bel-Air* (1990–1996)—constitute what Herman Gray terms the "*Cosby* moment."[21] The designation is not neutrally descriptive; it implies an ideological and industrial determinant tailored to the times. Ideologically, *The Cosby Show* patently served a capitalist realist agenda, in this case, the Reaganist discourse of color blindness. An affluent, well-adjusted African American family headed by a doctor father and lawyer mother proved that blacks had made it and needed no further government assistance. From an industrial perspective, the historical conjuncture (mid-1980s to early 1990s) during which *Cosby*'s cross-over success propelled an unprecedented black television trend corresponded to the paradigm shift in American network television from the classical to the postclassical era.

As the network triopoly (CBS, NBC, ABC), buffeted by new-media technologies, relinquished its thirty-plus-year stranglehold on U.S. audiences and the television institution, *Cosby* and company offered the old networks as well as the upstart new ones (Fox, later the WB and UPN) useful programming models. For the old networks, studies showed that blacks tended to watch about 50 percent more television than whites and, increasingly importantly, were less wired for cable, direct broadcast satellite, and other new media. For the upstarts, blackness connoted "cool," just the ticket for counterprogramming strategies geared toward a young, urban, hip (black and non-black) demographic. The opportunistic nature of the approach was clearly demonstrated by Fox, which gradually whitened its schedule as its audience share achieved parity with the majors.[22]

The "Lily White" Controversy

By the late 1990s, black-themed programming had congealed around the weak-sister networks, the WB and mainly UPN, or migrated to cable's Black Entertainment Network (BET, purchased by Viacom, UPN's parent, in 2001). Confirmation that the "*Cosby* moment" was history came in 1999, when media monitoring groups fomented what the press labeled the "Lily White" controversy. Already concerned about the deteriorating status in the mid-to-late 1990s of minority representation on television, these groups were appalled when the lineup of network shows announced for the fall 1999 season pointed to further reduction of people of color in lead or recurring roles. Kweisi Mfume, then president of the National Association for the Advancement of Colored People (NAACP), denounced the "virtual whiteout" of the upcoming season, front-page headlines derided the "all-white landscape" of primetime television, and an unprecedented multiethnic coalition of African American, Latino, Asian/Pacific Island American, and Native American groups converged to deal with the issue.[23] The uproar led to "a mad dash for diversity" on the part of the networks, but a Screen Actors Guild (SAG) study launched to monitor the newly negotiated standards showed mixed results (in both senses).[24] Released in February 2000, the SAG study, which focused on African Americans, revealed that blacks accounted for 16 percent of characters over the monitoring period, compared to the group's 12.2 percent proportion of the U.S. population. However, further analysis showed that "overrepresentation" was undermined by the shunting of African Americans to "ghetto" networks (the WB and UPN), "ghetto" genres (sitcoms rather than dramas), and "ghetto" scheduling (the less viewed Monday and Friday nights).[25] A study released by the San Francisco–based advocacy group Children Now pointed to an even more damaging aspect of the marginalization of minorities: programming that included people of color frequently did so in an "exclusionary manner," in other words, by depicting them disproportionately as vagrants, dopers, gang-bangers, and sundry predatory criminals.[26]

Divergence rather than convergence of ethnicity, it would seem, had become the order of the day at the dawn of the new millennium. Nor had much changed by the early 2000s. The post-network era, with its deregulated marketplace and multichannel universe, had promised a "'win-win' situation ... for industry as well as for multicultural diversity," as John Caldwell observed.[27] But in the new "culture of conglomeration," in which the single-medium monopolies of old were dwarfed by six multimedia giants (Time-Warner, Viacom, News Corp., Disney, GE, and Sony), something far less utopian had occurred. The same "caste system of genre and tiering" that the SAG and Children Now studies had documented in 2000 appeared just as entrenched in 2001. As Caldwell wrote:

> African Americans are diverted to the endless ethnic comedies and reality shows so prevalent on UPN, WB, and Fox. Diversity of representation now exists on television, but only because of generic and format "ghettoizations" that the multichannel conglomerates have established and profited from.... [T]his ghettoization allows the majors (CBS, NBC, ABC) to be no less appreciably white than they've been for two decades (and to be that way without regulatory pressure); ... the new conglomerates actually prosper by internalizing and mastering a regime of difference in the form of corporate affiliate tiering.[28]

A harsh assessment, certainly, but one that the media-monitoring coalition tended to affirm. Under a memorandum of understanding between the coalition and the networks in 1999, the various groups (black, Latino, Asian/Pacific Island American, and Native American) were to meet individually with the networks twice a year (since 2005, once a year) to discuss diversity in hiring and screen images, and each group was to issue a report card on each network's performance at the end of the year.[29] The reports assess not merely casting ("Actors: On-Air Prime-time Scripted Shows"; "On-Air Prime-time Reality Programming" was added as a category in 2004), but also behind-the-scenes hiring and other issues: "Writers and Producers: Prime-time," "Directors: Prime-time," "Program Development," "Procurement," "Entertainment Executives," and "Network Commitment to Diversity Issues."[30]

The reports have not been glowing, especially for Asian/Pacific Island Americans. "Overall" grades (an average of all the categories), which started off in the D or C– range for all the networks, have remained stuck in the C range since the mid-2000s, with only ABC finally moving up to a B– in 2008 (reports are issued at the end of each calendar year).[31] Grades for Latinos have fared somewhat better, with ABC and Fox rating overall Bs as early as 2004, NBC and CBS joining them at that level in 2006, and ABC shooting up to an A– in 2006 and 2007, although retreating to a B+ in 2008.[32] African Americans have remained, as they started, ahead of the field, despite a "stern warning" from the NAACP that much more needs to be done.[33] On the heels of Barack Obama's 2008 election victory, newly appointed NAACP president Benjamin Todd Jealous decried the level of minority hiring by the networks: "At a time when the country is excited about the election of the first African American president in U.S. history, it is unthinkable that minorities would be so grossly underrepresented on broadcast television."[34] Jealous's critique came shortly after a forty-page report by the NAACP charged that blacks and other minorities continued to be underrepresented in "nearly every aspect of the television and film businesses."[35]

Despite the frustrating lag in overall media diversity, unmistakable improvement in minority representation on primetime network television has occurred since the Lily White controversy erupted, for which the reinvigorated media monitoring movement must be given at least some credit. In all the areas of contention cited by Caldwell, for example, significant positive change has taken place. First, although the still ratings-challenged CW network (formed from a merger of the WB and UPN in 2006) remains the bastion of "black block" programming, the real "action" on the diversity front has migrated from the "ghettoized" to the *major* networks. Second, the center of gravity for diversity programming has shifted from "the endless ethnic comedies and reality shows" to the drama. Finally, in an area that neither Caldwell nor the image-analysis studies addressed, the ethno-racial discourse has expanded beyond the black/white binary to embrace Hispanics and Asians as well (Native Americans, except as historical relics in television movies or miniseries, remain almost invisible). And at the forefront of all these substantial changes have been the neo-platoon shows cited earlier: *Boston Public* (Fox), *Kevin Hill* (UPN), *Lost* (ABC), *Heroes* (NBC), *Grey's Anatomy* (ABC), and *Ugly Betty* (ABC).

The Neo-Platoon Shows

What sets apart the neo-platoon shows of most recent vintage from recent near-misses and their 1970s forbears is not merely the former's dramatic genre, soap structure, or televisuality. *Hill Street Blues* and *St. Elsewhere*, as previously described, pioneered these elements, and *Miami Vice*, *NYPD Blue*, *Homicide*, *New York Undercover*, and *ER* also engaged one or more of them. Two other medical dramas emerging from the Lily White controversy—*City of Angels* (2000) and *Gideon's Crossing* (2000–2001)—took additional steps in the neo-platoon direction, but the former was typecast (on the playbill and in the press) as an "African American show," and the latter was a single-protagonist show (albeit starring an African American, Andre Braugher).[36] What makes the latest crop of platoon shows stand out is their convergence of difference. First, the multiple-protagonist/interlocking-narrative structure creates a textual interdependence among the range of ethnicities represented. Second, this interdependence is reinforced by the egalitarian positioning of the main characters; that is, people of color are placed on a par with or even a notch above their white cohorts, both in screen time and social/occupational standing. Finally, and most distinctively, interracial romance is prominently displayed.

Although not directly connected to ethno-racial difference, gender egalitarianism is another notable, if not unique, aspect of several of the neo-platoon shows, namely, *Heroes*, *Grey's Anatomy*, and *Ugly Betty* (*Lost*, *Kevin Hill*, and *Boston Public* maintain more traditional patriarchal structures). *Taxi*, among the 1970s platoon shows, included women in its ranks, and 1980s/1990s shows upped the ante. However, the numerical ratio and power balance between the sexes has decidedly shifted in women's favor in the neo-platoon phase. This shift was precipitated partly by feminist political strides, since the platoon show period overlapped with women's increasing industrial clout as a "quality" demographic; but it also occurred by default, from a narrative perspective, to accommodate the interracial romance angle. The following overview of the main neo-platoon shows illustrates their generic attributes as delineated above.

Boston Public (2000–2004), though its premiere predated the neo-platoon show trend, established the pattern. Steven Harper, an African American, is the capable and compassionate principal of a problem-plagued inner-city high school. The pedagogical platoon is rounded out by a Jewish vice principal,

Scott Guber, and a rainbow faculty: two African Americans (Marla Hendricks and Marilyn Sudor); two Hispanics (Harry Senate and Carmen Torres); another Jew (Henry Lipschutz); and sundry nondenominational whites (Ronnie Cook, Lauren Davis, and Danny Hanson). Multicolored students and family members also play prominent roles, but the administration and faculty members are the main recurring players. As for interracial romance, extensive screen time is devoted to Guber/Sudor's and Harper/Davis's biracial intimacies, although not fully consummated, not to mention subplots involving students' mixed relationships. Capitalist realism, meanwhile, would appear a counterintuitive construct for such a seemingly dystopian show. However, the ability of *Boston Public* to redeem "the system" despite the titular school's apparent abject failure is precisely what makes the show an ingenious reinvention of the form. On first glance, the school's chaotic conditions make *Blackboard Jungle* (1955) look like *Pleasantville* (1998); but the fact that the institution makes it through each episode's descent into hell and comes up with its head (Principal Harper) bloodied but unbowed affirms the clichés that, despite the flaws of the United States as a society, it is still the best of all possible worlds, and that while all may not be well with America, the Dream is still alive.

The comparatively short-lived *Kevin Hill* (2004–2005) picked up where *Boston Public* left off, multiracially and romantically. Another African American male, former high-level attorney Kevin Hill, is again nominally "in charge," this time of a lower-level legal team he is forced to join when his cousin's death causes Hill to relocate in order to take custody of the cousin's ten-month-old baby girl. And again, the multicultural composition of the firm is foregrounded: Jessie Gray is black; Damien "Dame" Ruiz is Puerto Rican; George Weiss is Jewish; Veronica Carter and Nicolette Raye are white. Another Hispanic, Michele (no last name given), has a recurring role, as do two of Hill's love interests, Evelyn Cruz, a Latina, and Monroe McManus, an African American. Obviously, interracial romance is again at play here, with Ruiz's courting of Carter reinforcing the theme. Racial mixing of a more ambiguous, and ultimately somewhat controversial, sort resulted from the multi-casting of the role of baby Sarah. The little girl's portrayer was switched early on in the series to a lighter-skinned "black" infant, causing one viewer to complain on the show's website: "Why was the baby on the show switched to a different baby? The baby used for the premiere of *Kevin Hill*, better matched the star of the show. The new baby absolutely does not look as though it is remotely related to Taye Diggs. Was the first baby star, too dark-skinned? I for one thought she was absolutely beautiful. Taye Diggs and its producers should be ashamed of themselves. The new baby looks as though she would be perfectly matched, if the star of the show was LL Cool J or Terrence Howard, but certainly not Taye Diggs."[37] Sarah, of course, is not diegetically imputed to be

Kevin Hill's child but his cousin's, whose skin complexion is not referenced. When the birth mother, a light-skinned "black," does appear mid-season, in prison, both her previous absence and Sarah's lighter skin tone are explained (though doubtfully to the above viewer's satisfaction).

In subsequent neo-platoon shows, the ethno-racial spectrum is expanded and the interracial gloves are removed completely. In *X Files* meets *Survivor* style, *Lost* (2004–2010) revolves around airplane crash survivors on a mysterious tropical island.[38] The people of color in the enormous cast have (at various times and in various combinations) included seven blacks (Michael, Walt, and Susan Lloyd; Rose; Naomi Dorrit; Mr. Eko; and one of the mysterious "Others"), four Hispanics (Hurley, Ana-Lucia, Paulo, and Nikki), two Koreans (Jin and Sun Kwon), an unspecified Asian (Miles Straume, played by the Chinese Ken Leung), and an Iraqi (Sayid, played by the British-Indian Naveen Andrews). Several interracial relationships start up on the island or predate the crash: Sayid has an affair with the Paris Hiltonish Shannon; Hurley makes a play for Libby, who is white; Michael and Sun, in her husband's absence, veer toward intimacy; and Rose and Bernard, who is white, are married. The show also significantly extends the notion of multiculturalism from a purely American into a global phenomenon. Besides the two Koreans and the Iraqi, Alex and Daniel Rousseau are French, Claire is Australian, Charlie is British, Desmond is Scottish, and Mr. Eko is Nigerian. As for the "platoon leader," there are so many, ever-changing, interlocking combinations in constant flux that it's difficult to isolate, much less privilege, any one individual. Clearly, however, despite its overly determined multipolar structure, the show revolves most magnetically around the characters of Jack Shephard, John Locke, and Kate Austen (name symbolism duly noted by showrunners and fans), all of whom are white and whose love triangle thus provides the narrative(s) with the most traditional (and conservative) structural and ethno-racial anchor.

Heroes (2006–2010) makes global multiculturalism the veritable premise of its transnational tale of a geographically dispersed, karmicly connected group of men, women, teens, and children possessed of superhuman powers that may, or may not, save the world (from nuclear disaster in the first season, from a deadly virus in the second, from the unpredictable consequences of all humans' gaining special powers in the third). The main characters (also shifting over time) include seven African Americans (D. L. Hawkins, Micah Sanders, Simone and Charles Deveaux, Monica and Nana Dawson, and Benjamin "Knox" Washington), three Asian Indians (Mohinder Suresh, Chandra Suresh, and Niran), three Japanese (Ando Masahashi, and Hiro and Kaito Nakamura), three Hispanics (Isaac Mendez and Maya and Alejandro Herrera), and one West Indian ("the Haitian"). Besides the global scope (much of the musical score is Indian influenced and manga-style graphics permeate the visuals), as with *Lost*, several interethnic couplings have been featured or alluded to here: one Indian-white (Mohinder and Eden McCain); one Indian-Latina (Mohinder and Maya); one Latina-white (Maya and Gabriel Gray, aka Sylar); one Asian-"white" (Princess Yaeko and Adam Monroe—the latter actually Hiro/Takezo Kensei in disguise); one Latino-black (Isaac and Simone); and two black-white (Simone and Peter Petrelli, and D. L. Hawkins and his wife, Niki Sanders). The last pairing is of particular note, given its obvious play on the relationship of O. J. Simpson and Nicole Brown Simpson. Besides the name similarities (O.J./D.L., Nicole/Niki) and Niki's physical resemblance to her namesake (tall, slender, long blond hair), one plot has a frightened Niki warn the police about her fugitive husband, who, she fears, might be stalking her: "D. L. Hawkins is a killer!" This unsubtle commentary on the O.J. trial is ultimately reversed when Niki's evil alter-ego turns out to be the real killer. As for platoon hierarchies, they are perhaps even more difficult to establish here than in *Lost*; and if *Lost* is any indication, they are likely to grow even more ambiguous in the "hyper-diegetic" cyberspaces with which Internet-era shows have formed a feedback loop.[39]

Another notable biracial tryst, and the most high-profile (so far) among the neo-platoon shows, has unfolded on the medical drama *Grey's Anatomy*, the first American network television drama created by an African American woman, Shonda Rhimes.[40] No doubt partly owing to its showrunner's ethnicity, the storyline places an African American "in charge": Dr. Richard Webber, head of surgery at a Seattle hospital whose surgical staff includes (or included at one time) two other African Americans (Dr. Miranda Bailey and Dr. Preston Burke), a Chinese American (Dr. Cristina Yang), and a Latina (Callie Torres). Rhimes's gender likely also played a role in the show's partial tilt toward its female (not necessarily black) characters, as illustrated in the title's derivation from a (white) woman, Dr. Meredith Grey, and in the gender equality and even slight numerical dominance of women over men among the primary doctors featured on the show.[41] Biracial affairs, meanwhile, abound at the hormonally unchallenged medical facility. Torres has been the most adventurous of the bunch, having had affairs with three white doctors, one of whom was a woman. Burke and Yang, despite ups and downs, were a long-term item, briefly engaged, and almost married; Yang has since taken up with the Caucasian Dr. Owen Hunt. Another multicolored affair, between doctors Webber and Ellis Grey (mother of Meredith Grey), is part of the show's backstory.[42]

Whatever sympathy the rainbow romances may have engendered was overshadowed by the brouhaha surrounding seemingly homophobic remarks made by Burke portrayer Isaiah Washington at a 2007 Golden Globes backstage interview, directed at castmate T. R. Knight (Dr. George O'Malley), who is openly gay. The incident eventually led to Washington's being eased out of the show; however, ethno-racial tensions rather than issues of sexual orientation may have been the real problem behind the scenes of a show previously hailed for its multicultural harmony. In a pre-scandal *Nightline* segment, Washington had complained that ABC refused to give him the juicy part of Dr. Shepherd and had chosen Patrick Dempsey, who is white, "because casting a Black actor as a prime-time sex object for women was 'off limits.'"[43] The idealized multicultural formula of the neo-platoon shows, like the capitalist realism of the platoon films of old, obviously cannot hide American society's legacy of racism.

Ugly Betty (2006–2010), in some ways the most tenuous fit in the neo-platoon category, is in other ways its most quintessential example. More dramedy than drama and with a narrative premise heavily concentrated on its titular lead, the show demonstrates its neo-platoon credentials, first, by the eponymous Betty Suarez's Latina ethnicity and gender. Second, while the show's focus on Betty and her family seemingly makes it more a Hispanic than a multicultural show, the interracial affairs of Betty and other featured players and the textual interdependence among a range of ethnically (and sexually) diverse main characters redeem its neo-platoonism and then some. The show's power relations lean even more strongly toward women than in *Grey's Anatomy*. The show is not only named after but revolves around Betty. And while her job is a subservient one—she is the much-abused personal assistant to the editor-in-chief of a high-end New York fashion magazine—her nemesis (and boss, by the third season) is a domineering African American woman, Wilhelmina (note the Teutonism) Slater. The magazine's owner (following the early diegetic death of her husband), Claire Meade, is also a woman. And two other prominent cast members, Claire's "daughter" Alex/Alexis and staffer Marc St. James, are transsexual (male to female) and gay, respectively. Biracial liaisons, meanwhile, are as abundant and complex as on any primetime show: Claire's heterosexual son Daniel, an apparent serial inter-racialist, has affairs with the Latina Sofia Reyes, the Asian Grace Chin, and Wilhelmina's sister Nico; and Claire has a lesbian affair, in prison, with another inmate, Yoga, who is African American. Betty herself is romantically linked to three non-Latino whites: Walter, Henry Grubstick, and Giovanni "Gio" Rossi (the latter, though nominally Italian, is played by Freddy Rodriquez). Other recurring multicultural characters have included a Latina (Gina Gambarro) and two Asians (Kenny and Suzuki St. Pierre, who is also gay).

One other, short-lived show, *Six Degrees* (2006–2007), deserves honorable mention, if not actual inclusion, in the neo-platoon show trend. Its erstwhile unacquainted, ultimately enmeshed New Yorkers include an African American (Damian) and a Hispanic (Carlos) whose more than platonic attraction to one of the white women (Mae) drives much of the early action. In the brief 2007 revival of the show, Carlos has another interracial affair, graphically consummated this time, with his white real estate agent. Recurring characters significantly linked to the "core six" include Damian's black girlfriend, Regina, and an Asian American secretary, Melanie. While the show has no platoon "leader" per se (class distinctions provide the prime demarcation, especially in the upscale/downscale view of New York City), *Six Degrees* does mimic *Lost* and *Heroes* in thrusting a group of previously unconnected people together through chance or fate. As with *Lost*'s partly hard-luck, partly fortunate survivors and *Heroes*' reluctant "chosen ones," *Six Degrees*' half-dozen "intertwined destinies" remind us "that romance, success, peace or forgiveness might be right around the corner."[44]

The hybridization of *The X Files* (1993–2002) and *Survivor* (2000–), alluded to in relation to *Lost*, is more than superficially pertinent here. What the supernatural thriller and the so-called reality show share with the destiny-driven neo-platoon shows is a desperate diegetic attempt to wring connectedness and meaning (and for the reality show, its own youth-driven brand of neo-platoonism) from a steadily contracting yet increasingly disconnected and meaningless world.[45] Convergent ethnicity—and in the case of *Lost* and *Heroes*, convergent nationality—rides to the rescue, appearing to offer a textual solution to the postmodern dissolution of personal identity and the rupturing of the "really real," as well as an industrial solution to the postclassical breakdown in network hegemony. Viewers lost in a sea of psychological and socio-economic distress, and global capitalist networks fishing for crossover domestic and international hits, can find common cause in an ethnically convergent world where not only is everything interconnected but interconnectedness also provides the key to salvation.

What we have here is no longer capitalist realism but capitalist surrealism. As with the dysfunctional "families" that have dotted the contemporary television landscape from (on the networks) *Married with Children* (1987–1997), *The Simpsons* (1989–), *Seinfeld* (1989–1998), *Family Guy* (1999–), and *Arrested Development* (2003–2005) to (on cable) *The Sopranos* (1998–2006), *Curb Your Enthusiasm* (2000–), *Big Love* (2005–2011), *Lucky Louis* (2006), *Weeds*, *Dexter*, and *The Riches* (2007–2008), we're definitely not in Kansas (or even in 1950s Springfield) anymore. In a post–9/11, climate-changing, economy-tanking world where the American Dream has turned into a Global Nightmare, where "one nation under television" had become (until the financial meltdown) "one market under God," what we have come to value most—despite, or because of, our cellphones, iPods, MySpaces, and YouTubes—is survival.[46] As the British-accented narrator on *Heroes* unctuously proclaims: "Evolution is an imperfect and often violent process, a battle between what exists and is yet to be born. In the midst of these birth pains, morality loses its meaning, the question of good and evil reduced to one simple choice: survive or perish."

Conclusion

Commenting on the viewer fragmentation of the multichannel era, NBC West Coast president Don Ohlmeyer stated in 1998: "There is no 'audience' anymore. There are 200 different segments of the audience, and the goal is to try and pull together as many different segments and aggregate them at one time with something that they collectively want to experience. That's what programming is about today: providing a collective experience."[47] An obvious follow-up question to Ohlmeyer's remarks, in light of the neo-platoon trend, is whether these shows have filled the post-network programmer's bill. Has convergent ethnicity created, at least for the moment, an ideal "collective experience," what John Caldwell described as a "'win-win' situation ... for industry as well as for multicultural diversity" that ghetto shows on ghetto networks failed to achieve?[48]

First, the comparatively low grades given by media monitoring groups to the major networks for casting and hiring practices, up through 2008, preclude any triumphal assessments. Further disheartening developments, especially for African Americans, have resulted from the consolidation of the WB and UPN networks in the new CW. Shonda Rhimes and the neo-platoon shows notwithstanding, the situation for blacks has "actually gotten worse," according to veteran black writer/showrunner Felicia Henderson (*Moesha* [1996–2001], *Soul Food* [2000–2004], *Everybody Hates Chris* [2005–2009], *The Fringe* [2008–]). "Because we [blacks] basically work when black shows are on the air, and only the CW is airing black-themed shows, coupled with the dire outlook of half-hour comedies and therefore black comedies, what you have are the least number of black writers working since I started writing ten years ago."[49]

The NAACP concurred on the sitcom point, asserting that the 2006–2007 season marked "the first time in 'recent memory' that there is not a comedy on ABC, NBC, CBS, or Fox with an African American lead." "I feel we are losing ground," the civil rights group's then president Bruce Gordon averred. "The lack of African American leads in sitcoms is unconscionable. This is historically where many African actors, writers and show runners have honed their skills and found employment."[50] George Lopez, star of the *George Lopez Show* (2002–2007) and the lone Hispanic lead of a major network comedy since Freddy Prinz on *Chico and the Man* (1974–1978), went further, calling the television studios "modern-day plantation owners. They are only concerned with the problems white people have.... It's really sad for me to see, and I've seen it from the inside since 2001. There used to be these comedies with Bernie [Mac] and Cedric the Entertainer... but it seems like they got afraid and reverted back to what's safe for them."[51]

The smash, yet still token, success of the dramedy *Ugly Betty* is the rule-proving exception to the eulogies of Lopez, the NAACP, and Henderson for the ethno-racial sitcom. On the other hand, it is notable and more than a bit curious to find the comedy "of color," long maligned as a ghetto genre, bemoaned as an endangered species at the very moment when people of color—and people of various colors, not just blacks—are starring, and intermixing, in unprecedented numbers on the long-taboo network television drama.

Two other caveats obtain to this paean to post-network programming, however. First, the rise of the neo-platoon show has tilted toward one network, ABC. Of the seven neo-platoon shows (including *Six Degrees*) highlighted in this study, four air(ed) on ABC, with one each on NBC (*Heroes*), Fox (*Boston Public*), and UPN (*Kevin Hill*). *George Lopez* was also an ABC show. The unequal distribution of diversity is reflected, as it should be, in the higher marks ABC has garnered in the overall ratings from the media monitoring groups.[52] Thus, while it would be fatuous to label ratings-competitive ABC or current king CBS the new ghetto networks, it is also clear that convergent ethnicity, in the neo-platoon sense, has yet to achieve full programming acceptance across the network spectrum.

Second, convergent ethnicity, at least in its present capitalist realist (or surrealist) form, can hardly be considered an unqualified boon for people of color or for society as a whole. Color-blind casting may promise an end to "othering," but this potential benefit is compromised by a damaging cost: the dissolution of difference. The multiethnic members of the neo-platoon shows may look different, but they tend to act the same. Historical and cultural distinctions, not to mention persistent ethno-racial inequities, are ignored for the most part, if not denied altogether. While Reaganist color-blind ideology and the "common sense" status it has achieved (further "confirmed" by Obama's "postracial" presidency) is clearly at work in television's soft-pedaling of race, the commercial constraints of the culture industry also play a part. As Ann Oldenberg reported in *USA Today*, the showrunners of *Lost* ended up downplaying the racial "edge" displayed in one of the show's black-white liaisons due to fans' complaints.[53]

How to play the race card remains a hot-button issue within the industry. On one side of the debate are television producers and observers who claim that the color-blind approach represents "progress that should be celebrated . . . an evolution demonstrating that such [interracial] romances are no longer a big deal."[54] For example, Kari Lizer, creator of the CBS sitcom *The New Adventures of Old Christine* (2006–), felt that it was important to illustrate a loving relationship between the white protagonist (Christine) and a black schoolteacher (Daniel Harris) "where race was not the main focus." "We don't act like the race aspect is invisible," Lizer says. "We say it as a fact of life and move on."[55] Robert M. Entman, professor of the George Washington University School of Media and Public Affairs and author of *The Black Image in the*

White Mind, finds definite progressive elements in the current crop of interracial romances, color blindness and all: "It makes these couples more normal, and if they're more normal on TV, they might seem more normal out on the street."[56]

The other side argues that downplaying race oversimplifies race relations. Janette Dates, dean of the John H. Johnson School of Communications at Howard University, complains that the shows do not "allow the nuances of the reality of interracial couples to come through. I don't know of any interracial couple that doesn't have to work at it. Society forces them to deal with their situation."[57] Mara Rock Ali, creator of the black-cast shows *Girlfriends* (2000–2008) and *The Game* (2006–), emphasizes a positive aspect that's missing in most depictions: "I find it not only false but unfortunate that the very thing that defines the 'interracial couple' is not explored. And by not exploring race, not only do you miss the opportunity for great stories, you miss what is unique to their experience."[58] Henderson goes further: "The idea of multiculturalism [in these shows] is a sham to me. There is one culture, mainstream American culture. The characters have been stripped of all their cultural identity . . . as opposed to encouraging them to embrace and celebrate difference in a positive way."[59]

While sugar-coating difference may be the more prevalent problem of the neo-platoon show, exaggerating difference in stereotypical ways is an issue as well. In *Grey's Anatomy*, in particular, as Henderson observes, the "superspade and the mammy get a makeover." Dr. Preston Burke's updated version of the superspade role popularized in 1970s blaxploitation films is fulfilled not so much in his objective prowess as a heart surgeon as in his macho carriage and "overstated belief" in his own superiority.[60] Dr. Miranda Bailey, another African American doctor, addresses Burke's superiority complex head-on in "The First Cut Is the Deepest" (season 1, episode 2): "I think you're cocky, arrogant, pushy, and you have a God complex. And you never think about anybody but yourself." Bailey herself, however, often falls into the asexual, overweight, and crude characterizations associated with the stereotypical mammy. At once underscoring and overcompensating for the representational slight, Bailey is referred to by her cohorts on the show as "the Nazi."[61]

Traces of demeaning stereotypes relating to other ethno-racial groups include Hurley's quasi-Latino buffoon in *Lost*, the nerdy, asexualized Japanese duo Ando and Hiro in *Heroes*, and the Dragon Lady aggressiveness of Dr. Cristina Yang, also on *Grey's Anatomy*. The latter show's unflattering black images are the most perturbing, appearing as they do in a show created and run by an African American undoubtedly striving, and otherwise succeeding, in expanding multicultural horizons. The paradox is further proof, if any were needed, of the enduring power of media images to penetrate even the best defenses, ultimately serving to internalize such stereotypes among the groups

FIG. 9-2 *Grey's Anatomy* (ABC): Multicultural cast with an African American "in charge"—onscreen and behind the scenes. (DVD cover)

most damaged by them. The paradox also reinforces the sense that convergent ethnicity, both in front of and behind the camera, is no panacea for the problems of diversity in American television.

Convergent ethnicity is certainly not to be confused with David Hollinger's notion of "postethnicity," at least not yet. Postethnicity, while it resists the exclusionary aspects of identity politics, insists on acknowledging the historical reality and ongoing repercussions of difference."[62] A postethnic perspective encourages coalitioning among ethno-racial groups to achieve mutual political goals more effectively, much as the post-1999 media monitoring groups have succeeded (with mixed results) in doing.[63] Most ambitiously, postethnicity favors a "rooted cosmopolitanism" that "promotes multiple identities, emphasizes the dynamic and changing character of many groups, and is responsive to the potential for creating new cultural combinations."[64] Whether convergent ethnicity is capable of developing along postethnic lines within a commercial television system and a global capitalist economy is questionable. The color-blind ethos maintains a strong hold on American culture and society, for powerful psychological and ideological reasons: it assuages individual and collective guilt about past and ongoing inequities; it minimizes, if not eliminates, the need to address outstanding issues; and it provides a sense that "the system," with all its dysfunction, is still "the best of all possible worlds." In the face of such formidable obstacles to progressive change, perhaps the most we can expect from the post-network era is what Andreas Huyssen believes the postmodern age itself represents: "simultaneous delimitation and expanded prospects ... both our problem and our hope."[65]

Notes

Thanks to Michael Barry, Drew Morton, Jennifer Porst, Mallory Sussman, and Tim Wells for their invaluable commentary and feedback.

1. John Thornton Caldwell, *Televisuality: Style, Crisis, and Authority in American Television* (New Brunswick, NJ: Rutgers University Press, 1995); Eileen Meehan, "'Holy Commodity Fetish, Batman!': The Political Economy of a Commercial Intertext," in *The Many Lives of Batman: Critical Approaches to a Superhero and His Media*, ed. Roberta Pearson and William Uricchio (New York: Routledge, 1991), 47–65; John Fiske, *Understanding Popular Culture* (Boston: Unwin Hyman, 1989); Henry Jenkins, *Textual Poachers: Television Fans & Participatory Culture* (New York: Routledge, 1992); Lynn Spigel and Jan Olsson, eds., *Television after TV: Essays on a Medium in Transition* (Durham, NC: Duke University Press, 2004); Tasha G. Oren and Patrice Petro, eds., *Global Currents: Media Technology Now* (New Brunswick, NJ: Rutgers University Press, 2004); Lisa Parks and Shanti Kumar, eds., *Planet TV: A Global Television Reader* (New York: New York University Press, 2003).
2. Herman Gray, *Watching Race: Television and the Struggle for Blackness* (1995; Minneapolis: University of Minnesota Press, 2004); Darryl Y. Hamamoto, *Monitored Peril: Asian Americans and the Politics of TV Representation* (Minneapolis: University of Minnesota Press, 1994); Sasha Torres, ed., *Living Color: Race and Television in the United States* (Durham, NC: Duke University Press, 1998); Vincent Brook, *Something Ain't Kosher Here: The Rise of the "Jewish" Sitcom* (New Brunswick, NJ: Rutgers University Press, 2003); Kristal Brent Zook, *I See Black People: The Rise and Fall of African-Owned Television and Radio* (New York: Nation Books, 2008).
3. Amanda Lotz, "Textual (Im)Possibilities in the U.S. Post-Network Era: Negotiating Production and Promotion Processes on Lifetime's *Any Day Now*," in *Television: The Critical View*, ed. H. Newcomb, 7th ed. (New York: Oxford University Press, 2007), 223–244. *Desperate Housewives* (2004–) fails to make the neo-platoon grade, despite increased color consciousness in recent years, because of its still predominantly white cast. Other network shows that come close: *Whoopi* (2003–2004), a sitcom in which Whoopi Goldberg's hotel owner character, Mavis Rae, lords it over two quasi-white "servants" (an Iranian hotel doorman and a Bulgarian

maid) and Mavis's conservative Republican brother has an ongoing interracial romance with a female "wigger"; the sitcom *Luis* (2003), which foregrounds a Latino-white romance, disaggregates difference in its distinction between the Puerto Rican Luis and his Dominican Republican ex-wife, and might have qualified had it lasted more than a month; and *The West Wing* (1999–2006) and *24* (2001–2011), both of which broke major color barriers in their minority U.S. presidents—*West Wing*'s Josiah Bartlett (played by the Latino Martin Sheen) and his successor, Matthew Santos (played by Jimmy Smits), and *24*'s African American David Palmer (Dennis Hasbert) and his son Wayne (D. B. Woodside). Despite its white main character, Marin First (Ann Heche), the dramedy *Men in Trees* (2006–2008) exhibits the most unusual *consequence* to an interracial romance of any American show I know of. Although one of the main characters, Buzz, is clearly African American, and the woman he had sex with many years ago, Police Chief Celia Bachelor, is clearly Anglo, the fruit of their union, teenaged Patrick Bachelor, *looks* white—both in skin color and facial features (explanation: a recessive gene).

4   Quoted in www.comic-con.org/cci/cci07prog.shtml.
5   Quoted in Maria Elena Fernandez, "The TV Biz: Conceived Abroad, Born in the U.S.A.," *Los Angeles Times*, April 20, 2008, E1.
6   Jeanine Basinger, *The World War II Combat Film: Anatomy of a Genre* (New York: Columbia University Press, 1986), 62.
7   Quoted in C. R. Koppes, "Wartime Stars, Genres, and Production Trends," in *Boom and Bust: American Cinema in the 1940s*, ed. Thomas Schatz (Berkeley and Los Angeles: University of California Press, 1997), 203–261, 245, 246.
8   Lewis Jacobs, "World War II and American Film," *Cinema Journal* 7 (Winter 1967–1968): 1–21; Michael Schudson, *Advertising, the Uneasy Persuasion: Its Dubious Impact on American Society* (New York: Basic Books, 1984); Thomas H. Zynda, "*The Mary Tyler Moore Show* and the Transformation of Situation Comedy," in *Media, Myths, and Narratives: Television and the Press*, ed. James W. Carey (Newbury Park, CA: Sage, 1988), 126–145.
9   Zynda, "*Mary Tyler Moore Show*," 139.
10 J. R. Bowlt, *Russian and Soviet Painting*, exhib. cat. (New York: Metropolitan Museum of Art/Rizzoli, 1977), 13.
11 Patricia Erens, *The Jew in American Cinema* (Bloomington: Indiana University Press, 1984), 75.
12 Basinger, *World War II Combat Film*, 37.
13 Howard Suber, "Hollywood's Closet Jews," *Davka* (Fall 1975): 12–14.
14 Howard Suber, "Television's Interchangeable Ethnics: 'Funny, They Don't Look Jewish,'" *Television Quarterly* 12, 4 (Winter 1975): 49–56, 53.
15 The character of Fish is generally regarded as the lone explicitly Jewish character on the show. Barney himself, however, although the last name Miller was initially chosen for its ethnic ambiguity, comes out Jewish in a Christmas episode well into the series' run. Terry Barr, "Stars, Light, and Finding the Way: The Emergence of Jewish Characters on Contemporary Television," *Studies in Popular Culture* 15, 2 (1993): 87–89.
16 Suber, "Television's Interchangeable Ethnics," 50, 53.
17 Jane Feuer, Paul Kerr, and Tise Vahimagi, eds., *MTM: "Quality Television"* (London: BFI Publishing, 1984).
18 Zynda, "*Mary Tyler Moore Show*," 141.
19 Caldwell, *Televisuality*.

20 Although it lacked the full ensemble cast and other soap elements of *Hill Street Blues*, *The Mod Squad* (1968–1973) can be considered an adjunct of the platoon drama, with its triumvirate of countercultural undercover-cop protagonists—two male, one female; two white, one black. *I Spy* (1965–1968), meanwhile, with its black/white lead duo, also fits peripherally into the platoon show mold.

21 Gray, *Watching Race*.

22 Kristal Brent Zook, "The Fox Network and the Revolution in Black Television," in *Gender, Race, and Class in Media: A Text Reader*, ed. Gail Dines and Jean M. Humez, 2nd ed. (Thousand Oaks, CA.: Sage, 2003), 586–596. Whether the "*Cosby* moment" can be credited with the African Americanization of the *Star Trek* franchise is debatable, but the television series that presaged the platoon show trend of the 1970s tilted conspicuously toward blackness in its late 1980s sequel, *Star Trek: Next Generation* (1987–1994), and myriad spin-offs: *Star Trek: Deep Six Nine* (1993–1999), *Star Trek: Voyager* (1995–2001), and *Enterprise* (2001–2005).

23 National Association for the Advancement of Colored People, "Out of Focus—Out of Sync Take 4: A Report on the Television Industry" (2008): 3 (http:www.naacp.org/news/press/2008-12-18/naacp_ofos_take4.pdf); Greg Braxton, "A White, White World on TV's Fall Schedule," *Los Angeles Times*, May 28, 1999, A1; Elizabeth Jensen, "Groups Band Together to Press for Diversity Campaign," *Los Angeles Times*, September 11, 1999.

24 Greg Braxton, "A Mad Dash for Diversity," *Los Angeles Times*, August 9, 1999, F1.

25 Greg Braxton, "Study Finds Blacks Seen Most in Comedies, New Networks," *Los Angeles Times*, February 25, 2000, F24.

26 Greg Braxton, "TV's Color and Gender Lines, Examined Anew," *Los Angeles Times*, January 12, 2000, E1.

27 John Thornton Caldwell, "Convergence Television: Aggregating Form and Repurposing Content in the Culture of Conglomeration," in Spigel and Olsson, eds., *Television after TV*, 41–74, 67.

28 Ibid., 68.

29 The coalition comprises the National Association for the Advancement of Colored People (NAACP), the National Hispanic Media Coalition (NHMC), the Asian Pacific American Media Coalition, and American Indians in Film and Television.

30 www.nhmc.org; www.advancingequality.org.

31 www.advancingequality.org.

32 Ibid.

33 Greg Braxton, "NAACP Gives a Stern Warning to TV," *Los Angeles Times*, December 19, 2008, E26.

34 Quoted in ibid.

35 Quoted in ibid. Interestingly, the "much-maligned" reality show has actually fared better in overall diversity than the scripted shows, at least in recent years, according to the 2008 NAACP report. Whether this trend amounts to another "ghettoization" of diversity is another matter, but "just as the military and professional sports . . . became the unlikely vessels for breaking racial barriers decades ago," Greg Braxton observes, "reality programming may be a similarly transformational force in bringing greater diversity to television today" ("The Greater Reality of Minorities on TV," *Los Angeles Times*, February 17, 2009, A16). Matters certainly have progressed beyond the convergent ethnicity crisis generated in 2006 by the seminal reality show *Survivor* (2000–), when at the start of the season competing teams were divided by race—a controversial, ratings-driven ploy that was junked a few

weeks into the run. S. Zimmermann, "Jungle Fever: A Harmless Reality-TV Race Gimmick," *TNR Online*, September 14, 2006; Earl Ofari Hutchinson, "In Defense of 'Survivor' Segregation," *AlterNet*, September 13, 2006; Greg Braxton, "NAACP's Stance on 'Survivor,'" *Los Angeles Times*, September 13, 2006.

36   www.tv.com/city-of-angels/show/243/summary.html.
37   www.imdb.com/title/tt0412162.
38   Other of *Lost*'s conscious inspirations include the novel *Lord of the Flies*, the film *Cast Away* (2000), the sitcom *Gilligan's Island* (1964–1967), and the game *Myst*. David Bernstein, "Cast Away," *Chicago*, August 2007, http://www.chicagomag.com/Chicago-Magazine/August-2007/Cast-Away/.
39   Matt Hills, *Fan Cultures* (London: Routledge, 2002); Caldwell, "Convergence Television," 9. The web version of *Heroes* features two additional major characters of color: the African American Echo and the Latino Santiago.
40   Felicia Henderson was the first black woman to create a dramatic series for American television, network or cable: Showtime's *Soul Food* in 2000. Kevin Arkadie, a black male, preceded both Henderson and Rhimes in creating a network dramatic series (Fox's *New York Undercover*, in 1994), and Yvette Lee Bowser was the first African American showrunner altogether, albeit of a sitcom, Fox's *Living Single*, in 1993.
41   The moving force behind and star of the spin-off from *Grey's Anatomy*, *Private Practice* (2007–2013), also is a woman, Dr. Addison Montgomery. Of course, women's numerical parity cum superiority among the doctors in *Grey's Anatomy* also mirrors actual trends in medical school enrollment and graduates.
42   A non-hospital gay relationship between the recurring bartender character, Joe, and his male partner also has been limned, to the point of the couple's adopting a child.
43   Scott Collins, "Kiss & Make Up?" *Los Angeles Times*, February 5, 2007, E13.
44   abc.com; www.imdb.com/title/tt0801427.
45   The supernatural genre is worthy of a separate study, having undergone a major surge from the mid-1990s onward, including *The X-Files*, *Touched by an Angel* (1994–2003), *Buffy the Vampire Slayer* (1997–2003), *Charmed* (1998–2006), *Angel* (1999–2004), *Roswell* (1999–2002), *Dark Angel* (2000–2002), *Smallville* (2001–2011), *Medium* (2005–2011), *Ghost Whisperer* (2005–2010), *Reaper* (2007–2009), *Supernatural* (2005–), *Heroes*, *Pushing Daisies* (2007–2008), *Saving Grace* (2007–2010), *Life on Mars* (2008–2009), and *True Blood* (2008–).
46   J. Fred MacDonald, *One Nation under Television: The Rise and Decline of Network TV* (Chicago: Nelson-Hall, 1994); Thomas Franks, *One Market under God: Extreme Capitalism, Market Populism, and the End of Economic Democracy* (New York: Doubleday, 2000).
47   Quoted in Brian Lowry, "Change Is on the Air," *Los Angeles Times*, August 30, 1998, C10.
48   Caldwell, "Convergence Television." Convergent ethnicity, despite some initial promise, has neither solved nor is likely to solve network television's deeper systemic economic problems stemming from an inability to interface with the Internet and other new media more profitably.
49   Felicia Henderson, interviewed by Vincent Brook, 2006.
50   Quoted in Greg Braxton, "NAACP Says TV Still Lacks Color," *Los Angeles Times*, June 16, 2006, B2.
51   Quoted in ibid.
52   CBS, it should be noted, while lacking any neo-platoon shows, did receive, for its

multiracial *CSI* franchise, the highest grade among the networks for casting of Latinos in a scripted series, as well as a (somewhat backhanded) compliment from the NAACP for having two of the only three remaining shows in 2008—*CSI: Crime Scene Investigation* and *The Unit*—with minorities in leading roles (ABC's *Ugly Betty* was the third). Braxton, "NAACP Gives a Stern Warning."

53 Ann Oldenberg, "Love Is No Longer Color-Coded on TV," *USA Today*, December 20, 2005, http://www.usatoday.com/life/television/news/2005-12-20-interracial-couples_x.htm.
54 Greg Braxton, "The Hot Button of a Casual Embrace," *Los Angeles Times*, February 11, 2007, E21.
55 Quoted in ibid.
56 Quoted in ibid.
57 Quoted in ibid.
58 Quoted in ibid.
59 Henderson, interview.
60 Felicia Henderson, "Superspade and Mammy Get a Makeover: Race as a Floating Signifier in 'Grey's Anatomy,'" unpublished manuscript.
61 Ibid.
62 David Hollinger, *Postethnic America: Beyond Multiculturalism* (New York: Basic Books, 1995).
63 Henderson, in our interview, was skeptical about the ultimate efficacy of the media monitoring groups: "I think what [the process] does is allow both sides to feel more comfortable with the lack of progress. The NAACP can say, 'Look what we forced them to do,' and the networks can say, 'Look how open we are,' but in fact [little meaningful change has occurred]."
64 Hollinger, *Postethnic America*, 3–4.
65 Andreas Huyssen, "Mapping the Postmodern," *New German Critique* 33 (Fall 1984): 5–22.

10
Translating Telenovelas in a Neo-Network Era

●●●●●●●●●●●●●●●●●●●●

Finding an Online Home for MyNetwork Soaps

KATYNKA Z. MARTÍNEZ

The promotional video that introduced the press, network executives, and advertisers to MyNetwork TV at a 2006 upfront referenced what were perceived to be five factors behind the inevitable success of the soon to be launched network. Although it was touted as the "biggest change in primetime television," MyNetwork TV was also associated with the decades-old genre of the limited-run dramatic serial. The promo began with a rapid series of images from MyNetwork's original programming, then immediately moved to an image of a spinning globe with the names of countries like Brazil, the Czech Republic, and Greece scrolling behind the world. The voice of a male narrator announced: "Billions around the globe are watching the short dramatic series." The globe rolled off screen and was replaced by scenes from MyNetwork's dramatic series. The narrator resumed, "Millions tune in daily. Thousands of hours of nonstop drama. Already a rating sensation in hundreds of countries around the world. And now as one television network brings the genre to America, with highly promotable stars and fresh new faces that are destined for stardom, here are the five big reasons why you should make a play with MyNetwork TV." A female voice started the countdown with a breathy "Five." The male narrator continued, "Ninety-percent-plus U.S. clearance when MyNetwork TV premieres in September. Four: Addictive dramatic stories that have been proven hits around the globe. Three: Quality strip programming that employs breakthrough digital production techniques. Two: Compatible original web and mobile content that advances our storylines and extends the reach of your advertising message. One: Fifty-two weeks of original programming in high definition. It all adds up to the biggest change in primetime television you've ever seen. MyNetwork TV."

The programming format of MyNetwork TV did indeed mark a change in U.S. primetime television, and to associate the network with the limited-run dramatic series of countries such as Brazil was not unfounded. However, the orientation of the network was less a homage to the successful formats of "hundreds of countries around the world" than a cost-effective response to the loss of programming after a significant 2006 television network merger. MyNetwork TV was launched in September 2006 following the introduction of the CW Television Network, which was created through the merging of the WB and UPN. The creation of this network meant that local Fox Television stations that had been affiliated with UPN could no longer draw from UPN programs. MyNetwork was created as a hybrid between a national network and a syndicated programming block that would air on former WB and UPN affiliates. The CW positioned itself as a network for the eighteen-to-thirty-four-year-old demographic, especially young women and African Americans. The two flagship shows promoted through the launch of MyNetwork TV were remakes of Latin American soap operas.

At first glance it would appear that the genre of the limited-run serial was chosen as a way to draw in the bilingual Latino audience that watches telenovelas on Spanish-language networks. Such a reorientation of Fox Television's programming is possible in the neo-network era of satellite and cable media, which, according to Michael Curtin, have expanded the range of television characters and gender roles. Curtin points to shows like *Absolutely Fabulous*, which was picked up by Comedy Central after ABC passed on the British program, and suggests that the transnational circulation of multiple and alternative representations of feminine desire is an outcome of the neo-network era.[1] Channeling the Latina desire that fuels telenovelas could be a wise investment, considering that ratings for the telenovela-heavy networks Univision and Telemundo often surpassed those of the WB and UPN. However, with the exception of the video played at the 2006 upfronts, most of the promotional materials for MyNetwork were careful not to include too many references to the Latin American origins of its programming. Instead, a discourse of innovative television was promoted, the innovation being that the network would not air repeats and that its series would run in thirteen-week, sixty-five-episode arcs.

The name of the network, *My*Network TV, was emphasized to suggest that a special relationship would exist between audience and media. The viewer was promised "a new brand of television. It's your brand." The relationship between viewer and network is further encouraged via cast members' video confessionals, which are available online. This feature invokes the reality television genre, as do online voting contests to choose the band that will be featured in a future television episode, on the MyNetwork website, or on the network's MySpace profile. However, these marketing strategies were ultimately unsuccessful, and the network suffered from extremely low ratings, perhaps in part because the majority of English-language U.S. audiences were unfamiliar with the telenovela genre.

In adapting the telenovela format for U.S. audiences, MyNetwork neglected to investigate what makes the genre popular. While many Spanish-language telenovelas make efforts to draw attention to recognizable parks, buildings, or other cultural landmarks, and garner audience praise for doing so, MyNetwork's telenovelas were primarily shot on soundstages in San Diego, California. In this chapter I will present an industry analysis of MyNetwork and its ambivalent relationship with the Latin American telenovela and with U.S. Latina/o audiences. I focus on the series *Fashion House* and explore the irony inherent in MyNetwork's claim to be cutting-edge television while it simultaneously invokes the traditional genre of the Latin American telenovela. I consider the role that familiar physical locales have played in viewers' relationships with successful Spanish-language telenovelas and compare this reality to the online world offered to the viewers of MyNetwork's evening dramas.

Cutting Costs for a Cutting-Edge Format

In January 2006 CBS Corporation and Time Warner Inc.'s Warner Bros. closed their broadcasting companies UPN and the WB. They held on to the strongest performing shows from the two networks, such as *Everybody Hates Chris* and *Gilmore Girls*, and aired these on a new network, called the CW, along with new programs aimed at viewers aged eighteen to thirty-four. The creation of the CW and the shuffling of WB and UPN programming left the News Corporation without programming for those of its local television stations that had been affiliated with UPN. The News Corporation responded by putting together plans for a network that would function as a hybrid between a national network and a syndicated programming block. The plans for MyNetwork TV were completed in three weeks.[2]

MyNetwork tried to appeal to local stations' desire for control over costs and airtime by emphasizing that the stations would have nine minutes of local inventory per hour while MyNetwork would keep five minutes of national time. This deal was presented as an uncommon amount of control over airtime in comparison to the CW's offering of three minutes of local time in primetime. In addition, MyNetwork promised to give local stations 65 percent of advertising inventory in its two-hour primetime block.[3] Jack Abernethy, the CEO of Fox Television Stations, alluded to the economic benefits this model holds for local stations that "don't want to give up 30 hours."[4] He also acknowledged that relying on the local networks' already existing programming could save MyNetwork from the pitfalls that plagued many other networks: "What we are trying to avoid is the model that failed for 10 years with WB and UPN, where you're constantly failing, trying and developing 20 shows to keep 10 hours of programming on the air."[5] Instead, MyNetwork focused on launching and promoting only two original shows, *Fashion House* and *Desire*, both English-language adaptations of Latin American telenovelas.

The thirteen-week story arcs of these evening soaps and the fact that they aired five nights a week were presented as innovations that would be attractive to audiences and advertisers alike. However, ads sold for about $25,000 per thirty-second spot when MyNetwork debuted versus the $30,000 to $150,000 per spot received by the CW. In essence, MyNetwork was receiving syndicated rates rather than network ad rates.[6] Fortunately, in light of these meager advertising revenues, MyNetwork's signature programs were shot on a shoestring production budget of $100,000 to $120,000 per episode. In comparison, daytime soaps shoot for $200,000 to $400,000 per episode; primetime limited-run dramas usually shoot for $500,000 per episode until success drives the prices higher; and primetime dramas can cost as much as $3 million per episode.[7] By Abernathy's own admission, MyNetwork "managed to save costs by economies of scale." He added that "there's no sitting around. Everyone is acting, editing, writing, shooting at the same time."[8]

Four production crews worked simultaneously on MyNetwork's telenovelas. Each one shot about twenty pages per day and came out of production with about eighty pages of material that was given to editors at the end of the day.[9] The shows were shot on Panasonic High Definition Varicam using single-camera lighting that effectively made MyNetwork's programs look more like a primetime drama.[10] Taping was done on non-union soundstages in San Diego, which also kept costs low.[11] However, the most cost-effective measure may have been the adaptation of Latin American telenovelas to the English-language limited-run serial format.

McTelevision and the Adaptation Process

The act of bringing a format across international borders is emblematic of both the interconnectivity of global television industries and the reliance on a general system of "McTelevision." Silvio Waisbord defines McTelevision as "the selling of programming ideas with a track record that are sufficiently flexible to accommodate local cultures to maximize profitability."[12] Making a link between the McDonald's fast-food chain and the business model preferred by the contemporary television industry, Waisbord argues that the professional sensibilities of television executives worldwide have become homogenized, resulting in the globalization of television trends via the purchase of successful formats. The commercial success of these formats implies an already existing efficiency and predictability of the program. Indeed, Tony Cohen suggests that success and profitability reside in the format itself. Cohen, the head of Freemantle Media, an international production company that produces regional versions of shows such as *American Idol* and *Thank God You're Here*, maintains that a type of biological determinism can be applied to the branded approach to programming: "They're programs, but they can be so much more than that. We call it the global-formatting phenomenon. If a show has global-formatting DNA, it can be reconstituted."[13] However, the program must be altered sensitively to resonate with the tastes and culture of its new environment.

Television scholars Horace Newcomb and Paul Hirsch argue that all television producers, even those who are not adapting a series from a foreign country, create new meaning out of the combination of familiar signifiers. Jeffrey Miller builds off of the work of Newcomb and Hirsch and attempts to account for the success of the sitcom *All in the Family*, an adaptation of the British situation comedy *Till Death Do Us Part*. He compares the bigoted fathers in the two series and explains that the Archie Bunker character was not a simple clone of Alf Garnett. Instead, the Bunker character had to embody cultural meanings and realities that were significant for U.S. audiences debating Richard Nixon's bombing of Cambodia and witnessing violence on the campuses of Kent State and Jackson State. Miller argues that *All in the Family*'s engagement in discussions surrounding these events brought a level of familiarity to the characters and made the show relevant and topical to audiences. In addition, the show was adapted to be presented within the genre of what Miller calls "the most basic form of American television: the domestic family comedy."[14]

More recently, U.S. television screens have been filled with numerous adaptations of reality programs, contests, and games that were originally successful abroad. The short-form game show is a format that lends itself to adaptation because it does not offer many spaces in which national differences can manifest themselves. Games are also inexpensive to produce. For example, the average cost of ABC's *Who Wants to Be a Millionaire?* is $750,000 per show versus $1.2 million for a dramatic series like ABC's *The Practice*.[15] Although the American game show *The Price Is Right* did not appear overseas until fifty years after its premiere, it is no longer surprising to find that versions of shows like *American Idol* are able to reach thirty-nine countries within five years.[16] Yet, these quicker rates in global acquisitions and adaptations by the numerous companies now focusing on global entertainment brands do not ensure the smooth re-creation of a successful format. The process is especially problematic with dramatic series in which the adaptation and the original usually have very little in common.[17]

The work that went into adapting telenovelas for MyNetwork TV points to the challenges of adapting dramatic series even in an era dominated by McTelevision. *Desire* was originally a Colombian telenovela titled *Mesa para Tres*, and *Fashion House* aired in Cuba as *Salir de Noche*. When MyNetwork purchased these telenovelas, each came with a story arc that covered more than a hundred episodes, but the network planned to tell the stories in only sixty segments. Upon reviewing the original episodes, the head adaptor issued bullet points to a team that was expected to condense the drama.[18] Even more challenging was the translation. Patrick Perez was hired by MyNetwork as a translator and eventually became an adaptor when it became clear that simply translating the original telenovelas would not be sufficient for U.S. audiences.[19] For example, some of the main characters in the Latin American originals were not familiar protagonists of American serial dramas, much less sexy evening soaps. Perez was faced with scripts that featured priests as main characters and storylines in which it was not uncommon for adults to live at home with their parents. The depiction of Cuban day-to-day existence in *Salir de Noche*, for example, included an unsavory lead character who sells sandals in the town square and is the protagonist's cheating husband. Perez describes him as a man who "stayed out all night with the boys and . . . bought black market leather, so that's what made him bad. That doesn't translate. There is no black market leather in America, so of course everything was changed to make everybody way more wealthy, way more American, and basically almost nothing was salvaged from the original."[20] The husband in *Salir de Noche*'s English-language adaptation became an adulterous, alcoholic, money-laundering accountant.

Note to Audience: "Don't Be Afraid"

Even though the *Fashion House* and *Desire* series had been stripped of much of their original telenovela roots, they were still marked as different in relation to the rest of U.S. network television. A *Wall Street Journal* article announcing the network described it in these terms: "High-brow the endeavor is not. MyNetwork TV will kick off with two campy sex-drenched dramas in the style of Spanish-language telenovelas."[21] Words like "campy," "sex-drenched," and "Spanish-language telenovelas" mark the programming as something excessive and exotically different when compared to the usual English-language network television offerings—and thus something that highbrow audiences needed to justify watching. Even MyNetwork's early on-air promos referred to the possibility that shame could be associated with viewing its dramatic series. An illicit affair between audience member and network is hinted at in a promotional piece that ran soon after its launch. Superimposed on images from the station's two telenovelas was the message: "Six Nights a Week. You're Invited. Be there for the passion. Betrayal. And True Love. New episodes. Every Night. And there's never a repeat performance. Let's spend every night together. Guilty pleasure. Six nights a week. Don't be afraid."

References to an illicit affair and feelings of guilt continued throughout the run of the first two series and were used in the ad introducing two new series that debuted in December 2006. It begins with a close-up of a woman seductively blowing out a candle, followed by a scene in which a man and woman embrace in a swimming pool. The next image shows a man standing near the silhouette of a woman taking off her blouse. The following text cuts between additional clips from the new series: "Every affair starts with one seductive look. Now all eyes are on us. MyNetwork TV. New network with a radical idea. All new episodes all night long. All year long. MyNetwork TV. No repeats. No reruns. No kidding. Because who wants the same thing every night? MyNetwork TV. Dramas so intense they'll leave you . . . breathless. Stars so sultry you'll need a cold shower. MyNetwork TV. All in Hi-def. Because the devil is in the details. Had enough? . . . we didn't think so. Don't get caught . . . not watching. MyNetwork TV. Guilty Pleasure. Six Nights a Week. Don't Be Afraid."

The promotion campaign did not mention that MyNetwork was drawing from the Spanish-language telenovela programming model. Instead, the network continuously emphasized its high-definition format and six nights of original programming while highlighting the scandalous storylines of its series and the sex appeal of their stars. Twentieth Television President and Chief Operating Officer Bob Cook defined MyNetwork as "the first all-original, all-hi-def network." Cook oversaw MyNetwork TV's promotional strategy and hoped this description of the network would be sufficient to attract viewers. Early on, he realized that the all-original approach also meant that MyNetwork could not rely on familiar programming when introducing itself to new viewers. Reflecting on this reality before the launch, Cook said, "There was probably a greater confusion factor than we anticipated. It is going to take a little longer. It was easy for them [smaller, more rural markets] to understand their favorite shows on WB and UPN are now on the CW. For MyNetwork TV, it is a bigger challenge."[22]

The CW had positioned itself as a network for the eighteen-to-thirty-four demographic, especially young women and African Americans. Early press reports leading up to MyNetwork's launch claimed that it was targeting the eighteen-to-forty-nine and eighteen-to-thirty-four demographics and hoping to appeal especially to Latino viewers.[23] However, network executives and national advertising efforts rarely addressed Latinos directly, perhaps for fear of adding to the "confusion factor." The approach taken by Fox Television

Stations' CEO Jack Abernethy was more welcoming. He explained that the telenovela format was not specifically pointed toward Latino audiences and added, "We think that everyone will like watching them."[24] Rather than associate the network too closely with the programming common on Spanish-language networks, MyNetwork executives justified their pursuit of primetime short-run content by aligning the network more with reality programs that have been successful in the English-language U.S. context. According to Cook, viewers had become accustomed to "investing time and energy into a rooting interest for a character, then having that character go away."[25]

Referring to the global presence of telenovelas may have been appealing in an upfront, but such potentially controversial references did not have a place in viewers' living rooms. Cook invoked the "confusion factor" again when justifying his decision to avoid presenting MyNetwork's dramatic series as "Hispanic content." "It will be funneled right at the general market," he affirmed, "then a guerrilla aspect would be to get into the Latin pipeline to let them know [*Desire* and *Secrets* are] novela-based."[26] Getting into the "Latin pipeline" was a multi-pronged endeavor. The section that follows focuses on three measures: local media blitzes, a story on a nationally syndicated Latino entertainment news and culture television program, and infiltration of telenovela websites by MyNetwork bloggers.

Confusion, Guerrilla Marketing, and the "Latin Pipeline"

MyNetwork's name was chosen for synergistic reasons—News Corporation owns both MyNetwork TV and MySpace.com—and also to suggest that a special relationship exists between audience and media. Similar claims can be found in the names of artifacts and websites such as iPod, iTunes, YouTube, and so on. With MyNetwork, the viewer is promised "a new brand of television. It's your brand." A promotional spot from KXMN (Spokane, Washington) drives this point home. A middle-aged woman directly addresses television viewers and tells them to "get ready for a whole new generation of TV." Another woman says, "I want a network with no repeats." A third woman agrees, "No repeats!" A male voice is heard singing along with a rock song: "I want it my way! I want it now! My Network Television!" The last woman to speak to the camera announces the launch date of MyNetwork and suggests that it will serve her entertainment needs: "September 5th I get TV my way." This promotional spot, which features three women on camera, two Anglo and one African American, all wearing conservative, buttoned-up blouses, is quite different from the one that aired on the Phoenix, Arizona, MyNetwork channel. A young Latino male wearing a black fedora hat raps over the sounds of flamenco guitars as young Latino, African American, and Anglo dancers make their way across a Spanish-style patio. He intersperses Spanish and

FIG. 10-1 MyNetwork's poster at a Los Angeles casting call invokes the global audience of telenovelas without referencing the Cuban origins of *Fashion House*. (Photo taken by the author)

English words like "smoking" and "caliente," rhymes "MyNetwork TV" with "Phoenix AZ," drops the names of syndicated shows on the station, and identifies MyNetwork as the place "where the Suns play." A woman coded as Latina joins the bilingual rapper. Wearing a flowing white skirt and a red bikini top with tassels, she sings: "Get it on, feel the heat. My 45. My 45."

The ads that ran in Spokane and Phoenix exemplify MyNetwork's regional approach to promoting the network. According to Cook, "Networks are all about localism and about individual television stations."[27] It therefore makes sense that the Phoenix spot would present images of Latinidad, however limited and problematic they may be. Surprisingly, however, no mention is made of the telenovela-style programming on MyNetwork. Although the ad appears to be catering to the young Latino demographic, references to Latinidad within the text are limited to Spanish words, such as the translation of the station's channel number from "forty-five" to "*cuarenta cinco.*" Fears of the "confusion factor" seem to have contained all references to Latinidad within the sphere of familiar Spanish words, dance moves, and flamenco guitars.[28]

Meanwhile, the Puerto Rican salsa singer Marc Anthony provided a station identification on Los Angeles's MyNetwork 13. This station also aired a spot in which Natalie Martinez, the sympathetic protagonist of *Fashion House* and the face of Jennifer Lopez's clothing line, recalled watching telenovelas with her family. These spots ran only infrequently, however; for the most part, Latino audiences were courted via local media blitzes organized around events such as casting calls and parades. MyNetwork made efforts to reach the bicultural Latino audience by sponsoring a float in the 2006 New York City Puerto Rican Day parade. Prior to its launch, the network also staged a thirty-city promotional tour that featured autograph signings with the actors and giveaways of stickers and DVD promos for *Fashion House* and *Desire*.

The Los Angeles casting call drew many aspiring actors. More than two hundred people lined up early in the morning at the Glendale Galleria mall to audition for a chance to win a walk-on role in MyNetwork's dramatic series. By 11:30 A.M. people were being turned away. The casting call was hosted by the "hurban" format radio station Latino 96.3, which plays a mix of reggaeton, hip-hop, and R&B.[29] A bilingual radio personality entertained actors as they waited in line. Promotions Coordinator Claudine Guerrero did more to reach out to the Latino audience in the four hours that she was at the mall than the television station had done in all of its on-air advertising spots. She described the drama of Spanish-language telenovelas, joked about the storylines, and tried to generate excitement about English-language versions of the Latin American series. Televisions ran clips from *Fashion House* and *Desire*. Scenes of passionate embraces, scowling men, and women slapping each other were interspersed between taglines such as "*Sus novelas favoritas ahora en ingles*" (Your favorite telenovelas now in English) and "*¡Mi canal es su canal!* My 13 LA" (My channel is your channel!"). These slogans never aired on MyNetwork TV.

While one could bemoan MyNetwork's failure to employ on-air marketing that acknowledged Los Angeles's sizable Latino population, the slogans used during the casting call should be analyzed in terms of their misguided attempts at reaching a bilingual, bicultural Latino audience. For example, the use of the Spanish-language "*Sus novelas favoritas ahora en ingles*" makes very little linguistic sense. A more appropriate approach would be to use the type of Spanglish that young second- and third-generation Latinos are familiar with, that is, the type of Spanglish regularly heard on Latino 96.3 and among the young actors lined up to audition for MyNetwork TV.

Network executives from NBC, ABC, CBS, and MyNetwork TV have all adapted or plan to adapt Latin American telenovelas. They anticipate that the familiarity of the format will appeal to second-generation Latinos who grew up watching Spanish-language telenovelas with their parents and grandparents.[30] However, the loyalty of this audience is not a given. In their study of the sale and purchase of television programs and formats at international trade fairs, C. Lee Harrington and Denise Bielby observed that audiences are absent both physically and discursively from the proceedings.[31] This absence is obvious in MyNetwork's Spanish-language taglines, in its telenovela adaptations,

FIG. 10-2 Claudine Guerrero, promotions coordinator at Latino 96.3, interviews Robert Buckley, who plays Michael Bauer on *Fashion House*. (Photo taken by the author)

and even in its choices among original programs. The series that Twentieth Television purchased were chosen precisely because they were inexpensive, not because they were ever actually anyone's "*novelas favoritas*." The Colombian telenovela *Mesa para Tres*, which became *Desire*, had nowhere near the popularity of the wildly successful *Yo Soy Betty, la Fea*.[32] Claims that B-list telenovelas are U.S. Latinos' favorite soaps grossly underestimate the level of affection and commitment that Latino audiences actually have to their novelas.

The nationally syndicated Latino entertainment news and culture television program *American Latino TV* addressed the popularity of Spanish-language telenovelas in a short segment titled "English Language Novelas." The subjects of the segment were the ABC series *Ugly Betty*, the two series that MyNetwork debuted in 2006, and the NBC soap *Passions*, which features a Latino family among its cast of characters. The host of the show began by addressing a presumably Latino audience: "I'll admit it, I grew up watching telenovelas. You probably watched them at one point. Most Latinos have watched novelas at one point or another. But recently there's been an explosion of telenovela style programming on English-speaking networks. Apparently they're catching on."

The assumption that the audience members had a multigenerational connection and insider's view into what U.S. network executives were just beginning to recognize as a television trend turns the usual outsider status of Latinos vis-à-vis the entertainment industry on its head. Interviewed for the segment, the *Fashion House* star Natalie Martinez, who is identified as a "Cuban actress," recalls watching novelas as a kid and enjoying the cliffhangers at the end of episodes. Speaking about MyNetwork's telenovela-inspired series, she affirms, "These are just like novelas in Spanish—they have a beginning, middle and end. And the stories are just as crazy. This is a show that's only gonna happen within the span of thirteen weeks. As for daytime soap operas, as you know the world's been turning forever." Again, the privileged position of English-language television is challenged, albeit in a cheeky wordplay on the name of a successful U.S. soap.

Allowing a series star to engage in this type of discourse on *American Latino TV* was a safe strategy for MyNetwork. Because the majority of the show's audience is Latino, MyNetwork did not have to justify direct attention to this demographic. As discussed earlier, MyNetwork executives stated that they were committed to focusing on the general market first and using guerrilla marketing later to get into the "Latin pipeline." The most secretive aspect of the network's Latino marketing strategy was the hiring of bloggers to visit telenovela and soap opera websites to post favorable commentary regarding MyNetwork's shows. Patrick Perez was among the staff at MyNetwork who posted on these websites. Posing as a female soap fan, Perez asked, "Have you checked out MyNetwork TV?" He also gushed about the "cute" stars of the series.[33] MyNetwork was committed to reaching not only the soap fans who visited these sites but also the general audience that actively consumes online digital media. To do this, MyNetwork integrated interactive features within the programs that it was promoting.

Online Extratextual Content and MyNetwork's Claims of Difference

Even before its telenovela series began airing, MyNetwork TV introduced interactive features. Actors who auditioned for walk-on roles in *Fashion House* and *Desire* could view their audition tapes on MyNetwork's website a week later. This promotion was the network's first attempt to cultivate the type of convergence culture described by Henry Jenkins in his book of the same name. The idea of convergence culture rests on the premise that the flow of content across multiple media platforms can enable new forms of participation and collaboration.[34] However, the closest thing to participation that MyNetwork TV encouraged was simply inviting the public to vote for their favorite audition. The results of this poll were announced online, and a couple of actors won walk-on roles. This first effort to draw potential audience members to MyNetwork's web content was intended to make viewers feel as if they were a part of the network or that it was, indeed, *their* network. Afterward, the more common invitation to MyNetwork's web content came via the swipes that appeared in the lower corner of the television screen.

Swipes are text messages that appear during a television program to promote the show on iTunes or to inform viewers that extra footage of a television show, football highlights, and other content are available on the channel's website. MyNetwork made use of swipes and also ran advertisements directing viewers to its website. One such advertisement, which appeared during each episode of *Fashion House*, began by asking viewers, "Wanna fulfill your passion for fashion?" Clips from the show played, and viewers were instructed to log on to mynetworktv.com, where they could "Go inside the runways of *Fashion House* with the latest beauty tips, designer downloads, and character confessions." This ad included medium shots of the show's characters, a close-up of a woman applying lip gloss, and an extreme close-up of a bottle of Maybelline Great Lash mascara.

Although the ad featured the Maybelline mascara prominently, brand names were not present within MyNetwork's web content, largely because more than twenty-five products had already been written into the series.[35] For example, one female character confided to another female character that she uses Revlon hair color when she doesn't have time to make it to the salon. Thus, it was unnecessary for the "Gorgeous Glow" and "Designer Download" features on MyNetwork's website to reference specific brands when offering tips on how to choose a lip liner or how to wear leggings. Instead, in the short video clips the female vixen or two designers held up clothing and beauty products that did not show any visible brand names or labels. These short beauty tip and fashion videos were available online and also ran on television alongside episodes of *Fashion House*. On the other hand, the "Confessionals" and "Inside the Mind" features were available only as online content.

Viewers were encouraged to go online to gain insights into the motivations behind a character's actions. The aesthetics of the "Confessionals" borrowed heavily from reality shows. Characters were all seated in the same location, framed in a medium shot, and engaged in a direct address with the camera. One character confided his feelings upon learning that his love interest is pregnant by another man; another character offered a justification for drugging her son. The "Inside the Mind" features offered additional insights into the characters' psyches. In these videos a woman, self-identified as a marriage and family therapist, sat on a leather chair and offered commentary on the actions of particular *Fashion House* characters as scenes from an episode ran. Her comments ranged from "Anger is a powerful emotion. It can even drive a calm person to insane behavior" to "Jealousy comes from two feelings: fear and anger."

Although this form of analysis may have been enticing for audience members fond of the self-help discourse found on shows like *Dr. Phil*, it is questionable whether younger audiences would gravitate toward this aspect of

MyNetwork's online content. The general youth audience is more likely to engage with multiple forms of media at one time, for example, sending and receiving text messages and using a social networking site while watching television.[36] MyNetwork attempted to reach this audience by collaborating with MySpace.com to promote both the dramatic series and the social networking site's music feature.

An advertisement announcing this collaboration ran on MyNetwork TV at the end of September 2006. It began by asking viewers, "Are you ready to help new musical artists find fame?" They were then told that MySpace and MyNetwork TV would collaborate to find "new buzzworthy bands." Language drawn from reality shows like *American Idol* promised, "Each week, America votes on which songs are hot and which songs are not." In his detailed discussion of *American Idol* Jenkins emphasizes that audience participation is a key strategy that marketers used to stimulate viewer investment in the show. In the case of the MyNetwork band competition, viewers were told that they would have the power to decide which songs would be featured on an upcoming episode. The ad ended with instructions to "click on to mynetworktv.com for the 411 on how to enter your songs and how to vote on who will become a rock star." Most of the images used in this advertisement appear on what looks like the screen of a Mac iBook computer and on an iPod. However, the visual presence of such popular new media artifacts and the use of presumably youth-oriented colloquialisms like "the 411" and "rock star" were not enough to generate a sizable audience on MySpace. A month after announcing the band search collaboration, MyNetwork had acquired only 4,000 MySpace friends; in the meantime, the CW had accumulated 27,000.[37] By June 2007, the CW had 34,340 MySpace profiles linked to its page while MyNetwork had only 7,647.

The Nielsen ratings for MyNetwork's original dramatic series were just as dismal as the network's status on MySpace. *Desire* and *Fashion House* averaged 950,000 viewers and a 0.4 rating among adults eighteen to forty-nine. The second cycle of telenovela series averaged 700,000 viewers and a 0.2 rating.[38] Still, MyNetwork made one last-ditch effort to promote its new cycle of dramatic series. Surprisingly, it did so by comparing itself to other networks. An advertisement announcing the new dramatic series *Wicked Wicked Games* and *Watch Over Me* invoked such hit shows as NBC's *The Office*, ABC's *Grey's Anatomy*, and numerous crime investigation series. The ad included scenes of a janitor vacuuming an empty office, a man walking away from an abandoned crime scene, and a hospital room in which the sound of a flatline could be heard. A narrator intones, "This winter, on the other networks the office is closed, the crime scene goes cold, and the doctors have left the OR." Clips from *Wicked Wicked Games* and *Watch Over Me* play as a voiceover suggests that these series will "keep you toasty" while the other networks are showing reruns.

Although MyNetwork continued to present itself as different from other networks, it ultimately had to follow the trend toward reality programming in order to survive—even going so far as to bring back reality series that had previously aired on NBC.[39] In February 2007 MyNetwork announced that it would not produce additional scripted series. Instead, its programming hours would be filled with reality shows, International Fight League matches, and second-run movies. The debut of *International Fight League* drove up ratings 60 percent in its time period and scored a 0.8 rating.[40] The addition of this programming resulted in MyNetwork's ratings among men eighteen to thirty-four and eighteen to forty-nine going up between 200 percent and 500 percent.[41] The broadcast of second-run films also raised ratings. For example, *The Cider House Rules* nearly tripled the 0.4 rating that the Saturday telenovela recaps were averaging.[42] Celebrity biography specials produced by *Access Hollywood* also drove up ratings. A special that aired after the death of Anna Nicole Smith earned a 1.0 household rating on March 7, 2007. Although this number is low, it doubled the previous Wednesday night average of 0.5 and was significant enough to prompt MyNetwork to announce that it planned to air six more *Access Hollywood* specials.[43]

Telenovelas, Locality, and Networked Communities

The creation of MyNetwork's limited-run dramatic series did not mark the first time that U.S. English-language networks attempted to draw from the telenovela format. C. Lee Harrington and Denise D. Bielby trace the "telenovelaization of U.S. soap operas" back to March 2000, when ABC's *Port Charles* shifted to a twelve-week story arc format. Soon after, in May 2001, CBS began airing simulcasts of *The Bold and the Beautiful* in Spanish and English. By July 2001, NBC began Spanish closed-captioning of *Days of Our Lives* and *Passions*. The change in format and the inclusion of the Spanish language in the audio feeds of the major networks could be interpreted as an attempt to court Spanish-speaking Latino telenovela fans. However, Bielby and Harrington emphasize that these developments did not necessarily mark a moment in which the U.S. English-language soap genre was truly adapting telenovela conventions or honoring the cultural and literary traditions of Latin America. They argue that the never-ending storytelling typical of English-language soaps has become incompatible with viewers' needs. Thus, the change in *Port Charles* format was an attempt to engage distracted viewers with short attention spans.[44] In much the same way, MyNetwork's adaptations of Latin American telenovelas never signaled a commitment to Latino fans of telenovelas. On the other hand, NBC recognized the popularity and market success of telenovelas and sponsored a writing program with a Miami university.[45] The scripts that come out of this city, often referred to as the media capital of

Latin America, could be intriguing insofar as they are informed by both bicultural sensibilities and major network media economics.

One reason that telenovelas have garnered a strong audience among U.S. Latinos is that they often include references to the cultures of the country in which the story is set. Lucila Vargas's ethnographic study with Latina teens in the North Carolina towns of Durham and Carrboro demonstrated that viewers watch telenovelas to experience familiar aspects of their home countries that were left behind in the process of migration.[46] Vicki Mayer's discussions with San Antonio Mexican American girls about soaps and telenovelas revealed similar constructions of transnational femininity. These girls often made negative comments about English-language soaps' never-ending stories, slowly evolving plots, and unexciting locales. On the other hand, these young girls, many of whom had never visited their parents' Mexican hometowns, felt a connection to the locales featured in telenovelas and eagerly spoke of learning about Mexico by watching Mexican telenovelas.[47] Vivian Barrera and Denise Bielby interviewed older fans of telenovelas and found that the scenery, language, architecture, and even hairstyles featured in telenovelas made viewers feel like they "had a piece of Latin America in their homes."[48]

Of course, not every member of MyNetwork's audience has a connection to Latin America, so it would not necessarily make market sense for the network to present telenovela-inspired programming that is premised on making direct references to Latin American countries. However, lessons can be learned from the failures that resulted from MyNetwork's misguided efforts at adapting telenovelas. For example, the dramatic series rarely included visual cues regarding the geographic locales in which the stories took place. The fact that the series were shot on soundstages in San Diego meant that most scenes that referenced a city were aerial shots of the downtown Los Angeles skyline and the Griffith Park Observatory. Thus, viewers never saw the cast of *Fashion House* actually engage with the landscape and cultural markers of the city of Los Angeles.

In his discussion of McTelevision and the adaptation of programs into a new cultural context, Waisbord emphasizes that the locality of the original series needs to be removed during the adaptation process but reintroduced in the adapted series in such a way as to bolster the basic concept that made the original series a success in the first place.[49] Locality was indeed evicted from *Fashion House*. That is, we never saw the protagonist's husband selling sandals in the town square. However, locality was never reintroduced. The show made no effort to foster a connection between viewers and any specific geographic locale. The closest sense of a geographic locale that the series referenced was an online world. MyNetwork continuously invited viewers to its website, where they could receive more information about the characters, fashion, and music featured on the show. This emphasis on the digital world may have been pursued in an effort to infuse the traditional genre of the telenovela with a sense of connectivity to the networked age.[50]

Adapting, Dubbing, and Destilando

Perhaps MyNetwork and those networks considering telenovela adaptations in this neo-network age could take a lesson from the success of *Ugly Betty*. This remake of the Colombian telenovela *Yo Soy Betty, la Fea* retained its original storyline of a homely young woman working at a high-end fashion magazine, but other elements of the story were changed. Most notably, the protagonist in the original telenovela was in love with her philandering boss, whereas Betty Suarez does not pursue this type of relationship with her employer in the U.S. version. The slapstick and the catty dialogue of the original telenovela remain essential elements of the U.S. version, but other conventions of traditional telenovelas have been revised. For example, *Ugly Betty* makes use of wide-angle lenses to create a depth of field that is rare in most low-budget telenovelas.[51] The difference in the look of *Ugly Betty* becomes most apparent during scenes in which the Suarez family watches telenovelas, or "telenovelitas," in their living room. The Spanish-language telenovelitas may be over-the-top in terms of their storylines, characters, and acting, but they pay homage to the genre that gave birth to *Ugly Betty*. Salma Hayek, the Mexican-born actress and producer of *Ugly Betty*, says of her cameo appearances in the telenovelitas, "I started in telenovelas. It's a lot of fun to go back and do it, to make fun of myself."[52]

Hayek's comment regarding her early career in telenovelas should not be considered a self-deprecating swipe against the genre. The presence of the telenovelitas on *Ugly Betty* and on the show's website serves as a reminder of the complexities inherent in adapting successful television formats to a new context.[53] The Suarez television is often tuned to a Spanish-language telenovela while Betty's father cooks or while the family eats dinner. Although the audience catches only a few seconds of the telenovelita, longer clips are available on the *Ugly Betty* website, where the originally Spanish-language telenovelitas are dubbed into English. The quality of the dub is intentionally poor to highlight the campy characteristics of the genre. Such a self-reflexive engagement with telenovelas could have rescued MyNetwork's soaps, which never clearly confronted their relationship to their telenovela roots.

Dubbing *Ugly Betty*'s telenovelitas into English serves to make the foreign familiar to those who do not already have a connection to telenovelas. In addition, the telenovelita is placed within the familiar setting of the Suarez home. Upon clicking on the button for a telenovelita episode, visitors to the

Ugly Betty website are led through the family's kitchen, past the dining room, and into the living room, where the camera zooms into the television and finds that an episode of *Sins of the Heart* is already in progress. This process of accessing the telenovelita serves to situate the program within a context that is familiar to viewers. On the other hand, MyNetwork's "Confessionals," "Inside the Mind" features, and other online extra-textual content take place in sterile studio environments.

Although both *Ugly Betty* and the MyNetwork soaps were filmed on soundstages, *Ugly Betty* used computer-generated imagery to create a context that included markers of New York like the Brooklyn Bridge, Manhattan street vendors, Queens bridal shops, and elevated subway trains. The MyNetwork soaps did not include many references to city landmarks; rather, the series attempted to connect viewers to a landscape that they were already familiar with, a virtual online space. *Ugly Betty* recognized that even in this space it is important to refer to the physical markers that viewers associate with the show.

Upon the success of *Ugly Betty*, the show's producer, Ben Silverman, acquired rights to an English-language version of another Colombian telenovela by Fernando Gaitán, the creator of *Yo Soy Betty, la Fea*. *Café con Aroma de Mujer* has not yet been remade for an English-language U.S. audience. However, the finale of the Mexican remake of the telenovela, titled *Destilando Amor*, set records on December 3, 2007, when it drew in more viewers in all demographics than the four English-language major networks.[54] For the coffee plantation setting of *Café con Aroma de Mujer*, *Destilando Amor* substituted the agave fields surrounding a Mexican hacienda near the small town of Tequila within the Mexican state of Jalisco. Although the central story concerns the family business of distilling tequila, *Destilando Amor* pays homage to its Colombian roots through the character of Alonso Santoveña, who continuously recites the history and healing attributes of coffee as he pours a cup for those in crisis. Notwithstanding Alonso's references, *Destilando Amor* made a clear break from the original Colombian telenovela. The picturesque scenes shot in Tequila are considered responsible for fueling an 83 percent growth in tourism to the town during 2007, the year that the telenovela aired in Mexico and the United States.[55]

While *Destilando Amor* sparked a boost in tourism, MyNetwork could not persuade audiences to visit its website—even when it offered viewers an extended online connectivity with the characters of its dramatic series. Among the lessons learned through the launch of MyNetwork TV is that the content of "old media" must generate engagement with and excitement about a series that will prompt audiences to seek out extra-textual content via the highly extolled "new media." Sometimes this content must include actual geographic markers like the Brooklyn Bridge, even if they are created with a computer.

Notes

1. *Absolutely Fabulous* is an outlandish comedic series about two female middle-aged self-indulgent former flower children. Michael Curtin, "Feminine Desire in the Age of Satellite Television," *Journal of Communication* 49, 2 (1999): 55–70.
2. Brooks Barnes, "New Network Will Showcase Greed, Lust, Sex," *Wall Street Journal*, February 23, 2006, B1.
3. Allison Romano, "MyNetworkTV Counts on a September to Remember," *Broadcasting & Cable* 135, 19 (May 8, 2006): 28; Michele Greppi, "MyNetworkTV to Center on Stations," *Television Week*, February 27, 2006, 33.
4. Romano, "MyNetworkTV."
5. Greppi, "MyNetworkTV."
6. James Hibberd, "MyNet's Switch Attracts Viewers," *Television Week*, March 19, 2007, 3.
7. Jim Benson, "Hot in Any Language," *Broadcasting & Cable*, May 1, 2006, 20.
8. Jennifer Ordoñez, "A Turn for Telenovelas," *Newsweek*, September 25, 2006, 42.
9. Patrick Perez, personal communication, May 14, 2007.
10. Benson, "Hot in Any Language."
11. Brooks Barnes, "With Sexy Story Lines, Low Budgets, News Corp. Will Launch MyNetwork TV," *Wall Street Journal*, August 31, 2006, B1.
12. Silvio Waisbord, "McTV: Understanding the Global Popularity of Television Formats," *Television & New Media* 5 (2004): 378.
13. Quoted in Jacqueline Marmo, "Special Delivery," *Hollywood Reporter*, March 20, 2007, S20.
14. Jeffrey S. Miller, *Something Completely Different: British Television and American Culture* (Minneapolis: University of Minnesota Press, 2000), 156.
15. Waisbord, "McTV."
16. Marmo, "Special Delivery."
17. Albert Moran, *Copycat Television: Globalisation, Program Formats and Cultural Identity* (Luton: University of Luton Press, 1998); Marmo, "Special Delivery."
18. Despite such changes and rewrites, Twentieth Television designated the writers as non-union "adapters" rather than "writers," thereby making them ineligible for coverage by the Writers Guild of America.
19. Perez later worked translating the script of the 2012 film *Casa de Mi Padre* from English to Spanish. Unlike his MyNetwork translations, the goal was to produce intentionally awkward dialogue so as to parody the telenovela genre.
20. Perez, personal communication, 2007.
21. Barnes, "New Network Will Showcase Greed."
22. Quoted in Christopher Lisotta, "Promo Hurdles for MyNetwork TV," *Television Week*, August 21, 2006, 3.
23. Jon Lafayette, "Buyers Eye Their My TV," *Television Week*, February 27, 2006, 1.
24. Quoted in Greppi, "MyNetworkTV."
25. Quoted in Jeff Zbar, "Law and Disorder Are Hot," *Advertising Age*, March 27, 2006, S8.
26. Quoted in ibid.
27. Quoted in Lisotta, "Promo Hurdles."
28. The colors, clothing, emphasis on dance, and use of words like "hot" and "caliente" are representative of the tropical narrative often used to define Latinos. Frances R.

Aparicio and Susana Chávez-Silverman, *Tropicalizations: Transcultural Representations of Latinidad* (Hanover, NH: University Press of New England, 1997).
29 "Hurban" is literally the combination of the words "Hispanic" and "urban." It refers to a radio programming format that draws from reggaeton, hip-hop, R&B, and Latino dance music.
30 Jeff Zbar, "Telenovela Format Takes on a Decidedly Anglo Look," *Advertising Age*, May 8, 2006, S26.
31 C. Lee Harrington and Denise D. Bielby, "Global Television Distribution: Implications of TV 'Traveling' for Viewers, Fans, and Texts," *American Behavioral Scientist* 48, 7 (2005): 902–921.
32 ABC's successful *Ugly Betty* is based on *Yo Soy Betty, la Fea*.
33 Perez, personal communication, 2007.
34 Henry Jenkins, *Convergence Culture: Where Old and New Media Collide* (New York: New York University Press, 2006).
35 Andrew Hampp, "Road to the Upfront: MyNetwork TV," *Advertising Age*, May 7, 2007, 4.
36 Mizuko Ito et al., *Hanging Out, Messing Around, and Geeking Out: Living and Learning with New Media* (Cambridge, MA: MIT Press, 2010).
37 Claire Atkinson and Abbey Klaassen, "Wanting Conversation, TV Nets Beef Up Web Presence," *Advertising Age*, October 23, 2006, 35.
38 James Hibberd, "MyNet TV Changes Production Script," *Television Week*, March 5, 2007, 6.
39 James Hibberd, "MyNetwork Revives 'Folks,' 'Paradise,'" *Television Week*, May 14, 2007, 4.
40 Hibberd, "MyNet TV Changes Production Script."
41 Hampp, "Road to the Upfront."
42 Kimberly Nordyke, "MyNet TV Execs, Affils: Change Is Good," *Hollywood Reporter*, March 9, 2007, 1.
43 Hibberd, "MyNet's Switch Attracts Viewers"; Hibberd, "MyNetwork Revives 'Folks.'"
44 C. Lee Harrington and Denise D. Bielby, "Opening America? The Telenovelaization of U.S. Soap Operas," *Television and New Media* 6, 4 (2005): 383–399; Harrington and Bielby, "Flow, Home, and Media Pleasures," *Journal of Popular Culture* 38, 5 (2005): 834–855.
45 Joshua Chaffin and Adam Thomson, "NBC Looks Home for Support as Its Mexico Adventure Froths into Soap," *Financial Times*, February 20, 2007, 13.
46 Lucila Vargas, *Latina Teens, Migration, and Popular Culture* (New York: Peter Lang, 2009).
47 Vicki Mayer, "Living Telenovelas/Telenovelizing Life: Mexican American Girls' Identities and Transnational Telenovelas," *Journal of Communication* 53, 3 (September 2003): 479–495.
48 Vivian Barrera and Denise Bielby, "Places, Faces, and Other Familiar Things: The Cultural Experience of Telenovela Viewing among Latinos in the United States," *Journal of Popular Culture* 34, 4 (2001): 1–18.
49 Waisbord, "McTV."
50 As a contrast, see Courtney Brannon Donoghue's discussion of *Ugly Betty*'s online audience interactivity. "Importing and Translating Betty: Contemporary Telenovela Format Flow within the United States Television Industry," in *Soap Operas and*

Television in the Digital Age: Global Industries and New Audiences, ed. Diana I. Rios and Mari Castañeda (New York: Peter Lang, 2011).
51 Denise Mann, interview with Silvio Horta, Paris, February 16, 2007.
52 Alex Strachan, "Salma Hayek Has Put on Her Ugly Face," *The Gazette*, November 9, 2006, D4.
53 It is worth noting that early press releases and journalistic media coverage include many examples of *Ugly Betty* executives and producers distancing the production from its telenovela origins and instead labeling it a dramedy. This change occurred when Salma Hayek and Silvio Horta became associated with the production. Juan Piñon explains that the "intervention of Latina/o producers underscores cultural understandings placing the telenovela's value in perspective, particularly in relation to producer Hayek and writer Horta whose personal and family trajectories have been tied to the genre." Juan Piñon, "Ugly Betty and the Emergence of the Latina/o Producers as Cultural Translators," *Communication Theory* 21, 4 (2011): 392–412.
54 Rick Kissell, "Univision's on the Rise," *Variety*, December 17, 2007, 26.
55 Daniel Garibay, "Le Agregan Atractivos a Tequila," *Reforma*, September 16, 2007, 3.

11

The Reign of the "Mothership"

••••••••••••••••••••

Transmedia's Past, Present, and Possible Futures

HENRY JENKINS

Transmedia storytelling has always been a blue-sky concept, the idealized intersection between the hopes of production personnel to gain more respect for their creative contributions, of networks to intensify viewer engagement, and of fans for more "complex" forms of storytelling. The chapters in this book provide some sobering reality checks, suggesting ways that our transmedia aspirations may be blocked by the silo-ing of production decisions within contemporary media conglomerates, co-production and co-licensing agreements with outside parties, and contradictory expectations of producers and audiences. Adopting a production studies perspective, these contributors offer case studies of what worked and what failed as television producers sought to extend their series into new media platforms.

The term "transmedia" means simply "across media" and implies a structured or coordinated relationship among multiple media platforms and practices. Marsha Kinder's 1991 book, *Playing with Power in Movies, Television, and Video Games*, described characters such as the Super Mario Brothers, the Muppet Babies, and the Teenage Mutant Ninja Turtles, who appear in many different media contexts, as operating within "transmedia supersystems . . .

[that] position consumers as powerful players while disallowing commercial manipulation."¹ I adopted Kinder's term in my 2006 book, *Convergence Culture*, to track emergent storytelling practices represented by the extension of *The Matrix* franchise (beyond film to comics, games, and anime), the development of "Dawson's Desktop," the promotional campaign for *The Blair Witch Project*, and the emergence of alternate reality games (such as "The Beast" and "I Love Bees"): "Transmedia storytelling represents a process where integral elements of a fiction get dispersed systematically across multiple delivery channels for the purpose of creating a unified and coordinated entertainment experience. Ideally, each medium makes its own unique contribution to the unfolding of the story."²

This definition assumes two key elements: radical intertextuality and multimodality. Radical intertextuality is present, say, when DC or Marvel Comics unfolds a "major story event" across many issues and multiple comics series, complexly intertwining the lives of their superhero protagonists. Radical intertextuality also occurs when *Battlestar Galactica* begins with a mini-series, evolves into a continuing series, includes stand-alone made-for-television movies, evolves into a prequel series, *Caprica*—all of which may be connected through webisodes. "Multimodality" is borrowed from the semiotician Gunther Kress, who describes how the affordances of different media shape how communication takes place: "There are things you can do with sound that you cannot do with graphic substance, either easily or at all; not even imitate all that successfully graphically."³ Writers such as Christy Dena have extended Kress's pedagogical analysis to explore how stories get told across multiple media.⁴ Some discussions of transmedia, thus, stress the movement across texts, while others emphasize the movement across media.

Transmedia Today

Over the past decade, transmedia has been both theorized and operationalized through ongoing exchanges between academics, journalists, fans, and media makers (both mainstream and independent). A growing number of conferences, such as Futures of Entertainment, Transmedia Hollywood, Storyworld, Power to the Pixel, Ted X Transmedia, DIY Days, and the Telefonica Foundation's Transmedia Lab events, bring together diverse participants to share and critique new projects and to explore how transmedia stories are constructed, how they are funded, and especially what assumptions producers make about audience engagement and participation. These conversations continue across blogs, podcasts, social media, and regular meet-ups in key production hubs. More and more books, written by both academics and industry insiders, engage in what might be called speculative aesthetics, imagining the future of transmedia and offering advice to would-be creators.⁵

Transmedia exploits Hollywood's consolidation of media ownership across once separate entertainment industries, creating a strong incentive for content to be deployed across as many platforms as possible. The concept transforms an economic imperative into an aesthetic possibility, asking how transmedia approaches produce more meaningful entertainment experiences. These first two motives are closely aligned with a third: the desire to intensify audience loyalty and increase engagement by recognizing and rewarding the most fully committed viewers. Many transmedia projects are funded through promotional budgets and judged by their success in attracting audience attention. In the case of a television series, transmedia content often fills gaps in the program flow in the days, weeks, or months before the next installment airs. Contemporary U.S. television has shifted from an appointment-based model toward an engagement-based paradigm. Under the appointment model, committed viewers arranged their lives to watch programs when they were broadcast. By contrast, engagement-based models place a premium on audiences willing to pursue content across multiple channels, accessing shows on their own schedules, thanks to videocassette recorders and later digital video recorders (DVRs), digital downloads, mobile video devices, and DVD boxed sets. Television embraces transmedia within this logic of engagement.[6]

Academics initially responded to early transmedia projects primarily with skepticism about their marketing motives. For example, P. David Marshall spoke about the "new intertextual commodity," which he described more or less entirely through a logic of branding: "the audience is imbricated in an elaborate extratextual matrix . . . designed to encircle, entice, and deepen the significance of the film for the audience."[7] By contrast, Brenda Laurel's 2001 *Utopian Entrepreneur* offered "transmedia" design principles, which included: "Think in transmedia terms from the beginning. . . . Build worlds, not just stories. . . . Create a foundational narrative. . . . provide for rituals. . . . Support community formation." Earlier, Janet Murray discussed the hyper-seriality of early transmedia experiments (such as those around *Homicide*) as part of a larger exploration of the emerging aesthetics of digital storytelling.[8] Contemporary accounts of transmedia still struggle to balance these two competing polls: transmedia as promotion and transmedia as storytelling.

Although many transmedia advocates stress the newness or innovation of such approaches as responsive to dramatic changes in the media landscape, these practices have been assimilated into mainstream media strategy so quickly in part because they rework or reimagine a much older history of licensing and franchising. And, as the studios and networks have embraced transmedia, the shift has often come with a cost. *The Matrix* depended on strong integration across all participating media. More recently, Hollywood has shifted toward a "mothership" approach; that is, the focus is on one core property that may be extended into other platforms depending on market response. Above all, creators seek to drive audiences to the mothership—most often a feature film or a television series. Other media should deepen the audience's engagement without "cannibalizing" the market. The mothership should not depend for its dramatic pay-off on something that consumers have to track down elsewhere. The mothership must be perceived as self-contained, even if other media add new layers.

Elsewhere, I have documented the ways in which studio-era Hollywood responded to the technological disruptions (the coming of sound) and the economic risks (the Great Depression) of another transformative moment in its history—the early 1930s.[9] Here, we saw Hollywood recruit performers from vaudeville and Broadway revues and construct vehicles for their distinctive talents. This strategy reflected the tastes of early adapters; for example, the theaters wired for sound were mostly in urban areas where stage entertainers were more likely to be appreciated. Hollywood's musical comedies focused on spectacular performance, often at the cost of narrative integration and character motivation. Over time, the system restablized, exerting more conservative pressures to conform with classical Hollywood norms.

We might see the pull toward the mothership model as another example of the system seeking to incorporate a somewhat alien aesthetic—one that reflects the potentials of interactive media, networked consumption, and participatory culture—and to make it more compatible with established norms and practices. The push for transmedia storytelling emerges at a moment when net access has become more widespread, especially for eighteen-to-twenty-seven-year-old males. The studios and networks look toward the web as a distribution channel, long term for media content, short term for promotional materials. And this moment created an opening for new models for entertainment content to be marketed and stories told. As the Web 2.0 business model has become more widespread, the focus has started to shift from top-down canonical content and forms of limited interactivity toward more ambitious plans for audience participation. Despite grander ambitions, transmedia extensions are still often made to work as a mechanism for creating or sustaining audience interest in the program itself, an unavoidable situation as long as ratings for aired episodes count differently from viewer attention online or through other media outlets.

If transmedia has the possibility to transform what counts as mainstream entertainment, it will be through a gradual evolution rather than a dramatic revolution in storytelling practice. In the short term, most of the innovation occurs on the fringes—more on cable than broadcast, more in independent and public media than in commercial media, and perhaps, more outside rather than inside the U.S. media system. None of these caveats should detract from the very real creativity, within constraints, that characterizes contemporary transmedia production within the mainstream industry. Transmedia has never been "simply a marketing strategy," as critics have implied, but marketing logics still exert strong pulls, at every stage, in shaping what kinds of transmedia extensions are produced and distributed. As Starlight Runner's Jeff Gomez explains, "Transmedia doesn't replace marketing, it is infused into it, turning marketers into storytellers who are helping to enrich and expand the franchise."[10] Or more pessimistically, transmedia turns storytellers into marketers.

This chapter will first offer a quick overview of how current transmedia practices evolved from older histories of licensing and franchising. Then we will consider two examples of contemporary transmedia production, exploring how the goals of expanded storytelling were shaped by logics of engagement and promotion. And finally, we will explore and situate the West Coast mothership model against alternative approaches to transmedia production that are emerging in other contexts, including the East Coast School's focus on alternate reality games, the concentration on education and cultural enrichment that emerges from public sector media economies, and the emphasis on transcultural communication that is appearing as transmedia makes its way into Brazil. Taken as a whole, this discussion is intended to help us to better understand why the mothership model has come to dominate mainstream U.S. media production but also to suggest the ways that this approach limits the possibilities that a more robust model of transmedia storytelling might be able to achieve.

Transmedia: A Prehistory

Often, transmedia advocates draw parallels between contemporary entertainment "mythologies" and the interlocking and recurring characters or elaborate histories associated with older folk tales, myths, and fairy stories. Fantasy writers such as C. S. Lewis and J.R.R. Tolkien, both scholars of medieval and folk literatures, explicitly modeled their fictions after the structures of old myths, and George Lucas's fascination with mythologist Joseph Campbell paved the way for the "hero's journey" to become a core structure shaping Hollywood blockbusters over the past several decades.[11] Angela Ndalianis has described these same franchises as "neo-baroque," suggesting a connection with earlier artists who embedded stories into their physical surroundings, often in ways that made it difficult to distinguish real space and representations.[12] Frank Rose maps parallels between the serial publication and consumption of Victorian novels and contemporary forms of transmedia seriality.[13]

L. Frank Baum, author of *The Wizard of Oz* books, was an early prototype for today's transmedia storytellers. For most of us, *The Wizard of Oz* is a single story reduced from the twenty or so books Baum wrote and, from there, to only those characters and plot elements from the MGM musical. Baum understood Oz as a world and depicted himself as its Royal Geographer, offering mock travelogues to illustrate its different places and peoples. Oz was also elaborated through comic strip series, stage musicals, and films, each of which added new elements.[14] Many transmedia extensions work much like Baum's travel lectures, offering guided tours of a fictional setting. Baum would have understood the promotional value of these media extensions, having edited and published a trade journal and established a professional organization centered around the craft of constructing window displays for early department stores. Baum sought to move retailers from a cluttered display of goods toward show windows that captured consumers' imagination through the use of fantasy, spectacle, and storytelling, even as he acknowledged that the ultimate goal was to move merchandise.[15] But Baum also saw these other texts as part of his larger project of constructing fairy tales appropriate for American children.

Seeking creative control over his stories, Baum started his own film production company; but there were rival comic strips set in Oz, reflecting conflicting copyright claims by Baum and his illustrator, W. W. Denslow.[16] Over the first few decades of the twentieth century, as Ian Gordon documents, a whole industry emerged around the licensing and reproduction of the likenesses of characters from nationally syndicated comic strips, such as Buster Brown.[17] For example, Felix the Cat originated in the early 1920s as a cartoon character created by Pat Sullivan and Otto Messmer, but within a few years Felix moved from animated cartoons into comic strips and children's books, sheet music and sound recordings, and manufactured goods, such as ceramics, stuffed animals, board games, and other toys.[18] Buster, Felix, and the other characters offered engaging personalities, easily recognizable iconography, and strong hooks for young consumers, yet they lacked any core narrative that might be linked across media.

By the 1930s, such licensing arrangements had become much more common. In his case study of George Trendle, who controlled the fates of the Lone Ranger and the Green Hornet, Avi Santos gives us valuable insights into how such arrangements responded to shifts in industry structure across the mid-century:

> Licensors initially positioned themselves as intermediaries who managed the articulation of brand formulas across multiple but separate media and manufacturing industries. As the national television networks consolidated their power by the late 1950s, they began operating in-house licensing and merchandising divisions that effectively adapted strategies introduced by independent licensors like Trendle, undercutting his autonomy and authority and eventually rendering him anachronistic by the mid-1960s. By the time conglomeration began to take hold by the late 1960s, which centralized licensing and merchandising even further, the era of the independent licensor was effectively over. Yet, the brands formulated and managed by Trendle and other independent licensors would

persist. Many (including the Lone Ranger) were bought up by larger more diversified companies, like the Wrather Corporation, which stockpiled multiple licenses and engaged in both licensing and production activities.[19]

The Lone Ranger, Superman, and other action-adventure characters embodied a core mythology that might include origin stories, secondary characters, and recurring actions from venue to venue. The information conveyed across these different media tended to be repetitive rather than elaborative,[20] though there are some examples where the mythology emerged across media. For example, much of what we know about the destruction of Krypton comes from the radio drama; the cartoons and live action series moved Superman from "leaping tall buildings in a single bound" to flying; and the newspaper drama played a much more central role in the live action serial than in any other medium.[21] Our current understanding of Superman as a character, thus, emerged across different media, even if there was only limited effort to coordinate the generally episodic stories, especially given that the rights had been licensed to different parties.

We might see Walt Disney as the progenitor of transmedia practices on television. He coupled the launch of his Disneyland theme park, for example, with the development of a *Disneyland* television series, which organized materials—mostly drawn from his theatrical releases—into themed segments based around Fantasyland, Frontierland, Adventureland, and Tomorrowland.[22] In turn, Disney built rides and attractions based on elements from the films. He also used television to experiment with long-form storytelling: his *Zorro* series, for example, relied on a season-long story arc as well as a more episodic structure. He developed participatory formats around content through *The Mickey Mouse Club*, which gave viewers a chance to perform their affiliation with the series. Some of Disney's screen characters had extended adventures through his comics, with Carl Barks developing a whole mythology around Donald Duck and his various relatives.[23] Much like Baum, Disney used himself as the point of intersection for these various media extensions; the smiling "Uncle Walt" persona put a friendly face on his cross-promotion strategies. Part of the appeal of visiting a Disney theme park was the notion of returning to a fictional place familiar from childhood memories and capable of being shared with the coming generation. Disney's focus on familiarity, however, meant that the links across media were not especially demanding or complex. He was among the first artists to license his cartoon characters; but through the years, he brought more and more content under his own corporate control, modeling the set of practices we associate with today's media conglomerates.

With greater media consolidation, we entered into the era Justin Wyatt has described as "high concept," where entertainment properties are assessed on their capacity to be exploited across the multiple divisions of a production company.[24] Wyatt argues that decisions to greenlight a project in the Hollywood of the 1970s and 1980s were determined by "the book, the hook, and the look," that is, a pre-sold commodity adapted from another media, a compelling short description that tapped into a topical interest, and a recognizable style or iconography that would work well across media. As Wyatt describes it, the high concept approach required little to no elaboration of the core narrative as it moved between media.

But the 1970s also saw a key turning point when George Lucas famously forfeited payment for his role in directing the *Star Wars* films in favor of a larger portion of the revenue from ancillary products.[25] Suddenly, Lucas had a strong incentive to pour craftsmanship and storytelling into comic books, novels, animated series, and computer games, which helped to sustain public interest in *Star Wars* far beyond the release of the first three films. Over time, characters, plots, and elements of the world that originated in these other sources have taken on a life of their own as the focal points of fan interest and the drivers of their own narratives. The 1992–1993 television series *The Young Indiana Jones Chronicles* offers a glimpse of things to come in terms of how television fits within transmedia seriality. Lucas and Steven Spielberg hinted at the protagonist's backstory in *Indiana Jones and the Last Crusade* when they included a flashback to Indie's boyhood and introduced his father, Professor Henry Jones. These plot elements inspired a high-budget television series, which used this backstory to stitch together historical events across the early twentieth century. As aired, the series crisscrossed in time, showing the adventures of Indiana as a child, as a young adult, and in old age. When the episodes were released on DVD, they were re-sequenced into a chronological flow and coupled with original documentaries providing historical context.

Derek Johnson traces the movement of the concept of a "franchise" from discourse about fast food and hotel chains into discussions of entertainment media during this same period; practices of re-versioning and extension became a central way that the media conglomerates sought to extend the life span and diversify the market for their most successful intellectual properties.[26] The franchise model, for example, might result in expanding the time line of a series like *Star Trek* so that one could develop multiple television and film series based on different Star Fleet captains and their crews, refreshing the core formula for subsequent generations.

During this same period, deregulation of children's programming resulted in new partnerships between toy manufacturers and content producers, generating such series as *Transformers*, *Thundercats*, *G.I. Joe*, *She-Ra*, and *He-Man, Masters of the Universe*. (The television series that Marsha Kinder discusses in *Playing with Power* emerged as the second wave of this phase of co-production, driven as much by early video game production as by toy manufacturing; hence, her focus is on interactivity as a central aspect of this "transmedia supersystem.") Critics like Peggy Charren, founder of Action for Children's Television, saw *He-Man* and similar programs as simply "half-hour commercials" that "blur the distinction between program content and commercial speech."[27] In some ways, *Masters of the Universe* was already a transmedia story, at least as much as the technology of the day would allow. He-Man appeared not only in the Filmnation-produced cartoons but also in the mini comic books that came with each action figure and in all the collector cards, sticker books, coloring books, and other kids' books, each of which revealed a little something more about Eternia. The fact that these stories were shared through mass media with other kids and that they were so vividly embodied in the action figures meant that it was easy for children to have intersubjective fantasies, to share their play stories with each other, and to pool knowledge about the particulars of this fictional realm. The accessories were extensions of the characters, reflections of their personalities, artifacts of their stories, and signs of their capacities for action. Each character was connected to every other character through complex sets of antagonisms and alliances, and each character bore its own mythology, which could become the point of entry for a new, as-yet-unrealized story.

When I speak to the twenty- and thirty-somethings who are leading the charge for transmedia storytelling, many of them talk about childhood immersion in *Dungeons and Dragons* or *Star Wars*, playing with action figures or other franchise-related toys. From the beginning, they understood stories less in terms of plots than in terms of clusters of characters and in terms of world building. They saw stories as extending from the screen across platforms and into the physical realm, as resources for creating their own fantasies, as something that shifted into the hands of the audience once they had been produced, and, in turn, as something that was expanded and remixed on the grassroots level.

Marc Steinberg has traced the way similar trends gave rise to the "media mix" strategy that shaped production decisions within the Japanese entertainment industries, including "the convergence of commodity and advertisement or program and promotion," "the serial intercommunication of media texts and things," "the displacement of the text as a unified totality by the text as a series of transmedia fragments," "the expansion from media texts to media environment, entailing the wider circulation of the image," "the rise of the child as an increasingly important new consumer category," and "the reconceptualization of consumption as a form of productive activity or work."[28] Much like earlier models of marketing in the United States, the media mix model was character-centered rather than narrative-centered. Hiroki Azuma argues that anime characters function as a database of interchangeable elements, as figures for fan collection and exchange, often with little to no emphasis on narrative context.[29] But for many U.S. transmedia producers, anime was experienced as providing a much more elaborate mythology, complex plays with chronology, and serial connections across episodes than American counterparts. The Wachowskis' *Animatrix* project, which solicited short films by top Asian animators set in the world of *The Matrix*, or, more recently, the anime series focused on fleshing out the backstory of *Supernatural* suggest the degree to which American transmedia producers were inspired by Japanese media mix practices, although those approaches had to be redesigned to operate in a U.S. context.

Contemporary transmedia practice was also prefigured in the comics industry. Because of the lower costs of production and the smaller but more hardcore following, comics have often been a space within which to experiment with alternative genre strategies.[30] Over the past few decades, comics storytelling has become increasingly serialized, moving from self-contained episodes to story arcs across multiple issues, often intended to be collected and sold as graphic novels. The offerings of the two major comics publishers, Marvel and DC, have become interconnected universes, where characters pass regularly across titles, and several times a year company-wide events coordinate actions across all affiliated titles. Such trends were supported by the greater dependability of comics distribution in an era of specialty shops and online subscriptions and the increased availability of back issues. But such "radical intertextuality" was also supported by the growth of networked consumption, as online fan communities helped each other sort through confusions over the ever more extended mythologies. Today, DC and Marvel are owned by Warner Communications and Disney, respectively, which continue to create and test intellectual property that may ultimately play across other media platforms.

In the decade between *Playing with Power* and *Convergence Culture*, Hollywood's emphasis shifted from licensed characters and "high concept" style to a focus on complex narratives and the world-building process; many of the practices developed around children's programming were applied to adult dramas. The promotional departments at the networks and studios gained a certain freedom to experiment at a moment when it was widely understood that marketing needed a digital component but no one was certain yet what an effective model looked like. A generation that had grown up playing with action figures, running role-playing games, reading superhero comics, and watching Japanese anime entered the industry and started designing content for younger versions of themselves.

As *Convergence Culture* suggests, their visions of more expansive popular narratives came of age alongside a growing emphasis on creating multiple touch points with the consumer. Transmedia producers justified what they were doing as a kind of promotion, as expanding opportunities for branding, though increasingly they would also assert claims as content producers. Despite its box office success, *The Matrix* was regarded within the industry as a failed experiment, one that had expected too much of its audiences in terms of dispersing key elements of the story across multiple media and one that met with resistance because the public did not want to "do homework" before sitting down in front of a Hollywood blockbuster. The current mothership model, then, represented a way of constructing transmedia content that served the need to intensify and reward audience engagement without becoming a competing interest. Whatever transmedia storytellers might hope, the industry still understands these practices within a logic of promotion. But within that space there are still opportunities to use transmedia extensions as means to develop secondary characters, explore richer fictional worlds, provide backstory, or otherwise expand the time line of the narrative, so long as doing so will attract engaged consumers. In the next two sections, we will consider two high-profile examples of how transmedia producers have sought to drive audience engagement and, at the same time, justify their own interests in transmedia experimentation.[31]

Ghost Whisperer and "Total Engagement" Television

Kim Moses, who, along with her husband, Ian Sander, was the co-executive producer of *Ghost Whisperer* (2005–2010), describes the transmedia process as a "total engagement experience."[32] When CBS agreed to put her supernatural-themed series on the air, Moses recognized that her team faced an uphill battle in finding their desired audience. The program's star, Jennifer Love Hewitt, had built a youthful following, but CBS was noted for its maturing demographics. At the same time, the network had decided to air her series on Friday evenings. As Moses has often noted, this time slot has proven to be the kiss of death for most series, especially those targeting young people, who are often out on dates on Friday nights. "In this day and age," Moses has said, "it is not enough for a producer to just deliver a television show. I believe you are also responsible for delivering the audience."

Working with the network, Moses made sure that her production company maintained creative control over the content produced for the web and other platforms, seeing it not simply as a promotional device but also as a contribution tied to the "heart and soul" of the series: "The total engagement experience gives you a bridge experience in between each broadcast; it drives people to the show week after week through these experiences; and it also gets people to sample the show who have never seen the show before." She and her team closely monitored online fan response, trying to identify program elements that generated particular interest. One example was "the laughing man," a character introduced as a throwaway detail on one first-season episode. The character developed such intense audience interest that he became central to the online strategy and was ultimately reintroduced into the primary series

FIG. 11-1 *Ghost Whisperer* (CBS) "Internet Strategy" created by Sander-Moses. (Photo courtesy of Kim Moses, executive producer, *Ghost Whisperer*)

for the initial season's final episode. Building on audience fascination with the ghost world, the *Ghost Whisperer* team also developed a web serial, "The Other Side." The online series did not include any cast members or sets but instead explored the same issues and themes the series tackled from the perspective of the ghosts.

In *The Ghost Whisperer Spirit Guide*, Moses and Sander describe the total package:

> First and foremost, we are storytellers, so everything we do, from "The Other Side" webisodes to the interactive journey in Payne's Brain, tells a story relative to *Ghost Whisperer*. . . . We test assets we've created online, and as they get

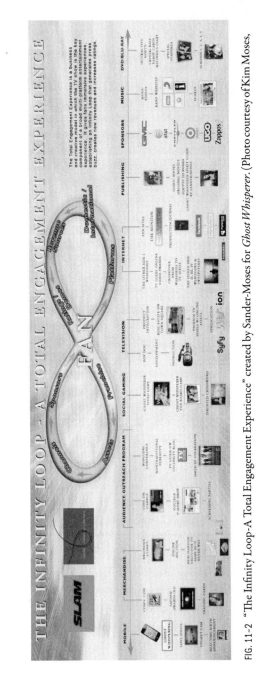

FIG. 11-2 "The Infinity Loop-A Total Engagement Experience" created by Sander-Moses for *Ghost Whisperer*. (Photo courtesy of Kim Moses, executive producer, *Ghost Whisperer*)

traction, we team with department heads at the studios and networks and roll them out into various platforms that service our fan base. The aim is to make *Ghost Whisperer* a multi-dimensional experience which the viewer can interact with in their own way on their own schedule. Go beyond what Melinda Gordon is experiencing in the spirit world on an episode by interacting online with all sorts of viral initiatives we've created.[33]

Some of what they created followed the story arc, while other elements were topical, responding to particular episodes. Some sought to expand the fiction; others offered behind-the-scenes looks at the production process. Much of what they did was accomplished on very low budgets, so they collaborated with amateur artists, offering greater visibility for them in return for their labor. Over time, the team expanded their efforts from the web to other media. When they saw how much traction Joss Whedon had gotten from extending *Buffy the Vampire Slayer* into a post-cancelation "eighth season" via comics, for instance, they began to produce *Ghost Whisperer*–themed comics and, from there, original novels.

By the time the series was canceled at the end of its 2010 season, *Ghost Whisperer* had been successfully picked up by the SyFy Channel, meeting Moses's goal of entering syndication. This approach, Moses argues, makes sense in the context of what scholars have described as an "experience economy,"[34] where hardcore fans want to become an active part of the world depicted on television, demanding the right to participate in a fiction that matters to them. "We had to collect them, bring them in," she says. "We had to court them and date them. It was like the Mickey Mouse Club for people who liked ghosts. As we did this, our relationships on the online world started to grow. And the audience started to feel enfranchised, and they started to do things virally for us.... [W]e really think that this total engagement experience is the silver bullet for diminishing audiences in the 21st century."[35]

Moses's courtship metaphor is especially helpful here in terms of thinking about "engagement" as the emotional connection that forms between viewers and desired content. A key point about transmedia, then, is that engagement requires building a relationship with fans, often through expanding fans' opportunities to participate in the unfolding story. Possibilities might include fans collecting and exchanging bits of information dispersed across the media environment, shaping central aspects of the story through voting mechanisms, adding their own content, or engaging through the interactive mechanics of a computer game. However, we must keep in mind that every act to structure interactivity and participation into a franchise supports some fan reading practices while excluding or marginalizing others. Some feminist critics have argued that, to date, transmedia content has been designed with a presumed masculine bias, supporting forms of participation associated with male audience members.[36] *Ghost Whisperer* may be a notable exception, where a husband-and-wife showrunner team worked to try to attract an older female following for their series and developed their engagement strategy accordingly.

Campfire and *True Blood*

Contemporary transmedia strategies assume that the gradual dispersal of material can reward a property's most ardent fans while inspiring others to become even more active in seeking and sharing new information. In this environment, events like the San Diego Comic-Con have become a key starting point for word-of-mouth campaigns around cult media properties. The event attracts more than 130,000 active fans each year, many of whom have blogs, Twitter accounts, and influential followings on social network sites focusing on popular culture.[37] Comic-Con was at the center of the campaign for HBO's 2008 launch of *True Blood*, developed by Campfire, a marketing agency formed, in part, by members of the Haxans, the team that pioneered "dispersed storytelling" with *The Blair Witch Project*.

Low-budget horror films have long been the object of hucksterism and hype, a history that goes back to the Universal horror films of the 1930s and perhaps reached a pinnacle in the strategies associated with producer William Castle in the 1950s and 1960s.[38] Horror walks a line between plausibility and implausibility, between the world as we know it and a world that follows radically different rules. Here, transmedia extensions often work to sustain that ambivalence and uncertainty by intensifying the reality of the depicted events. *The Blair Witch Project* created a wealth of fake but convincing materials, an approach inspired by the pseudo-documentary qualities of the film itself. Eduardo Sanchez, the film's director, described this successful strategy: "What we've learned from *Blair Witch* is that if you give people enough stuff to explore, they will explore. Not everyone but some of them will. The people who do explore and take advantage of the whole world will forever be your fans, will give you an energy you can't buy through advertising.... It's this web of information that is laid out in a way that keeps people interested and keeps people working for it. If people have to work at something, they devote more time to it. And they give it more emotional value."[39]

Many of these same theories of audience engagement continue to shape the projects launched by Campfire, a company that has helped to generate the engagement strategies for HBO's *True Blood* and *Game of Thrones*, USA's *Political Animals*, FX's *American Horror Story*, and A&E's *Bag of Bones*. Campfire is perhaps best known for developing the paratextual content that audiences encounter before they first watch the program, content that builds up their awareness and deepens their understanding of the fictional world. Increasingly, Campfire's approach also seeks to motivate these early fans to share their enthusiasm around this content through social media, where it can attract "earned" news coverage and reach more casual viewers.

As Michael Monello, executive creative director for the *True Blood* campaign, explained, "The strategy for season one was to begin small, excite core fans and then build to wider audiences as we got closer to the premiere."[40] During what Campfire described as the "discovery" phase, fan opinion leaders received personal mailings, including messages in dead languages or vials of fake blood intended to spark online discussions. Those who cracked the code were able to access hidden websites or secret telephone messages providing new leads. Their interests were fed by videos distributed via YouTube, "documenting" how humans responded to the discovery of vampires in their midst.

Having snagged these early adopters, Campfire extended the approach to advertisements in magazines and newspapers or billboards in major cities, shifting focus toward more casual fans. As they did so, Campfire and the *True Blood* team counted on those fans already engaged with the property to proselytize their friends and families. This phase began with the first public presentation by the series cast and crew—at a standing-room-only panel at San Diego Comic-Con—out of which came another immense wave of online buzz around the soon-to-start series. From there, Campfire engaged wider audiences through traditional publicity for the show and through more easily accessed content packages that ran on HBO's website, its video-on-demand platform, and elsewhere. Throughout the process, the fan buzz inspired news coverage, further increasing mainstream awareness. Ultimately, the online videos alone attracted more than 5.9 million viewers, according to Campfire's internal statistics, and 6.5 million people watched the first episode of the new series—high results for a program on a subscription cable channel. As Monello explained:

> Without social networks enabling fans to spread the word widely and quickly, our "start small" strategy wouldn't be nearly as viable. Those initial mailings worked because the people who received them had connections with other fans and could tell the stories to each other, and these stories could be discovered by other fans.... Freed from the need to make everything big and accessible, we can create elements within our stories for different types of fans, sparking passion and driving people to spread the story.... Hardcore fans act on their passion much faster, so the key is to create experiences that give them a piece of your story to tell.... Giving these dedicated fans early access to story elements and empowering them to help circulate the story became key to this highly successful launch strategy.[41]

Here, again, we can see Campfire making the case that the best way to engage and sustain fan interest is by creating compelling content that extends the story in meaningful ways. Without a compelling story, rich worlds, and intriguing characters, there is little to motivate these fans to help spread the word, and there's little to keep them coming back week after week when there are so many other options.

Alternative Models for the Future of Transmedia

Max Giovagnoli concludes his book *Transmedia Storytelling* with a series of interviews with leading transmedia producers in which he asked how they imagine the role of the audience in a transmedia experience. The responses hint at the complex interweaving of the logic of engagement with a range of other aesthetic and social goals:

> Transmedia storytelling invites the audience to become a part of a fictional world. It's more than just getting more interactively involved with a narrative, it's about getting immersed in a fictional world and feeling like you have agency within the world, that what you do matters and has an impact on the related story you experience as you travel across and between media to participate more fully in the story. (Drew Davidson)
>
> [Consuming transmedia] requires perceiving a world in all its guises, engaging with many artforms, and seeing them all as being part of some greater whole. (Christy Dena)
>
> For me personally transmedia asks people to collaborate and to co-create stories that can be jumping off points to social connections and if I do that, the stories will surely spread. (Lance Weiler)[42]

These comments reflect the kind of utopian discourse that has surrounded transmedia storytelling from its inception—a desire for "something more" than Hollywood has offered in the past, a desire by transmedia producers for more creative scope, a desire on the part of the audience for a deeper, more meaningful kind of engagement. These claims do not deny the marketing and branding logics of the Hollywood system, which always return to the idea that, ultimately, transmedia is a mode of promotion designed to intensify audience engagement; yet they are arguments for why transmedia producers might be able to use the creation of a more complexly layered fictional world as a means of building a new relationship with consumers.

In some cases, transmedia teams have been incorporated directly into the production process, as occurred with *Lost*, *Heroes*, *Smallville*, and, as described above, *Ghost Whisperer*. In other cases, transmedia consultants and subcontractors were commissioned to develop transmedia campaigns around television properties, as we saw in Campfire's role in promoting *True Blood*. Jeff Gomez, the head of Starlight Runner, a company that partnered with industry clients to develop transmedia frameworks around their properties, helped to lead a push within the Producers Guild of America that would solidify the status of transmedia through the creation of a Transmedia Producer job title. Here's how the official prose defines the concept:

> A Transmedia Narrative project or franchise must consist of three (or more) narrative storylines existing within the same fictional universe on any of the following platforms: Film, Television, Short Film, Broadband, Publishing, Comics, Animation, Mobile, Special Venues, DVD/Blu-ray/CD-ROM, Narrative Commercial and Marketing rollouts, and other technologies that may or may not currently exist. These narrative extensions are NOT the same as repurposing material from one platform to be cut or repurposed to different platforms.
>
> A Transmedia Producer credit is given to the person(s) responsible for a significant portion of a project's long-term planning, development, production, and/or maintenance of narrative continuity across multiple platforms, and creation of original storylines for new platforms. Transmedia producers also create and implement interactive endeavors to unite the audience of the property with the canonical narrative.[43]

Transmedia is defined here in terms of the linkage of platforms without necessarily integrating narratives (multiple storylines for multiple media versus a single story told across multiple media). The focus is on audience engagement as the motivation for transmedia production. The transmedia producer can be brought into the project at any point; "born transmedia" projects do not receive special emphasis. Here, transmedia is most often work for hire, developed by creative participants who have little or no ownership or, in many cases, direct control over the content they are helping to produce.

Not all potential "transmedia producers" embraced this approach. Two key transmedia figures interviewed in 2011 by Nick DeMartino for the Tribeca Film Festival's "Future of Film" blog capture some of the polarities in this debate. "Why do we have to define it yet?" asked indie filmmaker Lance Weiler. "Why can't we just continue to experiment?" Because, says television writer-producer Jesse Alexander (*Lost* and *Heroes*), "You have to give it a name so people can talk about it. Isaac Newton didn't discover gravity, he named it."[44] Most persistently, critics wonder whether the definition is biased toward the mothership model, as opposed to, say, alternative reality games, where no media element can be seen as autonomous from the whole. Responding to this debate, Brian Clark, the founder and CEO of GMD Studios, mapped what he saw as two distinctive models for transmedia production, identified in terms of their primary geographic concentration:

The West Coast transmedia tradition ... thinks more in terms of franchises, it has struggles with the relationships with the owners of the industry, and starts from the perspective that creators won't own the IP they are creating. They want to fix the studio system, or recreate a new kind of studio.

The East Coast transmedia tradition is quite different and emerges far more from the independent traditions of media. ... It thinks in terms of one story told across platforms, it has struggles with monetizing and financing, and starts from the perspective that creators own the IP they are creating. They want to extend an existing community into transmedia, or recreate a new kind of community.

Neither is wrong. Few practitioners or creators work exclusively in one sphere or the other. One is not more noble or pure or profitable than the other. But we're all guilty of conflating the two together in ways that lead to moments where it might sound like the community is, for example, telling documentary filmmakers that they need to think more like franchisers if they want to get on the transmedia bandwagon and not be left behind as storytelling changes forever.[45]

Clark's analysis suggests that transmedia strategy takes different shapes depending on the production context, the operative business models, and the dominant media and their relationship with their audiences.

If *The Matrix* and *Heroes* are among the defining projects for the West Coast mothership model, two projects created at about the same time laid the foundations for the East Coast model: "The Beast," developed by Sean Stewart for Steven Spielberg's 2001 feature film *A.I.*, and "I Love Bees," developed by 42 Entertainment for the release of the video game *Halo II*. "The Beast" was tied to a Hollywood feature film, and "I Love Bees" to a Microsoft game launch, suggesting that from the start, Clark's models were not as geographically located as his terminology may suggest. But, in both cases, these alternate reality games (ARGs) were only loosely linked to their media franchises, while there was strong integration across the different media pieces. Some subsequent ARGs have continued to be linked to Hollywood practices (such as "Why So Serious?" for *The Dark Knight* or "The *Lost* Experience"), but many others (such as *Perplex City*) have sought to be stand-alone experiences, and the latter most fully embody Clark's East Coast school.

These projects offered a model for transmedia that was grounded in the practices of independent media production (often developed through small teams of close collaborators) rather than those of studio- or network-based production. For Clark, this model ensures that the artists maintain a much tighter control over their work and that there is a higher chance for meaningful integration across media segments.[46] Here, the focus is on participatory and interactive experiences, especially collaborative game play and collective puzzle solving. Yet, Clark's East Coast model is only one of a series of alternative transmedia approaches emerging in other economic contexts.

In Canada, Great Britain, and members of the European Union, money for transmedia production comes from state-supported broadcasters or national and international funding agencies of the kind that have historically supported the production of animation, documentaries, and art cinema. For example, the nonprofit Canadian Media Fund created a Convergent Media track in 2010 with the goal of promoting the growth of transmedia production in Canada, and it specifically targeted "underrepresented genres: drama, documentary, children's and youth, & variety and performing arts."[47] Beyond focusing on genres of educational and high cultural value, the Canadian funders have also linked their "convergence" initiatives to content that creates bridges between cultural groups or amplifies minority voices, including aboriginal and Francophone peoples. Because of state mandates, transmedia production under this model often embraces broadly defined educational or civic goals. It is seen as a way of deepening the public's awareness of and involvement with social concerns, and in many cases it takes the form of serious games or links alternate reality games to various forms of transmedia activism and mobilization. Transmedia production under this model is more likely to be tied to museums, universities, and other public institutions, where a strong cultural agenda dominates and where transmedia is in some cases (for example, recent work by Peter Greenaway) understood as an avant-garde practice, carrying a very different aura than the general perception of transmedia extensions coming out of the Hollywood franchise model.

In the case of producers in the European Union, we see some tension between the historic goals of individual countries in building up their national cinemas as a source of prestige (and thus a strong emphasis on cultural heritage) and developing transmedia content as a way of promoting shared transnational identities. Consider, for example, the mission statement for Transmedia Europe, a nonprofit organization that seeks to "advocate" for transmedia as a "creative form and practice" ideally suited "for [Europe's] multi-cultural needs, languages, and understandings, in respect to the European Union and its politics, including international cooperation."[48] The organization has sought to position itself as an intermediary between the creative community and various national and international bodies that inform European cultural policy. The current European economic crisis is speeding up the movement from public to commercial broadcasting, which may result in forms of transmedia production having more in common with the West Coast model. In Great Britain, for example, we may see the BBC producing transmedia around cult television series such as *Doctor Who*, *Torchwood*, and *Sherlock*,[49] or we may see Channel 4, a commercial producer, developing serious games as a means of extending its reach to adolescent and school-aged consumers.

Brazil, like many other countries, provides substantial support to local arts, media, and cultural activities, with many of the most significant ones (in terms of both resources and track record) coming from the state and city of Rio de Janeiro.[50] The state has long recognized games and other digital media as part of its funding priorities, but in 2011 the state secretary of culture announced funding for transmedia, defined as "audiovisual projects in which content is developed, complementary, in multiple media, in order to explore and maximize the potential and characteristics of each one of them."[51] At the same time, transmedia production in Brazil is being informed by the example of public/private partnership set by RioFilme, a company funded by the City of Rio de Janeiro but run as a private firm, with the goal of sustaining the growth of the city's audiovisual industry through co-producing and co-financing films. By design, the company seeks to ensure that two out of every five films it produces will reach audiences of more than one million people inside Brazil, with all of the profits from successful productions reinvested into future film projects.

In 2012 the RioContent Market, the largest event for television, web, and mobile production in Latin America, launched its Transmedia Lab competition, attracting more than 160 entries from across the region. In the end, the event's jury chose to showcase twelve productions, mostly from Brazil, but also including finalists from Columbia and Argentina.[52] Many of the projects specifically targeted children as the most likely transmedia audience, reflecting the global perception that digitally literate youth are more apt to embrace products that flow across media platforms. Eleven of the twelve had a strong educational or documentary focus, with many of the producers explicitly identifying themselves as coming out of public media or web documentary traditions. For example, the producer's statement for *Tin Gods* explains: "As the material that we gathered in the process of research and preproduction was so vast, the message so positive, and Pre-Columbian art is a heritage of humanity, the project grew so much and was more than a feature film, limited only to the movie screens and home videos. Thus, given the advances in technology, the new devices, social networks and interactivity, I realized how important it is for the film's positive message to be transmitted by the greatest amount of possible means to children around the world, which are, in the short, the ones who can change it for good." *Stories of a New Life* documents some of the economic upheaval surrounding the growth of the middle class in Brazil: "On television, 5-minute episodes will show the faces of the new middle class. On the web, we will allow users to better understand the social and economic foundations that enabled this change.... The integration of both platforms will present the most thorough human and social scenario ever outlined about the new Brazilian middle class." These young media producers are groping toward a new model that accepts transmedia practices as the perfect vehicle and, perhaps, as an ideal metaphor for thinking about how to build bridges among multiple cultures, whether understood in terms of the cultural diversity within Brazil itself or, more generally, the ongoing negotiations between the global and the local within what they understand as an increasingly networked and transnational society. For them, transmedia offers a chance to fuse genres, combining, say, animation and documentary or games and narrative.

This all-too-brief global tour of different kinds of transmedia producers is intended to make two core points. First, transmedia has become a global phenomenon, as various kinds of media producers are drawn together by the shared project of imagining what kinds of media narratives and experiences make sense for an increasingly networked society. Second, transmedia production takes different shapes and pursues different goals within different modes of production. There is nothing inevitable or irreversible about the turn toward the mothership model within American film and television production. Given Hollywood's long history of licensing content, given the television industry's strong desire to deepen audience engagement, and given Madison Avenue's desire to increase touch points with its brands, this particular form of transmedia seems to serve the economic rationales of the current mode of production. Creative artists and fans, alike, have had to pursue their interests within these constraints. The focus on engagement is what enabled these transmedia projects in the first place, but that focus has perhaps made it harder for transmedia production to be driven by aesthetic impulses or to satisfy fully audience desires for richer content and greater opportunities for participation in the unfolding of their favorite stories. As these other examples suggest, transmedia is capable of doing much more than the dominant industry practices currently allow.

Notes

1 Marsha Kinder, *Playing with Power in Movies, Television, and Video Games: From Muppet Babies to Teenage Mutant Ninja Turtles* (Berkeley: University of California Press, 1991), 38, 119.
2 Henry Jenkins, "Transmedia Storytelling 101," *Confessions of an Aca-Fan*, March 22, 2007, http://www.henryjenkins.org/2007/03/transmedia_storytelling_101.html.
3 Gunther Kress, "Reading Images: Multimodality, Representation and New Media," *Information Design Journal & Document Design* 12, 2 (2004): 110–119.
4 Christy Dena, "Transmedia Practice: Theorizing the Practice of Expressing a Fictional World across Distinct Media and Environments" (PhD diss., University of Sydney, 2009).
5 Nuno Bernardo, *The Producer's Guide to Transmedia: How to Develop, Fund, Produce, and Distribute Compelling Stories across Multiple Platforms* (New York: Beactive Books, 2011); Drew Davidson, *Cross-Media Communications: An Introduction to the Art of Creating Integrated Media Experiences* (Pittsburgh: ETC Press, 2010); Max Giovagnoli, *Transmedia Storytelling: Imagery, Shapes, and Techniques*, trans. Feny Montesano and Piero Vaglioni (Pittsburgh: ETC Press, 2011); Frank Rose, *The Art of Immersion: How the Digital Generation Is Remaking Hollywood, Madison Avenue, and the Way We Tell Stories* (New York: W. W. Norton, 2010).
6 For a fuller discussion of these shifts, and especially shifts in audience measurement,

see Henry Jenkins, Sam Ford, and Joshua Green, *Spreadable Media: Creating Meaning and Value in a Networked Culture* (New York: New York University Press, 2013), 113–152.

7   P. David Marshall, "The New Intertextual Commodity," in *The New Media Book*, ed. Dan Harries (London: British Film Institute, 2002), 69–82. Also see Will Brooker, "Living on Dawson's Creek: Teen Viewers, Cultural Convergence, and Television Overflow," in *The Television Studies Reader*, ed. Robert Allen and Annette Hill (New York: Routledge, 2004), 569–580.

8   Brenda Laurel, *Utopian Entrepeneur* (Cambridge, MA: MIT Press, 2001), 85–86. Also see Janet Murray, *Hamlet on the Holodeck: The Future of Narrative in Cyberspace* (Cambridge, MA: MIT Press, 1998);

9   Henry Jenkins, *What Made Pistachio Nuts? Early Sound Comedy and the Vaudeville Aesthetic* (New York: Columbia University Press, 1992).

10 "Jeff Gomez on Transmedia Producing," Producer's Guild of America, http://www.producersguild.org/?jeff_gomez.

11 Michael Saler, *As If: Modern Enchantment and the Literary Prehistory of Virtual Reality* (Oxford: Oxford University Press, 2012); Joseph Campbell and Bill Moyers, *The Power of Myth* (New York: Anchor, 1991).

12 Angela Ndalianis, *Neo-Baroque Aesthetics and Contemporary Entertainment* (Cambridge, MA: MIT Press, 2005).

13 Frank Rose, *The Art of Immersion: How the Digital Generation Is Remaking Hollywood, Madison Avenue, and the Way We Tell Stories* (New York: W. W. Norton, 2010).

14 Mark Evan Swartz, *Oz Before the Rainbow: L. Frank Baum's The Wonderful Wizard of Oz on Stage and Screen to 1939* (Baltimore: Johns Hopkins University Press, 2002); Frank Kelleter, "'Toto, I Think We're in Oz Again' (and Again and Again): Remakes and Popular Seriality," in *Film Remakes, Adaptations, and Fan Productions*, ed. Kathleen Loock and Constantine Vervis (London: Palgrave Macmillian, 2012).

15 Patricia Krueger, "Wizard of Oz Writer's Wonderful Window Display," Popart, November 2008, http://blogs.popart.com/2008/11/wizard-of-oz-writer-s-wonderful-window-display/.

16 Peter Maresca, ed., *Queer Visitors from the Marvelous Land of Oz* (San Francisco: Sunday Press, 2009).

17 Ian Gordon, *Comic Strips and Consumer Culture, 1890–1945* (Washington, DC: Smithsonian Institution, 2002).

18 Donald Crafton, *Before Mickey: The Animated Film, 1898–1928* (Chicago: University of Chicago Press, 1993).

19 Avi Santos, "Transmedia Brand Licensing Prior to Conglomeration: George Trendle and the Lone Ranger and Green Hornet Brands, 1933–1966" (PhD diss., University of Texas at Austin, 2006), www.lib.utexas.edu/etd/d/2006/santod77474/santod77474.pdf.

20 William Uricchio and Roberta E. Pearson, "I'm Not Fooled by That Cheap Disguise," in *The Many Lives of the Batman: Critical Approaches to a Superhero and His Media*, ed. Roberta E. Pearson (New York: Routledge, 1991), 182–213; Will Brooker, *Batman Unmasked: Analyzing a Cultural Icon* (London: Continuum, 2001).

21 Gerard Jones, *Men of Tomorrow: Geeks, Gangsters, and the Birth of the Comic Book* (New York: Basic, 2005).

22 J. P. Telotte, *Disney TV* (Detroit: Wayne State University Press, 2004), 61–79.

23 Andreas Platthaus, "Calisota or Bust: Duckburg vs. Entenhausen in the Comics of

Carl Barks," in *Comics and the City: Urban Space in Print, Picture and Sequence*, ed. Jorn Ahrens and Arno Metelling (New York: Continuum, 2010), 247–264.

24 Justin Wyatt, *High Concept: Movies and Marketing in Hollywood* (Austin: University of Texas Press, 1994). See also Eileen Meehan, "'Holy Commodity Fetish, Batman': The Political Economy of the Commercial Intertext," in Pearson, ed., *Many Lives of the Batman*, 47–65.

25 Will Brooker, *Using the Force: Creativity, Community, and Star Wars Fans* (London: Continuum, 2003).

26 Derek Johnson, *Media Franchising: Creative License and Collaboration in the Cultural Industries* (New York: New York University Press, 2013).

27 The Mysteries of Greyskull, "The History of He-Man and Masters of the Universe," http://mysteriesofgreyskull.tripod.com/Articles/the_he-m.htm. See also Tom Englehardt, "Children's Television: The Shortcake Strategy," in *Watching Television*, ed. Todd Gitlin (New York: Pantheon, 1986), 70–85.

28 Marc Steinberg, *Anime's Media Mix: Franchising Toys and Characters in Japan* (Minneapolis: University of Minnesota Press, 2012). See also Mizuko Ito, Daisuke Okabe, and Izumi Tsuji, eds., *Fandom Unbound: Otaku Culture in a Connected World* (New Haven: Yale University Press, 2012); Ian Condry, *The Soul of Anime: Collaborative Creativity and Japan's Media Success Story* (Durham, NC: Duke University Press, 2013).

29 Hiroki Azuma, *Otaku: Japan's Database Animals* (Minneapolis: University of Minnesota Press, 2009).

30 Shawna Kidman, "Five Lessons for New Media from the History of Comics Culture," *International Journal of Learning and Media* 3, 4 (2012): 41–54. See also Tyler Weaver, *Using Comics to Construct Your Transmedia Storyworld* (New York: Focal, 2013).

31 Selections from the following two sections are taken from Jenkins, Ford, and Green, *Spreadable Media*.

32 Kim Moses, interview by Henry Jenkins, Los Angeles, October 28, 2009. Unless otherwise noted, comments by and information about Kim Moses come from this interview.

33 Kim Moses and Ian Sander, *The Ghost Whisperer Spirit Guide* (New York: Titan, 2008), 146–147.

34 B. Joseph Pine II and James H. Gilmore, *The Experience Economy: Work Is Theatre and Every Business a Stage* (Cambridge, MA: Harvard Business School Press, 1999).

35 Kim Moses, interview by Henry Jenkins.

36 Julie Levin Russo, "User-Penetrated Content: Fan Video in the Age of Convergence," *Cinema Journal* 48, 4 (2009): 125–130; Suzanne Scott, "Repackaging Fan Culture: The Regifting Economy of Ancillary Content Models," *Transformative Works and Cultures* 3 (2009), http://journal.transformativeworks.org/index.php/twc/article/view/150/122.

37 Henry Jenkins, "Super-Powered Fans: The Many Worlds of San Diego Comic-Con," *Boom: The Journal of California* 2, 2 (2012): 22–36.

38 Kevin Heffernan, *Ghouls, Gimmicks, and Gold: Horror Films and the American Movie Business, 1953–1968* (Durham, NC: Duke University Press, 2004); Rhona J. Berenstein, *Attack of the Leading Ladies: Gender, Sexuality, and Spectatorship in Classic Horror Cinema* (New York: Columbia University Press, 1996).

39 Quoted in Jenkins, *Convergence Culture*, 105.

40 Mike Monello, online interview by Henry Jenkins, June 25 2010.

41 Ibid.
42 Giovagnoli, *Transmedia Storytelling*, 129.
43 Producers Guild of America, "Code of Credits—New Media, Transmedia Producer," http://www.producersguild.org/?page=coc_nm#transmedia.
44 Nick DeMartino, "Why Transmedia Is Catching On, Part One," *The Future of Film*, July 5 2011, http://www.tribecafilm.com/tribecaonline/future-of-film/Why-Transmedia-Is-Catching-On-Part-1.html#.UP30N2guo1s.
45 Brian Clark, "Reclaiming Transmedia Storyteller," Facebook, May 2, 2011, http://www.facebook.com/note.php?note_id=10150246236508993.
46 Henry Jenkins, "Brian Clark on Transmedia Business Models," *Confessions of an Aca-Fan*, November 7, 2011, http://henryjenkins.org/2011/11/installment_1_transmedia_busin.html.
47 Canada Media Fund, "Convergent Stream," http://www.cmf-fmc.ca/funding-programs/convergent-stream/.
48 Transmedia Europe, "The Making of Transmedia Europe and Alliance," June 8, 2012, www.transmediaeurope.org/Transmedia-Europe.pdf.
49 Matt Hills, *Triumph of a Time Lord: Doctor Who in the Twenty-first Century* (London: I. B. Tauris, 2010); Louisa Ellen Stein and Kristina Busse, eds., *Sherlock and Transmedia Fandom: Essays on the BBC Series* (Jefferson, NC: McFarland, 2012).
50 Thanks for the assistance provided by Mauricio Mota and Barbara Mota from the Alchemists in informing my understanding of the transmedia context in Brazil.
51 "Secretaria de Estado de Cultura anuncia mais de R$ 40 milhões em editais," Cultura.RJ, November 8, 2011, http://www.cultura.rj.gov.br/materias/secretaria-de-estado-de-cultura-anuncia-mais-de-r-40-milhoes-em-editais.
52 All subsequent descriptions and details about these projects are drawn from Rio Content Market, Transmedia Lab Program, 2012, http://www.riocontentmarket.com.br/app/webroot/catalogo/labtransmidia/.

Notes on Contributors

VINCENT BROOK teaches at UCLA, USC, Cal-State LA, and Pierce College. He has written or edited five books, most recently *Land of Smoke and Mirrors: A Cultural History of Los Angeles* (2013) and *Woody on Rye: Jewishness in the Films and Plays of Woody Allen* (co-editor, 2013).

WILL BROOKER is currently director of research in film and television at Kingston University, London. He is the author/editor of nine books on popular narratives and their audiences, including *Batman Unmasked*, *Using the Force*, *Alice's Adventures*, *The Blade Runner Experience*, the BFI Film Classics volume on *Star Wars*, and *Hunting the Dark Knight* (2012). He is the first British editor of *Cinema Journal*, the publication of the Society for Cinema and Media Studies.

JOHN T. CALDWELL is a professor of cinema and media studies at UCLA and the author of *Production Culture: Industrial Reflexivity and Critical Practice in Film and Television* (2008) and *Televisuality: Style, Crisis, and Authority in American Television* (1995). He has edited or co-edited several books, including: *Production Studies: Cultural Studies of Film/Television Work Worlds* (2009, with Vicki Mayer and Miranda Banks), *New Media: Theories and Practices of Digitextuality* (2003, with Anna Everett), *Theories of the New Media: A Historical Perspective* (2000), and *Electronic Media and Technoculture* (2000). His current book project is entitled *Para-Industry*. Caldwell is also a producer/director whose award-winning films include *Freak Street to Goa: Immigrants on the Rajpath* (1989) and *Rancho California (por favor)* (2002). His current documentary film/media archaeology project is entitled *Highway 58: Boron to Buttonwillow* (2013).

M. J. CLARKE received his doctorate in film and television from UCLA. He is the author of *Transmedia Television: New Trends in Network Serial Production*

(2012). His articles on television have appeared in *Television & New Media* and *Communication, Culture & Critique*.

JONATHAN GRAY is a professor at the University of Wisconsin–Madison. He is the author of *Show Sold Separately: Promos, Spoilers, and Other Media Paratexts* (2009), *Television Entertainment* (2008), *Watching with The Simpsons: Television, Parody, and Intertextuality* (2006), and (with Amanda D. Lotz) *Television Studies* (2012). He has also co-edited numerous books, including *A Companion to Media Authorship* (2013) and *Satire TV: Politics and Comedy in the Post-Network Era* (2009).

HENRY JENKINS is a Provost's Professor of Communication, Journalism, and Cinematic Arts, Annenberg School for Communications and Journalism at USC. He co-authored *Spreadable Media: Creating Value and Meaning in a Networked Culture* (2013), *Reading in a Participatory Culture: Remixing Moby-Dick for the Literature Classroom* (2013), and *Science Fiction Audiences: Dr. Who, Star Trek and Their Followers* (1995). He is the author of *Convergence Culture: Where Old and New Media Collide* (2008), *The Wow Climax* (2006), *Fans, Gamers, and Bloggers* (2006), *Textual Poachers: Television Fans and Participatory Culture* (1992), and *What Made Pistachio Nuts? Early Sound Comedy and the Vaudeville Aesthetic* (1992). He co-edited *Classical Hollywood Comedy* (1994), *Democracy and New Media* (2003), *Rethinking Media Change: The Aesthetics of Transition* (2003), *Hop on Pop: The Politics and Pleasures of Popular Culture* (2003), *From Barbie to Mortal Kombat: Gender and Computer Games* (1998), and edited *The Children's Culture Reader* (1998).

DEREK JOHNSON is an assistant professor of media and cultural studies at the University of Wisconsin–Madison. He is the author of *Media Franchising: Creative License and Collaboration in the Culture Industries* (2013) and the co-editor of *A Companion to Media Authorship* (2013).

ROBERT V. KOZINETS has authored or co-authored more than 100 research publications, including a textbook and three books: *Consumer Tribes* (2007), *Netnography* (2010), and *Qualitative Consumer and Marketing Research* (2013). An anthropologist by training, he is a professor of marketing at York University and also chair of the Marketing Department.

DENISE MANN is an associate professor and chair, Producers Program, UCLA. She is the author of *Hollywood Independents: The Postwar Talent Takeover* (2008). She co-edited *Private Screenings: Television and the Female Consumer* (1992) and served as associate editor of *Camera Obscura* for six years. She has published essays in the *Journal of Popular Film and Television*, *Camera Obscura*, and the *Quarterly Review of Film and Video*.

KATYNKA Z. MARTÍNEZ is an associate professor of Latina/o studies at San Francisco State University. Her publications include "Pac-Man Meets the Minutemen: Video Games by Los Angeles Latino Youth," in *National Civic Review* (2011); "The Garcia Family," "Sharing Snapshots of Teen Friendship and Love," and "Being More Than 'Just a Banker,'" in *Hanging Out, Messing Around, Geeking Out: Living and Learning with New Media* (2008); "Real Women and Their Curves: Letters to the Editor and a Magazine's Celebration of the 'Latina Body,'" in *Latina/o Communication Studies Today* (2008); "Monolingualism, Biculturalism, and Cable TV: HBO Latino and the Promise of the Multiplex," in *Cable Visions: Television Beyond Broadcasting* (2007); and "American Idols with Caribbean Soul: Cubanidad and the Latin Grammys," in *Latino Studies* (2006).

JULIE LEVIN RUSSO is a faculty member at Evergreen State College. She received her doctorate in modern culture and media at Brown University (2010) and subsequently served as an acting assistant professor of film and media studies at Stanford University. Her forthcoming book, *Indiscrete Media: Queer Economies of Convergence*, was supported by a research associateship at the Five College Women's Studies Research Center at Mount Holyoke College (2012–2013). She has published in *Camera Obscura*, *Cinema Journal*, and many online journals and forums, and she co-edited a special issue of *Transformative Works and Cultures* on Fan/Remix Video (2012). In addition to numerous conference presentations, she has been invited to speak about her research, teaching, and video practice at events from "24/7 2011: The State of the Art in DIY Video" at California College of the Arts to THATCamp Feminisms East at Barnard College (2013).

Index

Page numbers in italics refer to figures. Page numbers in boldface refer to authors of chapters.

A&E, 18, 258
ABC, 4, 11–12, 14–16, 19, 155; "ABC Start Here," 136n2; and *Absolutely Fabulous*, 224, 241n1; and *Flashforward*, 92; and global market, 187, 227; and "The Lost Experience," 4, 12, 119–124, 127–129, 133–135, 136n2; and MyNetwork TV, 224, 232–233, 236–237, 242n32; and neo-platoon shows, 205–208, 211–212, 214–215; and post-network reflexivity, 148, 158n22; and video games, 60. *See also* Disney-ABC; *entries beginning with* ABC; *titles of ABC programs*
ABC.com, 19, 92–93, 118, 122
ABC News Close-up, 158n22
Abercrombie, Nicholas, 85, 88
Abernethy, Jack, 226, 230
Abrams, J. J., 10–13
Absolutely Fabulous, 224, 241n1
Acceptable TV, 149; "TV pilots," 149
Access Hollywood, 143, 237
"The Accountants," 122
action figures, 5, 35, 252–253
Action for Children's Television, 252
actors, 7, 20, 41; and *Ghost Whisperer*, 254; and "The Lost Experience," 121–122, 125; and *The L Word*, 108, 116n17; and MyNetwork TV, 223, 229, 231–234, *233*; and post-network reflexivity, 148; and transmedia, 254
Ad Age (periodical), 127
adaptations: and global market, 188, 226–228; and MyNetwork TV, 226–228, 232–233, 238–239, 241n18; and transmedia, 247; and video games, 54–59, 63, 67–68

Adorno, Theodor, 1, 178
advertising, 2–3, 5–6, 12, 19, 21–23, 26; ad-supported networks, 2, 15, 17–18, 21; and capitalist realism, 201; and consumer-as-fan, 164–165, 167, 169–170, 172; and downloads, 121–122; and FanLib, 78–79, 112; and global market, 177, 181–184; grassroots advertising, 14–15; hard-sell advertising, 141; and Kellogg's, 90–91, 93, 97n81; and "The Lost Experience," 119, 122, 124–134, *131, 132*, 138n46, 139n55; and *The L Word*, 107–108, 110; and MyNetwork TV, 223–224, 226, 229–232, 235–236; and neo-platoon shows, 199; and "Oxo Factor," 92; paying for digital downloads, 121; and post-network reflexivity, 141, 145, 154, 159n27; and propaganda, 169; "subliminal advertising," 130; and Sublymonal.com, 119, 126, 130–133, *131, 132*, 138n46, 139n55; thirty-second ads, 5, 8, 25, 127, 131–133; T-Mobile ad, 72–74; and transmedia, 252, 258–259; and *True Blood*, 259; and user-generated content, 76, 78–80, 90–94; and Videomaker, 80. *See also names of advertising agencies*
affiliates, 6–7, 19, 23, 26
African Americans: and MyNetwork TV, 224, 229–230; and neo-platoon shows, 198, 202–216, *217*, 218n3, 220n20, 220n22, 221n39
After Film School, 155–156
A.I. (film), 262
Air Force (film), 200
Alba, Jessica, 143
Alcatraz, 13, 135

273

Alchemists, 268n50
Alexander, Jesse, 12, 16, 40–41, 134, 261
Ali, Mara Rock, 216
Alias, 11–12
Aliens, 58
Aliens Versus Predator 2, 58
Allen & Co. (media banking firm), 19
All in the Family, 227
alternate reality games. *See* ARGs
amateurs, 4–5; and *Battlestar Galactica*, 34–35, 42; and *Ghost Whisperer*, 257; and licensing, 34–35, 42; and "The Lost Experience," 121, 135; and post-network reflexivity, 147, 149, 156, 158n24; and T-Mobile ad, 73; and transmedia, 257; and user-generated content, 74, 82, 84–85, 92, 110; and Videomaker, 82, 84; and YouTube, 20; and "You Write It," 110. *See also* nonprofessional production
Amazon, 5, 11, 20–22; Amazon Unbox, 20; gift certificates, 78; and user-generated content, 78, 85
AMC, 16, 155
Amen, 205
American Dream, 201, 209, 213
American Horror Story, 16–17, 258
American Idol, 25, 227, 236
American Indians in Film and Television, 220n29
American Latino TV, 233–234; "English Language Novelas," 233
America's Got Talent, 25, 137n17
AMPTP (Alliance of Motion Picture and Television Producers), 98, 119–121
AMT, 36
Anderson, Chris, 17
Andrews, Naveen, 210
Ang, Ien, 115n10
Angel, 11–12, 221n45
Angelo, Carole Panick, 29n33
Angry Birds, 20
Animatrix, 253
anime, 162, 245, 252–253
Anthony, Marc, 231
AOL, 19, 21
Apple, 5, 153, 165, 170; iLife, 153; iPads, 24; iPhone app, 13; iPods, 148, 213, 230, 236; iTunes, 11, 15, 19, 118, 121, 182, 230, 235
Archive of Our Own (Ao3), 96n59, 114–115
Argentina, 264
ARGs (alternate reality games), 3, 8, 12, 24, 26, 245; as "advergaming," 127; and *Dark Knight*, 90, 262; and fourth wall, 125–126, 128–129, 131; and "The Lost Experience," 4, 12, 25, 118–136; and puppet masters, 125, 129; and

suspension of disbelief, 128; and transmedia, 248, 261–262
Arkadie, Kevin, 221n40
Arnold, Mark, 62
Arnold, Richard, 37
Arrested Development, 213
Artisan Entertainment, 128
Asian Americans, 199, 202–203, 206–208, 210, 212–213
Asian Indians, 210–211
Asian Pacific American Media Coalition, 220n29
Askwith, Ivan, 124–125, 129, 131, 137n31
Atari, 54
Atkin, Douglas, 165
AtomFilms, 77–78
AtomicGamer.com, 45
audiences: black audiences, 205; and cable networks, 16–18, 141; and consumer-as-fan, 173; and *Destilando Amor*, 240; eighteen- to forty-nine-year-old, 14, 229, 236–237; eighteen- to thirty-four-year-old, 224–225, 229, 237; eighteen- to twenty-seven-year-old males, 247; engagement of, 245–248, 254–261, *256*; and *Entourage*, 141–143; ethnically diverse, 24; and fan fiction (fanfic), 86–87, 111; and fan filmmaking, 77; and FanLib, 78–79, 111; fourteen- to twenty-four-year-old, 130; and fourth wall, 125; and *Ghost Whisperer*, 254–258, *256*; and global market, 188, 193, 227; highbrow, 228; and Kellogg's, 90; and "The Lost Experience," 120, 122–123, 130–132, 135; and *The L Word*, 99–101, 104, 107–108, 110, 112; mass audiences, 14, 16, 76, 123, 135; and MyNetwork TV, 224–226, 228–230, 232–238, 240; and neo-platoon shows, 197, 199, 203, 205, 210, 213–214; older audience, 16–17, 254, 258; online audience as "hive," 152; and post-network reflexivity, 140–143, 145–149, 152–154, 156, 158n21, 158nn21–24; and producers-as-audiences, 152–154, 158nn21–24; and "quality" demographic, 178, 203, 208; and *Star Wars*, 77; and T-Mobile ad, 73; and transmedia, 244–248, 250, 252–255, 257–262; and *True Blood*, 259–260; twelve- to thirty-four-year-old females, 11; and *Ugly Betty*, 239; unruly audiences, 146–149; and user-generated content, 73, 76–79, 86–87, 91, 94; and video games, 54, 57–60, 66; and Videomaker, 76; younger audience, 17, 205, 254. *See also* consumers; fans/fan labor; niche culture
Australia/Australians, 119, 182, 185, 210
authorship/authority, 24, 32–41, 44, 46–47; and authorial unity, 33; and *Battlestar*

Index • 275

Galactica, 32–35, 39–41, 44, 46–49; and *Beyond the Red Line*, 46; and canonicity, 36–39, 47–48; charismatic, 190; and *Diaspora*, 46–49; and fan fiction (fanfic), 79, 87, 112; and global market, 184–185, 190, 192, 194; and *Heroes*, 40–41; and *The Office*, 179, 190, 192, 194; and platoon films, 201; and post-network reflexivity, 141, 145, 158n16; singular, 33–35; and *Star Trek*, 37–38; and transmedia, 249
Autonomist Marxism, 101–103
autonomy: and fan fiction (fanfic), 79, 86, 112–115; of licensed vendors, 24; and *The L Word*, 99, 102, 108–110, 113–115; and post-network reflexivity, 151; and transmedia, 249, 261
Avatar (film), 7
Avengers (film), 9
awards, 14–15; awards shows, 5; Emmy awards, 141; Golden Globe awards, 211; Webby awards, 15
Azuma, Hiroki, 252

Babylon 5, 11
The Bachelorette, 137n17
Back to You, 122
Bacon-Smith, Camille, 74
Bad Robot production company, 12–13
Bag of Bones, 258
Bakhtin, Mikhail, 63
Ball, Alan, *18*
Barks, Carl, 250
Barney Miller, 198, 203, 219n15
Barnum, P. T., 170
Barrabee, Linda, 159n25
Barrera, Vivian, 238
Barthes, Roland, 172
Basinger, Jeanine, 200–201
Bataan (film), 200–201
Batman: The Brave and the Bold, 92
Battleground (film), 202
Battlestar Galactica, 24–25, 32–49, 90; and action figures, 35; and *Beyond the Red Line*, 44–46, 51n48; canon of, 36–39, 47–48; and comics, 33, 35, 39–40, 51nn32–33; and decentralization, 32–35, 42, 44, 48–49; and *Diaspora*, 35, 44–49, *45*, 51n48; and fans, 25, 35, 43–44; and "Join the Fight," 43–44, 46, 89–91; and licensing, 32–49; as neo-platoon show, 198; and nonprofessional production, 34, 41–44, 46; promotional apparatus for, 24, 45–46, 48–49; and "Resistance" webisode, 41; and *Star Trek*, 34, 36–38, 41, 48; and transmedia, 245; and unauthorized work, 44–49, *45*; and user-generated content, 74; and video games, 24, 33, 35, 39, *39*, 42–49, *45*;

and Videomaker, 43, 74–76, 80–84, *81*, 89, 91–93, 95n24; and webisodes, 17, 41
Battlestar Galactica (1978), 33
Battlestar Galactica: Ghosts, 51n33
Battlestar Galactica: Origins, 40
Battlestar Galactica: Season Zero, 40
Baum, L. Frank, 248–250
Bauman, Zygmunt, 177, 188, 192
BBC, 186–187, 263; BBC2, 141, 149, 179, 182; BBC Wales, 85
Beals, Jennifer, 108, 115n12
"The Beast," 245, 262
Beckingham, Leah, 116n17
Bell, Daniel, 101
Bell, Jeff, 12
Benjamin, Walter, 97n86, 171
Benkler, Yochai, 42, 49
Bennett, James, 24
Benson, Mike, 12, 15–16, 119, 127, 129, 135
BET (Black Entertainment Network), 205
Bewitched, 166
Beyond the Box (Ross), 24
Beyond the Red Line, 44–46, 51n48; as "Mod of the Year" (2007), 44
Bibb, Robert, 11
Bielby, Denise D., 169, 232, 237–238
Big Brother, 25, 176, 181
Bigelow, Gloria, 116n17
Big Love, 213
Bilson, Danny, 35
bisexuals, 25, 104, 108
Black, Jack, 149
Blackboard Jungle (film), 209
The Black Image in the White Mind (Entman), 215–216
Blackshaw, Peter, 102
The Blair Witch Project (film), 8, 128–129, 245, 258
Blanchard, Ken, 163–164, 166
Blast! Entertainment Ltd., 55
Blish, James, 37
blockbusters, 9, 17, 53, 145, 155, 248, 254
BMW, 10
Boeing, 10
Bogost, Ian, 56
Bogusky, Alex, 138n46
Boker, Gabe, 61
The Bold and the Beautiful, 237
Bones, 198
Booz-Allen & Hamilton, 164
Born, Georgina, 149
Born of Hope (film), 73, 75
Boston Public, 197, 208–209, 215
The Bourne Conspiracy (film), 68

Bowlt, J. R., 201
Bowser, Yvette Lee, 221n40
Boyd, Betty, 116n17
brands/branding, 3, 11, 13, 20, 25, 31n69; and affective commitment, 170; and "brand fandom," 163, 166, 170–171; and brand tribes, 166, 170–171; and community superstructure, 170, 172–173; and consumer-as-fan, 163–173; and engagement, 167–172; and *Entourage*, 143; and fourth wall, 125–126; and "The Lost Experience," 119, 122, 125–126, 130; as lovemarks, 165–166; and *The L Word*, 99, 102–103, 106, 111; and MyNetwork TV, 225, 227, 230, 235; and open brands, 172; and post-network reflexivity, 143, 151, 183; and private label brands, 165–166; and storytelling, 170–171, 173; and transmedia, 246, 249, 253, 260, 265; and user-generated content, 73, 76, 90–91, 93; and video games, 60, 68; and "writerly texts," 172. *See also individual brand names*
Braugher, Andre, 208
Bravo, 149, 155
Braxton, Greg, 220n35
Brazil, 223–224, 268n50; and transmedia, 248, 263–264
Break.com, 147, 149
Brecht, Bertolt, 125, 143
Bridget Loves Bernie, 202
broadband, 2, 15, 18, 23; FCC policy statement (2005), 23–24
broadcast industries. *See* television industries
Broderick, Peter, 9
Brook, Vincent, 25, 197–222, 269
Brooker, Charlie, 93
Brooker, Will, 24–25, 72–97, 269
Brown, Stephen, 171
Bruckheimer, Jerry, 140–141
Bruns, Axel, 34, 42, 49
Buckland, Jeff, 45
Buckley, Robert, 233
Buena Vista Games, 55
Buffy the Vampire Slayer, 11, 221n45, 257
Bunuel, Luis, 130
bureaucratic structure, 1–2, 5, 7, 10–11, 14; and *Battlestar Galactica*, 34, 41; and global market, 25, 190–192; and licensing, 34, 36–38, 41; and "The Lost Experience," 119–120, 135; and *The Office*, 25, 190–192; and *Star Trek*, 34, 36–38
Burger King, 119, 126; and "Subservient Chicken," 119, 126, 132
Burnett, Mark, 141
Burns, Caitlin, 9
Busey, Gary, 143

Busse, Kristina, 86–87, 172
Buster Brown, 249

cable networks, 16–18, 20–23; basic cable networks, 18; and e-rentals, 121; and five hundred channels, 20–21; and global market, 180, 182, 186–187; and MyNetwork TV, 224; and neo-platoon shows, 198–199, 205, 213; and policy wars, 23; and post-network reflexivity, 141; premium networks, 16–18, 22, 141; subscription model for, 17–18, 22, 141, 183, 187, 259; and transmedia, 247, 259. *See also names of cable networks*
Cabral, Matt, 62–64
Café con Aroma de Mujer, 240
Cage, Diana, 116n17
Caise, Yule, 29n36
Caldwell, John T., 1–2, 25, 130, **140–160**, 269; and ethno/racial diversity, 206–207, 214; and "industrial self-reflexivity," 7–8
Call of Duty: Modern Warfare 2, 53
Cameron, James, 143
Campbell, Joseph, 248
Campfire, 8–9, 16–17, *18*, 26, 258–261
Canadian Media Fund, 263
canonicity: and authorship/authority, 36–39, 47–48; and *Battlestar Galactica*, 36–39, 47–48; and FanLib, 79; and *Star Trek*, 36–38; and transmedia, 247, 261; and user-generated content, 79; and video games, 47–48, 62
capitalism, 1; and Autonomist Marxism, 101–103; and *Battlestar Galactica*, 43, 49; capitalist realism, 201, 203, 205, 209, 212–213, 215; capitalist surrealism, 213, 215; convergent capitalism, 49, 99, 114; and global market, 188, 191–192; late capitalism, 102–106, 109; and *The L Word*, 99–106, 109–110, 113–114; "mercantile capitalism," 188; and neo-platoon shows, 201, 203, 205, 209, 212–213, 215, 218; and venture capitalists, 5, 77, 84, 88, 94, 112
Caprica, 33, 245
Carell, Steve, 192
Cartoon Network, 92
Casa de Mi Padre, 241n19
Cashmore, Pete, 108
Cast Away (film), 221n38
Castells, Manuel, 180–181, 185, 187–188, 193
Castle, 124
Castle, William, 258
casts/casting: and casting calls, 225, *231*, 232, 234; and "color-blind" casting, 199, 205, 215–216, 218; and *Ghost Whisperer*, 254–255; and *The L Word*, 100, *113*; multiracial/multi-gendered, 25; and MyNetwork TV, 225, *231*, 232, 234,

238; and neo-platoon shows, 199, 204–205, 207, 209, 215–216; and transmedia, 259; and *True Blood*, 259
CBS, 12, 19–20; and global market, 178; and MyNetwork TV, 232, 237; and neo-platoon shows, 205–207, 214–215, 221n52; and post-network reflexivity, 145–146; purge of hayseed comedies (1971), 178; and transmedia, 254. *See also* entries beginning with CBS; *titles of CBS programs*
CBS Audience Network, 19–20
CBS.com, 19
CBS Interactive, 14
celebrities, 5, 9–10, 14; "Celebrity-Sighting" fan chat forum, 143; and *Entourage*, 142–143; and fan fiction (fanfic), 87; and *The L Word*, 100; and MyNetwork TV, 237; and post-network reflexivity, 142–143, 149, 155–156; taking down of, 149; and telenovelas, 237
cellphones, 159n25, 161, 213
Certeau, Michel de, 127–128
CGI animation, 80, 84, 157n1
Chaiken, Ilene, 100, 106, 108–111, 114, 115n5
Channel101.com, 149
Channel 4 (U.K.), 181, 263
Charlie's Angels, 111
Charmed, 221n45
Charren, Peggy, 252
Cheng, Albert, 16, 19, 119–120, 135
Chicanos, 202
Chico and the Man, 203, 214
Un Chien Andalou (film), 130
Children Now, 206
Chin, Staceyann, 116n17
Chinese Americans, 211
Ching, Leo, 188
"The Chloe Chronicles," 12, 14
The Cider House Rules (film), 237
Cinema Journal (periodical), 55, 57–59
City of Angels, 198, 208
Clark, Brian, 261–262
Clarke, M. J., 6–7, 25, 36, 133, **176–196**, 269–270
class/educational privilege, 96n55, 213
Clayton, Damion, 138n46
Clearspring, 19
Clone Wars, 84–85
Cloverfield (film), 13
CNN, 159n25
Coca-Cola, 7, 9, 129, 171; "Happiness Factory," 7, 171
Cohen, Tony, 227
Cold Case, 198
Cold War, 202
collaborations, 3–4, 6–7, 10–11, 13–14, 24;

and *Battlestar Galactica*, 32–35, 40–42, 44, 46–49; and *Beyond the Red Line*, 44–46; and "Dawson's Desktop," 77; and *Diaspora*, 44, 46–49; and fan fiction (fanfic), 79, 81, 86–87; and FanLib, 79, 81; and *Ghost Whisperer*, 257; and global market, 188; and *Heroes*, 40–41; and licensing, 32–36, 38, 40–42, 44, 46–49, 133–136; and "The Lost Experience," 119–120, 122–123, 133–136; and MyNetwork TV, 234, 236; and produsage, 34–35, 42; and *Star Trek*, 38; and transmedia, 257, 260, 262; and user-generated content, 77, 79, 81–82, 86–87; and Videomaker, 82; and wikinomics, 10, 34–35
collectivity, 7, 10, 49; and consumer-as-fan, 172; and *The L Word*, 102, 113; and neo-platoon shows, 214; and post-network reflexivity, 150
Collinson, Phil, 85
Colombia, 182, 228, 233, 239–240
colonialism: and fan fiction (fanfic), 84, 88; and global market, 184; predatory colonialism, 84
Columbia, 264
Columbia Tri-Star, 160n35
Comcast, 19, 22–23, 121
Comeback, 141
Comedy Central, 23, 224
Comic-Con, 258–259
comics, 6, 8; and *Battlestar Galactica*, 33, 35, 39–40, 51nn32–33; and Buster Brown, 249; *Comic Shop News*, 40; and consumer-as-fan, 166; and *Ghost Whisperer*, 257; and *Heroes*, 40–41; and *Kick-Ass*, 92; and licensing, 33–35, 37–41, 51n33; manga, 189; and *The Matrix*, 245; and *Oz*, 249; and *Star Trek*, 38; and *Star Wars*, 84; subscription model for, 253; and transmedia, 249–253, 257; and *24*, 35, 39
computer games. *See* video games
conglomeration, 36–37; "culture of conglomeration," 206; and global market, 183–184; and neo-platoon shows, 197, 206; and post-network reflexivity, 143–145, 149, 154; and transmedia, 244, 249–251
consumers, 1–3, 5, 7, 12, 22–27; and affective commitment, 170; and *Battlestar Galactica*, 33, 42–44; and "brand fandom," 163, 166, 170–171; and community superstructure, 170, 172–173; and consumer-as-fan, 161–173; and consumer behavior, 2–3, 23–25, 162–163; and consumer culture, 154, 162–163, 166, 168–170, 172; and consumerism-as-production, 152, 154–156, *155*; direct-to-consumer marketing, 145; and distributed cognition, 152–153, 158n16, 158nn18–19; engagement of, 26–27, 167–172; and fourth wall, 125–126;

consumers (*continued*)
and *FreeSpace 2*, 44; and hyper-consumerism, 154; and ideological symbologies, 168–170; and licensing, 33, 36, 42–44; and "The Lost Experience," 123, 126–127, 130, 138n39; and *The L Word*, 98, 102–103, 108, 112; and neo-platoon shows, 199; as "networked externalities," 152–153, 158n16; and nonprofessional production, 34, 42; paying for digital downloads, 121; and post-network reflexivity, 144–146, 149, 152, 154, 156; and *Star Trek*, 36, 44; and storytelling, 170–171, 173; and transmedia, 247, 249, 251–254, 260; and user-generated content, 76, 85; and video games, 42–44, 57; and "writerly texts," 172. *See also* audiences; fans/fan labor

contests online, 11, 15, 24; AtomFilms, 77–78; and FanLib, 74, 76–79, 111; "Join the Fight," 43–44, 46, 89; MyEntourage contest, 143; and MyNetwork TV, 225; and prizes, 77–79, 83, 89, 109–110, 132; and Sublymonal.com, 132; and "Ultimate *Lost* Fan Promo Contest," 92–93; Videomaker, 43, 74–76, 80–84, *81*; and "You Write It," 25, 100, 109–115, *113*

Convergence Culture (Jenkins), 24, 253

convergence in mass media: and convergence culture, 234; convergent capitalism, 49, 99; convergent ethnicity, 199–200, 204–206, 208, 213–218, 220n35, 221n48; convergent nationality, 213; and *Diaspora*, 48–49; and fourth wall, 100; and global market, 189; and *The L Word*, 98–100, 102–103, 106, 114; and MyNetwork TV, 234; and neo-platoon shows, 197, 199, 204–206, 208, 213–218, 220n35, 221n48; and *The Office*, 189; and transmedia, 252, 263; and video games, 48–49, 54, 56, 58, 67–68

Convergent Media Track (Canadian Media Fund), 263

Cook, Bob, 229–231

Coppa, Francesca, 74–75

copyright, 5; and *Beyond the Red Line*, 45; copyright infringement, 45; and fan fiction (fanfic), 88; and "The Lost Experience," 132; and *The L Word*, 99; and Oz, 249; and user-generated content, 74; and YouTube, 20–21

Cornell, Paul, 85–86

corporate system, 3, 7, 10, 25; and *Battlestar Galactica*, 34; and consumer-as-fan, 169, 172; and Disney, 250; and global market, 177, 183, 185, 189, 193; and "The Lost Experience," 119, 123–124, 126–128, 132–133; and *The L Word*, 98, 100, 102–103, 108–114; and McDonald's, 6, 133; and moral purpose/altruism, 145; and neo-platoon shows, 206, 218; and *The Office*, 189, 193; and post-Fordism business, 145, 192; and post-network reflexivity, 141, 143–150, 156, 157n6; and transmedia, 250; and transparency, 140, 146–147; and user-generated content, 76, 93–94, 111–114. *See also names of corporations*

The Cosby Show, 198, 205; and the "*Cosby* moment," 205, 220n22

cosmopolitanism, 217

Coupling, 181

Cova, Bernard, 127–128

Crackle, 135

Crazy Taxi, 67

creativity, 6–9, 11–12, 17, 24, 26; and *Alias*, 12; and *Battlestar Galactica*, 32–43, 46–49; and *Beyond the Red Line*, 46; constraints on, 35–36; and consumer-as-fan, 167; and "Dawson's Desktop," 76–77; and *Diaspora*, 46–49; and FanLib, 78, 112; and *Ghost Whisperer*, 254; and global market, 179, 181, 183, 187, 191; and *Heroes*, 40–41; and licensing, 32–43, 46–49; and "The Lost Experience," 118–119, 124–125; and *The L Word*, 102, 109, 114; and neo-platoon shows, 199; and *The Office*, 181, 183, 191; and post-network reflexivity, 144, 152–153; and *Star Trek*, 34, 36–38, 48; and T-Mobile ad, 73; and transmedia, 244, 247, 249, 252, 254, 260–263, 265; and user-generated content, 73–74, 76–78, 82–84; and video games, 46–49, 55, 68; and Videomaker, 74, 82–84

creeps, 161–162; and fan creep, 161–162; and feature creep, 161–162

Criminal Minds, 198

Crispin Porter + Bogusky, 119, 122, 126, 129–133, *131*, *132*, 138n46, 139n55

critics, 16, 22; and consumer-as-fan, 166; and *Entourage*, 141; lay critics, 149; and licensing, 35; and post-network reflexivity, 141, 149, 154–156, 158n24; and transmedia, 248, 252, 257–258, 261; and video games, 56–57, 59–60, 63. *See also* reviews

crowdsourcing, 9, 145, 152

CSI: Crime Scene Investigation, 221n52

CSI: New York: The Game, 55, 61–63, 65, 67

Cuba/Cubans, 228, *231*, 234

The Culting of Brands (Atkin), 165

cult media, 11–12, 98–99; and consumer-as-fan, 165; and "cult branding," 165; and "cult TV," 141; and *Entourage*, 141; and transmedia, 258, 263; and user-generated content, 85, 91–92

Curb Your Enthusiasm, 25, 141, 213

Curtin, Michael, 224

Cuse, Carlton, 4, *4*, 10, 12, 28n27; and "The Lost Experience," 4, 12, 118–124, 126–127, 129, 131, 133–135, 137n16, 137n31, 139n61
CW Television Network, 2, 26, 27n7, 148, 207; and MyNetwork TV, 224–226, 229, 236
CylonJefferson, 89
Czech Republic, 223

Dahlgren, Glen, 38
Dallas, 204
Dancing with the Stars, 25, 187
Daniels, Erin, 115n12
Daniels, Greg, 122, 182–183
Dark Angel, 221n45
The Dark Knight (film), 90, 262; "Why So Serious?," 90, 262
Dates, Janette, 216
DAVE (Digital Animation and Visual Effects) School, 84
Davidson, Drew, 260
Davies, Russell T., 85
Davka (periodical), 202
Dawson-Hollis, James, 138n46
Dawson's Creek, 12, 76–77
"Dawson's Desktop," 12, 14, 76–78, 245
Day Break, 12
Days of Our Lives, 237
DC Comics, 36, 245, 253
DCI (Digital Cinema Initiatives), 158n24
The Decision Makers, 158n22
Deep Space Nine, 38
DeFamer.com, 149
De Kosnik, Abigal, 97n95, 117n34
DeMartino, Nick, 261
democracy, 145, 201–202
demographic. *See* audiences
Dempsey, Patrick, 212
Dena, Christy, 245, 260
Denslow, W. W., 249
derivatives, 3, 7, 16; and "The Lost Experience," 119–120, 123–125
Desilu studio, 36–37
Desire, 226, 228, 230, 232–234, 236
Desperate Housewives, 14–15; free streaming of, 19, 121; and neo-platoon shows, 218n3
Desperate Housewives: The Game, 55, 61–64, 66–67
Destilando Amor, 240
Dewey, John, 191
Dexter, 7, 17, 198, 213
DGA (Directors Guild of America), 121
Dialectic of Enlightenment (Horkheimer and Adorno), 178
Diamond Select Toys, 35

Diaspora, 35, 44–49, *45*, 51n48; "Shattered Armistice," 47
A Different World, 205
Diggs, Taye, 209–210
"digital Darwinism," 167
digital distribution, 3, 17–23; and global market, 176–177, 180, 182–183, 187; and Hulu, 20–22, 121, 124; and "The Lost Experience," 121–122; and *The L Word*, 99; and neo-platoon shows, 199; and *The Office*, 180; and post-network reflexivity, 144–145, 148; and "Resistance" webisode, 41; and transmedia, 247; and "TV everywhere," 20; and YouTube, 21
digital economy, 42, 48, 99
digital laborers, 7, 12–13, 15–16. *See also* workers
Digital Millennium Copyright Act (1998), 21, 24
digital natives, 16–17, 26; hardwired to share content, 26, 31n69; living online, 16, 123; and "The Lost Experience," 123. *See also* fans/fan labor
digital technologies, 1, 3, 5, 11, 13, 19, 22, 24; and "The Lost Experience," 135; and neo-platoon shows, 205; and nonprofessional production, 34, 42; and *The Office*, 193–194; and post-network reflexivity, 140, 146–148, 151; and unruly technologies, 146–148; and video games, 59
Dinner for Five, 155
directors, 7, 31n69, 121, 125, 153, 156. *See also* names of directors
Discovery, 18, 155
Dish Network, 23
Disney, Walt, 250
Disney-ABC, 2, 16, 19, 118–121
Disney Corporation, 7, 9, 19, 112, 206, 253
Disneyland, 250
Disneyland television series, 250
The Division, 198
DIY (do-it-yourself) digital media makers, 7, 9, 121, 135
DIY Days, 245
Doctor Who, 85–86, 263
Dog Eat Dog, 181
domination, 184–185, 190, 192
Donald Duck, 250
Dornfield, Barry, 178
downloads, 24, 121–122; ad-supported, 121–122; consumer-paid, 121; and global market, 177, 193; illegal, 20–21; and post-network reflexivity, 145, 148; and transmedia, 246
Dragnet, 204
Dramatica Pro, 153, 158n19
Dreamwidth, 96n59
Dreamworks, 92

Dr. Phil, 235
Dr. Strangelove (film), 202
"Dunder Mifflin Infinity" (*The Office*), 90–91, 189, 193, 196n59
Dungeons and Dragons, 252
DVDs (digital videos), 7, 11–12, 24; and global market, 187; and "The Lost Experience," 12, 121, 123; and MyNetwork TV, 232; and post-network reflexivity, 143, 145, 154, 158n23, 160n34; and transmedia, 246, 251; and video games, 24, 60; and WGA strike, 121–122
DVRs (digital video recorders), 2, 24–25, 130–132, 246; and fast-forwarding, 130–131
Dynamite Entertainment, 35, 39–40
Dynasty, 204

eBay, 11
Eick, David, 81, 83
electronic press kits (EPKs), 143
Elkington, Trevor, 44–49, 56–57
Ellen DeGeneres Show, 78–79
Ellin, Doug, 141
Elliott, Kamilla, 57
Endemol, 147, 176–177
"The End of Television (As You Know It)" (2005 trade article), 1
Engage! (Solis), 167
Enlightened, 17
Enterprise, 220n22
The Entertainment Economy (Wolf), 164–165
Entertainment Weekly (periodical), 143, 154
Enter the Matrix, 59
Entman, Robert M., 215–216
Entourage, 25, 140–143, *142*, 149, 152, 156; "Celebrity-Sighting" fan chat forum, 143; MyEntourage contest, 143
entrepreneurs, 3, 5–6, 21–22; and consumer-as-fan, 164–165, 169; and "The Lost Experience," 133–136; and McDonald's, 6, 133; and neo-platoon shows, 199; and platoon films, 201–202
Eqal, 126
ER, 198, 208
Erens, Patricia, 201–202
Eskimos, 202
EST (electronic sell-through), 121
ethno/racial diversity, 15, 24–26, 29n36, 197–218; and "accidental minorities," 203; and assimilation, 199; and blaxploitation films, 216; and Canada, 263; and color blindness, 199, 205, 215–216, 218; and convergent ethnicity, 199–200, 204–206, 208, 213–218, 220n35, 221n48; and cultural pluralism, 203; and difference, 199, 202–204, 206, 208, 215–217; 218n3; and ghettoization, 199, 206–207, 214–215, 220n35; and *Hill Street Blues*, 25, 197, 204; and identity politics, 203, 217; and integration in military, 201–202, 220n35; and interracial romance, 25, 197–198, *200*, 208–213, 215–216, 218n3; and Lily White controversy, 199, 205–208; and marginalization, 206; and minority representation, 198–199, 201–203, 206–207, 216; and multiculturalism, 25, 197, 199, 206, 208–212, 214, 216, *217*; and platoon films, 198, 200–203; and postethnicity, 217–218; and post-network reflexivity, 150, 160n33; and public good, 25; and *Star Trek*, 202; and stereotypes, 216–217. *See also* neo-platoon shows
E.T. The Extra-Terrestrial (video game), 54, 56
European Union, 185–187, 263; and Americanization, 186; TVWF ("Television Without Frontiers"), 186
Evans, Emma Jayne, 90, 93
Everybody Hates Chris, 214, 225
exclusivity, 101, 148–150
executive producers. *See* showrunners
expansions, 11–13; and video games, 54–55, 57–58, 68
experiential core, 58–59, 63, 66, 68
experimentation, 3–4, *4*, 9–10, 12–17, 19–20, 26; and "The Lost Experience," 123; and *The L Word*, 99; and Nielsen ratings, 102; and post-network reflexivity, 147; and transmedia, 253–254, 261
extensions, 3; licensed extensions, 36, 40–41; and "The Lost Experience," 125; and transmedia, 247–252, 254, 258, 261, 263; and video games, 57
Extras, 25, 141
Extreme Makeover: Home Edition, 155; *How'd They Do That*, 155

Facebook, 2, 5–6, 10–11, 24; and consumer-as-fan, 167–168, 172; and "The Lost Experience," 123–124, 127, 135; and *The L Word*, 105, 116n17; and user-generated content, 73, 90–93
Fahey, Kevin, 40
Family Guy, 213
Fanboys (film), 162
fan fiction (fanfic), 33, 74, 76–89, 93–94; and Big Name Fans, 87–88; and FanLib, 74, 76–81, 84, 86, 88–89, 93–94, 95n16, 109–112; and gift economy, 79, 86–88, 112–113, 117n34; and "going pro," 76, 79, 84–88, 96n47, 97n67; and slash fiction, 75, 87, 97n67; and "terms of service" contracts, 77–79; and "You Write It," 109–110, 115

Index • 281

FanFiction.net, 79
FanLib: and effeminacy, 78, 112; and fan fiction (fanfic), 74, 76–81, 84, 86, 88–89, 93–94, 95n16, 109–114, *113*; and "Life Without FanLib," 111; and "Naomi," 78; websites of, 78, 116n22; and "You Write It," 109–114, *113*
fans/fan labor, 3–4, 10–12, 16–17, 24–26; and affective commitment, 170; and *The Blair Witch Project*, 258; and "brand fandom," 163, 166, 170–171; broad sense of, 93; and community superstructure, 170, 172–173; and consumer-as-fan, 161–173; and customer service, 163–164; engagement of, 167–172, 259–260; and *Entourage*, 141–143; and evangelism, 163–165, 167; exploitation of, 111; and fanaticus, 164–166; and fan creep, 161–162; and fan fiction (fanfic), 33, 74, 76–89, 93–94, 97n67, 109–114, *113*; and fan filmmaking, 73–75, 77, 80–85, *81*, 89, 92–93, 95n24; and fanisodes, 109–110, 112, *113*; and *Ghost Whisperer*, 254, 257; and "going pro," 76, 79, 84–88, 97n67; and *Hey! Nielsen*, 102; and ideological symbologies, 168–170; and letter-writing campaigns, 11; and "The Lost Experience," 119, 122–123, 125–133, 136; and *The L Word*, 25, 98–100, 102, 106, 108–115, *113*; and mash-ups, 4, 11–12, 92–93, 111, 132, 153, 156; and opinion leaders, 259; and OTW, 114; and post-network reflexivity, 141–143, 145, 147–149, 153, 156, 159n25; "raving fans," 163–166, 173; as spoilers, 149; sports fans, 163–164, 166; and *Star Wars*, 251; and storytelling, 170–171, 173; and Sublymonal.com, 131–133; super-fans, 3, 120, 136; and transmedia, 244–245, 251–254, 257–260, 265; and *True Blood*, 258–260; and video games, 33–35, 41–49, *45*, 54, 57–61, 63–65, 68; and "writerly texts," 172; and "You Write It," 25, 100, 109–111. *See also* user-generated content
The Farm, 115n5
Farmer, Richard, 131, *132*, 138n46
Farmville, 91, 93
Fashion House, 225–226, 228, 231–236, *231*, *233*, 238
Favreau, Jon, 155
FCC (Federal Communications Commission), 23–24, 198; broadband policy statement (2005), 23–24; and WLBT case (1969), 203
Fear Itself, 182
Felicity, 11
Felix the Cat, 249
feminism, 88, 110, 112, 115, 208, 257–258
Fictionalley, 79
Film Festival, 155

Filmnation, 252
films. *See* Hollywood; *titles of films*
Film School, 155, *155*
Final Cut Pro (FCP), 146
Fincher, David, 9–10, 146
Fine brothers, 5
Fiske, John, 169
"500 Reasons to Love *Jericho*," 14
Flashforward, 92, 135; and "Mosaic Collective" website, 92
flash mob, 73
Foer, Franklin, 169
Fontana, D. C., 37
foreign markets. *See* global market for television
format market: and global market, 180–189, *191*; and MyNetwork TV, 226–228, 232, 239
42 Entertainment, 262
forums, 9; "Celebrity-Sighting" fan chat forum, 143; and fan fiction (fanfic), 87; and "Join the Fight," 89; and user-generated content, 80–83, 87, 89, 95n26; and video games, 54; and Videomaker, 80–84, 95n26
Foucault, Michel, 192–193
fourth wall, 100; and "The Lost Experience," 125–126, 128–129, 131
Fourth Wall Studios, 125
Fox Entertainment Group, 26
Fox Interactive, 67
Fox Movie Classics (FMC), 155
Fox Television, 14, 39; and MyNetwork TV, 224–226, 229–230; and neo-platoon shows, 205–208, 214–215, 221n40; and post-network reflexivity, 146, 148, 159n25
France/French, 183, 185, 210
franchises, 2, 4, 6–7, 9–12, 14–16, 23–26; *Battlestar Galactica* franchise, 32–38, 40–49, 90; *CSI* franchise, 198, 221n52; and global market, 25; *Law and Order* franchise, 198; and licensing, 33–35, 40–41, 133; Lonelygirl15.com franchise, 126; *Lost* franchise, 14, 118, 120, 124, 128–129, 133; *The Matrix* franchise, 245; and McDonald's, 6, 133; and neo-platoon shows, 198; and post-network reflexivity, 145; *The Simpsons* franchise, 59–60; *Star Trek* franchise, 34, 36–38, 220n22; and transmedia, 246, 248, 251–252, 257, 261–263; and video games, 35, 42–49, 53–54
Frank, Ted, 15
Frankfurt School, 1, 178
Frank's Place, 205
Freemantle Media, 227
FreeSpace 2, 44, 46–47; and "fredding," 46–47; source code for, 44
The Fresh Prince of Bel-Air, 205

Friends, 181
Fringe, 12–13, 214
Fuselage, 12
Futures of Entertainment, 245
FX, 16–18, 258; "No Box," 17

Gaitán, Fernando, 240
The Game, 216
Game of Thrones, 17, 258
Gamers' Temple (online periodical), 63
Gatiss, Mark, 85–86
gays, 9, 25, 74–75, 108, 211–212, 221n42
GE, 206
Geertz, Clifford, 178–179, 194
gender, 9; and egalitarianism, 208, 211, 221n41; and fan fiction (fanfic), 74–89, 111–112; and fan filmmaking, 73–75, 77, 80–85, 89; and FanLib, 74, 76–81, 84, 111–112; gender binary, 75–76; genderqueer, 75–76; and heteronormativity, 112; and *The L Word*, 107, 110; and MyNetwork TV, 224, 229, 238; and neo-platoon shows, 202–205, 208, 211–212; and patriarchy, 88, 112, 208; and post-network reflexivity, 150; and slash fiction, 12, 74–75, 77–78; and transmedia, 257–258; and user-generated content, 73–90, 95nn24–25, 96n47; and vidding, 75, 95n24; and video games, 59; and Videomaker, 74–75, 80
Gen-X, 131, 133
Gen-Y, 131, 133
The George Lopez Show, 214–215
George Lucas in Love, 85
Gervais, Ricky, 182
"Get Down On It" (pop song), 72
Ghost Whisperer, 3, 12, 221n45, 254–258, *255*, *256*, 260; and FanLib, 77–78; "The Infinity Loop," *256*; "Internet Strategy," *255*; and "the laughing man," 254–255; "The Other Side," 255; thousand montages of, 14
The Ghost Whisperer Spirit Guide (Moses and Sander), 255, 257
Gideon's Crossing, 198, 208
gift economy, 90, 99, 110; and fan fiction (fanfic), 79, 86–88, 112–113, 117n34
G. I. Joe, 251
Gillan, Jennifer, 16, 24
Gilligan's Island, 221n38
Gilmore Girls, 225
Giovagnoli, Max, 260
Girlfriends, 216
GirlwGuns, 82–83
The Girl with the Dragon Tattoo (film), 10
global economy, 25, 164, 166–167, 183–185, 193, 197, 218, 263

Global Hollywood 2 (Miller et al.), 184
global market for television, 25, 176–194; and alternative paradigms, 183; and "cultural discount," 188; and European Union, 185–187; and format market, 180–189, 191; and ideas thingified, 180–181, 183; and MyNetwork TV, 226–227, 230, *231*; and neo-platoon shows, 199, 213; and obsolescence, 188–194; and *The Office*, 25, 176–183, *180*, 188–194; and outsourcing, 182, 187, 192–193; and Reveille, 177–178, 181–183, 194; and *The Simpsons*, 176–177, 183, 194; and speed/fluidity, 177–179, 182–183, 189–190, 192, 194; statistics of, 185–186; and surveillance, 179, *180*, 192–194; and transmedia, 263–265; and United Kingdom, 179–189; and waste/waste disposal, 177–179, 183, 194
GMD Studios, 261
Godard, Jean-Luc, 143
The Godfather (film franchise), 61
Goldberg, Whoopi, 218n3
Goldeneye 007, 53
GoldPocket Interactive, 158n24
Goldstein, Hilary, 60
Goldstein, Lew, 11
Gomez, Jeff, 7–9, *8*, 17, 135, 171, 248, 261
Good Times, 203
Google, 5, 11, 19, 21, 23
Google-YouTube, 19, 21
Gordon, Bruce, 214
Gordon, Ian, 249
Gossip Girl Second Life (*GGSL*), 59
governmental policies, 23–24; and Canada, 263; and ethno/racial diversity, 198, 202, 205; Truman's Executive Order No. 9981 (1948), 202
Grand Theft Auto, 55; *Grand Theft Auto III*, 63, 66–67
graphic novels, 7–8, 15, 29n34, 253
grassroots level, 14–15, 26, 33–35; and consumer-as-fan, 172; and "The Lost Experience," 123; and transmedia, 252, 259; and user-generated content, 73, 78, 85, 93, 111–113
Gray, Herman, 205
Gray, Jonathan, 24, **53–71**, 133, 270
Great Depression, 247
Greenaway, Peter, 263
Green Hornet, 249
Grey's Anatomy, 14, 25–26, 121–122; free streaming of, 19, 121–122; and MyNetwork TV, 236; as neo-platoon show, 198, 208, 211–212, 216, *217*, 221n41
Griffiths, Nick, 85
Grillo-Marxuach, Javier, 119, 124, 126–128, 130, 134, 137n31; as DJ Dan, 126

"Grimm Webisodes," 29n34
Guerrero, Claudine, 232, *233*
Gulf + Western, 37
Guzman, Pato, 36

Hager, Chris, 45
Hailey, Leisha, 104, 108
Hall, Haines, 138n46
Halo II, 262
"The Happiness Factory" (Coca-Cola), 7, 171
Harley-Davidson, 164–165, 170–171; and Sturgis (S.D.), 164–165
HarperCollins Publishers, 111; HarperCollins e-books, 78
Harrington, C. Lee, 169, 232, 237
Harris, Andrew, 35, 38–39
Harvey, David, 101
Hasbert, Dennis, 218n3
Hasbro, 7, 9
Haxans, 258
Hayek, Salma, 239, 243n53
Haynes, Jeff, 62
HBO, 16–18, *18*, 155; and "Celebrity-Sighting" fan chat forum, 143; and *Curb Your Enthusiasm*, 25, 141; and *Entourage*, 25, 140–143, *142*, 152, 156; HBO-Go, 17; "It's Not TV, It's HBO," 16–17; MyEntourage contest, 143; and transmedia, 258–259
Heche, Ann, 218n3
Hegeman, John, 128
Hellekson, Karen, 77, 79, 86–88, 112, 172; and gift economy, 79, 86–88, 112
He-Man, Masters of the Universe, 251–252
Henderson, Felicia, 214–216, 221n40, 222n63
Heroes, 3, 12, 14–16, 29n34, 29n36, 125, 134; "Heroes 360," 15; and licensing, 40–41; as neo-platoon show, 198–199, *200*, 208, 210–211, 213, 215–216, 221n39, 221n45; parody of, 73; and post-network reflexivity, 148; and transmedia, 260–262
Hewitt, Jennifer Love, 254
Hey! Nielsen, 102, 116n15; and Q Ratings, 102
Hills, Matt, 85
Hill Street Blues, 25, 198, 204, 208, 220n20
Hirsch, Paul, 227
Hispanics, 26, 199, 208–210, 212, 214, 230
Hit and Run. See *The Simpsons Hit and Run Game*
Hobson, Karen, 121
Hollinger, David, 217
Holloman, Laurel, 115n12
Hollywood, 3–10, 13, 19–21, 26, 121; "Alumni in Hollywood," 154; and children's programming, 251–253; and *Entourage*, 25, 140–143, *142*, 152, 156; and erosion of job classifications, 144–148; and Great Depression, 247; and licensing, 39; and "The Lost Experience," 131; and platoon films, 201; and post-network reflexivity, 141–156, 158n21, 160n33; and transmedia, 246–248, 251–254, 260, 262–263, 265; and video games, 53, 55, 68. See also studios; *titles of films*
Holt, Jennifer, 23
Holzhauer, Grant, 65
Homicide: Life on the Street, 198, 208, 246
homoeroticism, 74–75
homophobia, 64, 211
Hope, Ted, 9
Horkheimer, Max, 1, 178
horror films, 58, 258
Horta, Silvio, 243n53
Hotel City, 91
hot-rodding, 85
Hot Wheels (Mattel), 7
How to Train Your Dragon (film), 92
The Hulk (film), 160n31
Hulu, 20–22, 121, 124, 135; Hulu Plus, 22
Hurley, Chad, 28n27, 134
Hutcheon, Linda, 57, 59
Huyssen, Andreas, 218
Hyperion Press, 138n40

IATSE (International Alliance of Theatrical Stage Employees), 147–148
IBM, 10
Iceland, 90–91, 93
ideological symbologies, 168–170
IDW Comics, 35, 38–39
IFC, 155
Iger, Robert, 19
"I Love Bees," 245, 262
"Immaterial Labor" (Lazzarato), 101–102
Indiana Jones and the Last Crusade (film), 251
In Living Color, 205
innovations, 10–12, 15–17; and *Battlestar Galactica*, 41; and consumer-as-fan, 161, 167, 172; and global market, 25, 178, 183, 192–193; and licensing, 35, 41; and "The Lost Experience," 118–119, 127; and *The L Word*, 102, 108; and MyNetwork TV, 224, 226; and neo-platoon shows, 204; and *The Office*, 25, 192–193; and post-network reflexivity, 140, 144, 146; and transmedia, 246–247; and video games, 56
intellectual property, 6–7; and *Battlestar Galactica*, 32–35, 40–43, 49; and global market, 184, 186; and licensing, 32–35, 40–43, 49;

intellectual property (*continued*)
and *The L Word*, 102; and post-network reflexivity, 152; and transmedia, 251, 253; and video games, 42–43, 49, 53
interactivity, 7, 11–14, 17; and consumer-as-fan, 172; and *Entourage*, 141; and *Ghost Whisperer*, 255, 257; and "The Lost Experience," 25, 119, 126, 131–132, 134; and *The L Word*, 100, 105–106, 111; and MyNetwork TV, 234–235; and post-network reflexivity, 141, 144, 152; and Sho.com, 114; and Sublymonal.com, 119, 131–132; and "Subservient Chicken," 126; and transmedia, 26, 247, 251–252, 255, 257, 260–262; and video games, 57, 59, 63
International Fight League, 237
international market. *See* global market for television
Internet, 1, 7, 10–11, 23; and *Battlestar Galactica*, 34; and consumer-as-fan, 164, 166; and global market, 183; and "The Lost Experience," 118–119; and *The L Word*, 98, 100, 105–108; and Napster, 19; and neo-platoon shows, 211, 221n48; and *The Office*, 182–183; and post-network reflexivity, 144, 147, 149, 154; and user-generated content, 74, 90; and WGA strike, 122. *See also names of online sites*
Isbin, Sharon, 101
I Spy, 220n20

Jackson, Michael, 181
Jackson, Peter, 56, 68
Jacobs, Lewis, 201
James, William, 191
Jameson, Fredric, 101
Japanese, 189, 210, 216, 252–253
Jealous, Benjamin Todd, 207
Jeep, 124, 126–127
The Jeffersons, 203
Jenkins, Henry, 24, 26, **244–268**, 253, 270; and *American Idol*, 236; and convergence culture, 234; and fan culture, 138n39, 162, 172; and fan fiction (fanfic), 74, 77, 111; and fan filmmaking, 80, 85; and MyNetwork TV, 234; and "spreadable media," 123, 137n18; and video games, 58–59
Jerwa, Brandon, 33, 39–40, 51n33
Jews, 202–205, 208–209, 219n15
Jimenez, Angela, 116n17
The Jimmy Kimmel Show, 123, 128–129
"Jinky Williams," 130–131
Johnson, Derek, 6–7, 24, **32–52**, 56, 80, 89, 133, 251, 270
"Join the Fight," 43–44, 46, 89–91; and "terms of service" contracts, 43–44, 46

Jones, Chris, *132*
Joost, 19
Jowell, Tessa, 185

Kaliban, Karel, *132*
Kath & Kim, 15, 182
Kazmi, Hassan "Karajorma," 46–48, 51n48
Kellogg's, 90–91, 93, 97n81
Kevin Hill, 197–198, 208–210, 215
Kick-Ass (film), 92–93
Kickstarter, 9, 135
Kilar, Jason, 21–22
Kilmer, Val, 143
Kinder, Marsha, 244–245, 251–253
King Kong (DVD boxed set), 158n23
Kinsella, Sharon, 189
Kirshner, Mia, 115n12
Klepek, Patrick, 60–61
Knight, T. R., 211
Kompare, Derek, 33
Koreans, 210
Korean War, 202
Kozinets, Robert V., 25, 127–128, **161–175**, 270
Krave cereal, 90–91, 93
Kress, Gunther, 245
Kring, Tim, 10, 12, 15, 199
Krinsky, Tamara, 119, 123–124
Krzywinska, Tanya, 58
Kurzman, Alex, 12
KXMN (Spokane, Wash.), 230–231

labor, 1, 3; and Autonomist Marxism, 101–103; cheap labor, 184; division of, 178, 180–181, 183–184, 187, 189; and global market, 177–178, 180–181, 183–184, 187, 189, 192–193; immaterial labor, 101–103, 106; and "The Lost Experience," 118, 133; low-level digital, 5, 7, 12–13, 15–16, 122; and *The L Word*, 25, 98–102, 106–109, 114, 115n12; and "NBC 2.0" agenda of layoffs, 177; NICL (New International Division of Cultural Labor), 184, 187; and *The Office*, 180–181, 183, 189, 192–193; and post-network reflexivity, 144–150, 157n6; runaway labor, 184; and sweatshops, 5, 121, 184. *See also* fans/fan labor; workers
labor guilds, 7, 19, 26, 120–123; and post-network reflexivity, 147–148, 150–151, 154; rituals of, 145, 150–152; and "scientific management," 150. *See also names of labor guilds*
L.A. Law, 100
Lash, Scott, 180, 183
Latin America, 224–226, 228, 237–238. *See also names of Latin American countries*
Latinas/os: and "Latin pipeline," 230–234; and

MyNetwork TV, 224–225, 229–234, 237–238, 241n28, 243n53; and neo-platoon shows, 204, 206–207, 209, 211–212, 216, 218n3, 221n39, 221n52; second-generation, 232; and Spanglish, 232
Latino 96.3, 232–233, *233*; and "hurban" format, 232, 242n29
Laurel, Brenda, 246
Law and Order, 198
lawsuits, 67, 78, 145, 147
"Lazy Sunday" rap-parody, 20
Lazzarato, Maurizio, 101–102
Lee, Garnett, 62–63, 66
Legacy, 55
lesbians, 25, 99–112, 116n15; commodity lesbianism, 102, 107–112; and first television kiss, 100; and ghettoization, 103; identity of, 99, 101, 103–104, 108, 110; and neo-platoon shows, 212; and OurChart.com, 25, 100, 104–109, *105*, 116n21; representation of, 100–101, 106–110; and "You Write It," 109–112. See also *The L Word*
Leung, Ken, 210
Levitan, Steve, 122
Levitz, Paul, 36
Levy, Sidney J., 162–163
Lewis, C. S., 248
licensing, 5–7, 10–11, 24, 32–49, 53–68; and approval, 33, 35–36, 38–39; and *Battlestar Galactica*, 32–49, 51n33; and *Beyond the Red Line*, 46; and constraints, 35–36; and copyright infringement, 45; creative tradition of, 35–41; and *Diaspora*, 35, 44–49; and Disney, 250; and *E.T. The Extra-Terrestrial* (video game), 56; and global market, 176, 181–182, 187; and *Heroes*, 40–41; licensed extensions, 36, 40–41; licensed vendors, 6–7, 24, 133; and "The Lost Experience," 120, 122, 133–136; management of, 34, 40–41; marginalization of creativity, 38, 49; and neo-platoon shows, 198, 203; and nonprofessional production, 34, 41–44, 46; and *The Office*, 181; and post-network reflexivity, 147; public input on licensing decisions, 203; and *Star Trek*, 34, 36–38, 44, 48; and subsidiaries, 36–37; and "terms of service" contracts, 43–44, 46; and transmedia, 246, 248–250, 253, 265; and unauthorized work, 44–49; and video games, 24, 33, 35, 38–39, 42–49, 53–68, 133
Life on Mars, 221n45
"Life Without FanLib," 111
Lindelof, Damon, 4, *4*, 10, 12; and "The Lost Experience," 4, 12, 119–122, 124, 127, 129, 131, 133–135, 137n31

Linux, 172
Liquid Entertainment, 55
"liquid modernity," 177, 188, 192
Little Britain: The Video Game, 55, 61, 63–64, 66–67
Live365.com, 147
LiveJournal, 79, 81, 86–89, 96n59, 111, 113; "fanarchive"/"otw_news," 113
Living Single, 221n40
Lizer, Kari, 215
Loeb, Jeph, 12, 16
London, 72, 181–182, 188; Liverpool Street station, 72–74
The Lonely Crowd (Reisman), 1, 190
Lonely Girl15, 20, 126
Lone Ranger, 249–250
Longhurst, Brian, 85, 88
long-tail effect, 11, 17, 21, 166
Longworth, James, 33
Lopez, George, 214–215
Lopez, Jennifer, 231
Lopez, Miguel, 65
Lord of the Flies (Golding), 221n38
Lord of the Rings (film franchise), 73
Lord of the Rings (video game), 56, 68
Los Angeles: and *The L Word*, 106; and MyNetwork TV, 231–232, *231*, 238; and post-network reflexivity, 149–150, 154, 156, 158n23; "Silicon Beach," 6
Lost, 3–5, *4*, 11–16, 51n32, 98, 136n2; "The Dharma Initiative," 12, 123; free streaming of, 19, 121; interracial romance in, 25; "The Lost Diaries," 4, 120, 134; "The Lost Parodies," 5; "Lost Untangled," 4–5, 12, 134; and mobisodes, 118–119; as neo-platoon show, 198–199, 208, 210, 213, 215–216, 221n38; Oceanic Airlines, 14, 138n40; and transmedia, 260–261; and "Ultimate *Lost* Fan Promo Contest," 92–93; and user-generated content, 92–93; and video games, 54, 60–67, *64*, *65*; as watercooler event, 123. See also entries beginning with *Lost*
"The Lost Experience," 4, 12, 25, 118–136; *The Bad Twin* novel in, 128–129, 138n40; and Crispin Porter + Bogusky, 119, 122, 126, 129–133, *131*, *132*; and fourth wall, 125–126, 128–129, 131; Hanso Foundation in, 125–126, 128–130; Hugh McIntyre in, 128–129; and immersion, 125–129; and licensing, 122, 130, 133–136; and marketing, 4, 12, 119–120, 122–124, 127–129; and mass collaboration, 122–123; and publicity stunt, 128–129; Rachel Blake in, 125–126, 128; sponsors for, 119, 122–128, 130; and "spreadable media," 123;

"The Lost Experience" (*continued*)
and transmedia, 262; and WGA strike, 119–125, 133; white paper on, 124–125; and writers, *4*, 12, 25, 119–126, 129–131, 133–135, 139n61
Lostpedia, 91
Lost: The Final Season, 92–93
Lost: Via Domus, 54–55, 61–67, *64*, *65*
Lotz, Amanda, 24
lovemarks, 165–166
Lovemarks (Roberts), 165
Lucas, George, 85, 248, 251
Lucas, John Meredyth, 37
Lucas, Matt, 61
Lucasfilm, 77, 84
Lucky Louis, 213
Luddites, 193
Luis, 218n3
Lury, Celia, 180, 183
The L Word, 25, 98–117; and Alice Pieszecki, 104–107, *105*, 111, 116n21; and Autonomist Marxism, 101–103; and costume design, 101; and decentralization, 99; and fanisode e-zine, 109–110, 112, *113*; and FanLib, 77–79, 109–115, *113*; and lesbian identity, 99, 101, 103–104, 108, 110; and lesbian representation, 100–101, 106–110; and "Lez Girls," 110–111; as neo-platoon show, 198; and OurChart.com, 25, 100, 104–109, *105*, 111, 114–115, 116nn15–17, 116n21; and Sho.com, 114; and special created for finale, 100–101, 115n5; and unauthorized work, 99; and "You Write It," 25, 100, 109–115, *113*

MacBook Pro, 146
MacDermid, Susan, 102
Mac iBook, 236
Madison, Kate, 73, 75
Madison Avenue, 8, 13, 21, 131, 265. *See also* advertising
Mad Men, 73, 166
Mafia Wars, 91
Magrs, Paul, 85–86
Making of the Band, 155
Mancuso, Frank, 37
manga, 189, 210; "adult manga," 189
Mann, Denise, **1–31**, 25, **118–139**, 270–271
Mapplebeck, Linda, 91
marketing, 3–19, *18*, 23–27, 31n69, 33; and cable networks, 17–18; and consumer-as-fan, 161–170, 172–173; direct-to-consumer marketing, 145; and *Entourage*, 141, 143; and feature creep, 161; and fourth wall, 125–126, 131; and guerrilla marketing, 230; and licensing, 40;

and "The Lost Experience," 4, 12, 119–120, 122–129, 135; and *The L Word*, 98, 100, 104, 107–111, 114–115, 116n15; and MyNetwork TV, 225, 230–234, 236, 238; and post-network reflexivity, 140–141, 143–146, 148, 152, 156, 160n35; and transmedia, 246–248, 252–253, 258, 260; and user-generated content, 83, 90, 111; and Videomaker, 83; viral marketing, 13, 73, 137n18, 141–143, *142*, 145, 147, 156
Married with Children, 213
Marshall, P. David, 246
Martínez, Katynka Z., 25–26, **223–243**, 271
Martinez, Natalie, 231–232, 234
Marvel Comics, 9, 38, 245, 253
Marxism, Autonomist, 101–103
Marxist theory, 115, 191–192
The Mary Tyler Moore Show, 201, 203
M*A*S*H, 25
Mashable.com, 108, 168
mash-ups, 4, 11–12, 24, 92–93, 111, 132, 153, 156
"Master and Apprentice," 75
Masters of the Universe, 251–252
Mastertronic, 55
Mastrapa, Gus, 62
The Matrix (film franchise), 59; *Matrix Reloaded*, 160n31; and transmedia, 245–246, 253–254, 262
Mattel, 7
Maybelline mascara, 235
Mayer, Vicki, 238
McDonald's, 6, 73, 133; "Avatarise Yourself," 73
McGonigal, Jane, 125
McPherson, Steve, 16
McRobbie, Angela, 157n15
McTelevision, 188, 226–228, 238
Media Franchising (Johnson), 6–7, 133
media monitoring groups, 198–200, 203, 205–208, 214–215, 217, 222n63; and Lily White controversy, 199, 205–208; and report cards, 207, 214–215; and WLBT case (1969), 203
Medium, 198, 221n45
Men in Trees, 218n3
mentors/mentoring, 9, 12, 110, 143–144, 150
merchandising, 36, 40, 56, 69, 249; and "The Lost Experience," 120, 123, 134; and post-network reflexivity, 145, 154, 158n23; and transmedia, 249; and window displays, 249. *See also types of merchandising*
Merchant, Stephen, 182
Merchant-Ivory period pieces, 188
Mesa para Tres, 228, 233
Messmer, Otto, 249
Metacritic.com, 54, 154
Mexico/Mexicans, 238–240

Mfume, Kweisi, 206
MGM, 248
Miami Vice, 198, 204, 208
The Mickey Mouse Club, 250, 257
Microsoft, 5, 19, 21, 262; Powerpoint, 194
military, U.S., 7; "don't ask don't tell," 100–101; integration in, 201, 220n35; and Office of War directives, 201; and World War II combat films, 198, 200–203
millennials, 5, 14, 128
Miller, Jeffrey, 227
mobisodes, 118–120, 134, 159n25. *See also* webisodes
Mod Database, 44
mods, 44–49, *45*
The Mod Squad, 220n20
Moennig, Katherine, 108, 115n12
Moesha, 214
Moffat, Steven, 85
Moise, Lenelle, 116n17
Monello, Michael, 259
monetization, 5, 11; and fan fiction (fanfic), 78, 86, 88, 94, 111–112; and "The Lost Experience," 123; and *The L Word*, 104, 114; and transmedia, 262
Monster.com, 127
Montfort, Nick, 55–56
Monty Python, 188
Moonlighting, 204
Moore, Anne Elizabeth, 166
Moore, Candace, 116n15
Moore, Chris, 138n46
Moore, Ronald D., 33–34, 36, 38, 41, 47, 49
Moses, Kim, 12, 254–258, *255*, *256*
Mota, Barbara, 268n50
Mota, Mauricio, 268n50
mothership model, 26, 246–248, 255, 261–262, 265
movies. *See* Hollywood; *titles of movies*
MTV, 23, 155
multiculturalism, 25, 197, 199, 206, 208–212, 214, 216, *217*
multitasking, 20, 24, 145, 154, 159n27
Muppet Babies, 244
Murray, Janet, 246
Murray, Simone, 36
music, 11; and ideological symbologies, 168; music industry, 19; music swapping, 18–19; and MyNetwork TV, 225, 236, 238–239
Mussolini, Benito, 191
MyDamnChannel, 135
MyNetwork 13 (Los Angeles), 231
MyNetwork TV, 223–240; and adaptations, 226–228; bloggers for, 230, 234; and casting calls, 225, *231*, 232; and "confusion factor," 229–231; control over airtime, 225–226; cutting edge claim of, 225–226; and *Desire*, 226, 228, 230, 232–234, 236; and economies of scale, 226; and *Fashion House*, 225–226, 228, 231–236, *231*, *233*, 238; and geographic locales, 225, 238, 240; and guerrilla marketing, 230, 234; and high-definition format, 229; and "Latin pipeline," 230–234; and limited-run dramatic series, 223–224, 226, 237; and local media blitzes, 230, 232; and McTelevision, 226–228, 238; and neo-network era, 224, 239; promotions for, 223–224, 228–232, 234, 236; repeats not aired on, 224, 229–230; and second-run movies, 237; and *Secrets*, 230; six nights a week, 229; and story arc, 224, 226, 228, 234, 237; and swipes, 235; and telenovelas, 26, 224–226, 228–240, 241n19; and translations, 228, 241n19; and *Ugly Betty*, 233, 239–240, 242n32, 243n53
mynetworktv.com, 234–236, 238–240; and band competitions, 236; "Confessionals," 235, 240; "Inside the Mind," 235, 240
MySpace.com: and fan fiction (fanfic), 110; and "The Lost Experience," 127; and MyNetwork TV, 225, 230, 236; and neo-platoon shows, 213; and post-network reflexivity, 143, 147, 156
Myst, 127, 138n34, 221n38

NAACP (National Association for the Advancement of Colored People), 206–207, 214–215, 220n29, 220n35, 221n52, 222n63; NAACP/NBC Fellowship in Screenwriting, 29n36
Napster, 18–19
Native Americans, 206–208
NBC/NBC-Universal, 11–16, 20, 22, 29nn33–34, 73; and *Battlestar Galactica*, 34–35, 42–45, 48, 51n34; and global market, 177, 181–183; and Hulu, 20–22, 121; and licensing, 34–35, 40, 42–46, 48, 51n34; and MyNetwork TV, 232–233, 236–237; "NBC 2.0" agenda, 177; and neo-platoon shows, 199, 205–208, 214–215; and *The Office*, 89–90, 179–183, *180*; and post-network reflexivity, 146–149, 159n25; Sci-Fi Network, 42–43, 80–83, 89, 95n26; and *Studio 60*, 141; and TV "360," 15. *See also entries beginning with* NBC; *titles of NBC programs*
NBC-Reveille, 177–178, 181, 183, 194
NBC-Universal Digital Entertainment and New Media, 15, 19
NCIS, 198
Ndalianis, Angela, 248

Negri, Antonio, 101–102
neo-platoon shows, 197–218, 218n3, 219n15, 220n20; *Boston Public* as, 197, 208–209, 215; and capitalist realism, 201, 203, 205, 209, 212–213; and *The Cosby Show*, 198, 205; and egalitarianism, 208–210, 213, *217*; and gender, 202–205, 208, 211–212; *Grey's Anatomy* as, 198, 208, 211–212, 216, *217*, 221n41; *Heroes* as, 198–199, *200*, 208, 210–211, 213, 215–216, 221n39, 221n45; and *Hill Street Blues*, 25, 197, 204, 208, 220n20; and interracial romance, 25, 197–198, *200*, 208–213, 215–216, 218n3; *Kevin Hill* as, 197–198, 208–210, 215; and Lily White controversy, 199, 205–208; *Lost* as, 198–199, 208, 210, 213, 215–216, 221n38; and media monitoring groups, 198–200, 203, 205–208, 214–215, 217, 222n63; and platoon films, 198, 200–203; proto neo-platoon shows, 198, 204–208, 220n22; *Six Degrees* as, 212–213, 215; and *St. Elsewhere*, 25, 204–205, 208; *Ugly Betty* as, 25, 198, 208, 212, 214–215, 221n52. *See also* ethno/racial diversity
Netflix, 5, 11, 20–22, 121; and e-rentals, 121
Netherlands, 176–177
networks, 1–7, 10–18, 20–27; and analog past, 14, 22, 135–136; and digital distribution, 18–23, 121–122; and gatekeepers, 2–5, 187; and global market, 25, 176–179, 182–184, 187, 194; and "The Lost Experience," 118–120, 123, 133–136; and neo-platoon shows, 197–199, 205–207, 213–215, 222n63; and policy wars, 23; and post-network reflexivity, 141, 143, 147–156; Spanish-language, 224–225, 228–234, 237, 239; and transmedia, 244, 246, 249, 253, 257, 262; weak-sister networks, 205. *See also* television industries; *names of networks*
The New Adventures of Old Christine, 215
Newcomb, Horace, 227
New Economy, 189, 191–194
New International Division of Labor, 184
Newman, Kyle, 162
Newman, Michael, 33
new media, 2, 24, 29n36, 41; and Kellogg's, 97n81; and "The Lost Experience," 120–121, 123–125, 135; and *The L Word*, 114; and MyNetwork TV, 236, 240; and neo-platoon shows, 205; and WGA strike, 120–121, 123–124
News Corp., 147, 206, 225, 230
New York, 6, 8, 21, 41, 154, 212–213; and Brooklyn Bridge, 240; Puerto Rican Day Parade, 232; and *Ugly Betty*, 240
New York Times, 13, 116n17, 123
New York Undercover, 198, 208, 221n40

Next Action Hero, 155
Next Action Star, 141
NHMC (National Hispanic Media Coalition), 220n29
Nice Work if You Can Get It (Ross), 5
niche culture: and bicultural Hispanics, 25–26; and cable networks, 17, 198; and consumer-as-fan, 172; and *Entourage*, 141; and fan filmmaking, 92; and *The L Word*, 101, 107; and neo-platoon shows, 198; and post-network reflexivity, 141, 145, 159n27; and science fiction, 98; and Videomaker, 91; and The WB, 11
Nicholson, Neville, 59
Nickelodeon, 23
NICL (New International Division of Cultural Labor), 184, 187
Nielsen ratings, 102, 115n10, 116n15, 193, 236
Nigerians, 210
Nightline, 212
90210, 125
1984 (Orwell), 167
Ning, 19
Nintendo 64, 53
9th Wonders, 12
Nissan, 15
Nixon, Richard, 227
NLRB (National Labor Relations Board), 148
No-Collar (Ross), 5
No Hay Paraiso, 182
nonprofessional production, 26; and licensing, 34, 41–44, 46, 48
Novák, Franta, *132*
novelizations, 6, 10, 41, 123
Nussbaum, Joe, 85
NYPD Blue, 198, 208

Obama, Barack, 207, 215
obsolescence, 184, 188–194
The Office, 3, 15, 89–90, 122; "Back from Vacation," 194; "Basketball," 190; "Beach Games," 190; "Christmas Party," 190; "Diversity Day," 190; and "documentary cameraperson," 191–193; and Dunder Mifflin, 90–91, 189, 193, 196n59; "The Fight," 190; and gift economy, 90; and global market, 176–183, *180*, 188–194; "Health Care," 192; "Launch Party," 190; "Money," 194; and MyNetwork TV, 236; as parody, 191, 193; as "show of the future," 182; and Social Ethic, 190–191; and surveillance, 179, *180*, 193–194; and upward mobility, 191
"The Office Is Closed," 122
Ohlmeyer, Don, 214
Oldenberg, Ann, 215

Oliver & Ohlbaum Associates, 187–188
Olsson, Ian, 24
"The Only Way Is Up" (pop song), 72
On the Lot, 141, 152, 156
open-source architectures, 35, 42, 44, 46, 171–172
Orci, Roberto, 12
The Organization Man (Whyte), 1, 190–191
Orry, Tom, 62–63
Orwell, George, 167
OTW (Organization for Transformative Works), 113–114
OurChart.com, 25, 100, 104–109, *105*, 111, 114–115, 116nn15–17, 116n21; and "friends plus," 116n16
outsiders, 2–3, 5–7, 9–10, 17–18, 21; and *Entourage*, 142–143; and fan fiction (fanfic), 79, 87; and FanLib, 79, 111; and "The Lost Experience," 134, 136; and MyNetwork TV, 233–234; and post-network reflexivity, 142–143, 149–150; and transmedia, 26, 247
outsourcing, 144–146; and global market, 182, 187, 192–193
"Oxo Factor," 92
Oz, 198

Pacific Island Americans, 206–207
Pacino, Al, 61
Paramount, 37–38; Merchandising and Licensing, 38
Parnian, Parisa, 116n17
participatory culture, 14, 99, 113, 247, 250, 257–258, 262
Passions, 233, 237
PBS, 178, 187
PC software applications, 53, 138n34, 153, 159n25
Pedowitz, Mark, 2, 27n7, 120
Pensavalle, Mark S., 9
Perez, Patrick, 228, 234, 241n19
Perlstein, Ed, 37
Perplex City, 262
PGA (Producers Guild of America), 7, 134–136, 261; and transmedia producers, 7, 135–136, 139n64, 261
PGUs (producer-generated users), 152, 154–156, *155*, 160nn31–35
Phenomenon, 182
Phoenix (Ariz.), 230–231
Pinkner, Jeff, 12, 134, 139n61
Piñon, Juan, 243n53
Pirates of the Caribbean (film franchise), 9
plagiarism, 66, 78
platoon films, 198, 200–203, 212; and integration in military, 201–202; and Office of War directives, 201; and propaganda, 201–202; resurgence of, 202
Playing with Power (Kinder), 244–245, 251–253
Pleasantville (film), 209
poaching, 11, 77–78, 91, 94, 177, 197
Political Animals, 258
Popular Science (periodical), 44
Port Charles, 237
post-industrial economy, 99, 102–103, 203
postmodernism: and consumer-as-fan, 162, 166, 173; and "The Lost Experience," 128, 130; and neo-platoon shows, 197–198, 204, 213, 218; and Sublymonal.com, 130
post-network industry, 1–3, 24, 26; and erosion of job classifications, 144–148; and *Lost*, 118; and neo-platoon shows, 198, 205–206, 213–215, 218; and reflexivity, 8–9, 140–156
Powell, Jim, 60
Powell, John, 62–63
Power to the Pixel, 245
The Practice, 227
The Price Is Right, 227
Primetime TV, 158n22
Prinz, Freddy, 214
Prison Break, 148
Private Practice, 122, 221n41
Proctor & Gamble, 10
producers/production, 2–3, 6–9, 12–13, 15, 17, 23–24, 26, 34; and *Battlestar Galactica*, 33–38, 40–44, 46–49; and consumer-as-fan, 169, 171; and "Dawson's Desktop," 76–77; and DeFamer.com, 149; and distributed cognition, 152–153, 158n16, 158nn18–19; and *Doctor Who*, 85; and FanLib, 77–79; and *Ghost Whisperer*, 254–258, *255*, *256*; and global market, 176–184, 186–189, 191–194; and *Heroes*, 40–41; and licensing, 33–38, 40–44, 46–49; and "The Lost Experience," 118, 120, 124, 134; and *The L Word*, 100–101, 110; and MyNetwork TV, 224, 226–227; and neo-platoon shows, 215; and *The Office*, 177–182, 189, 191–194; and PGUs (producer-generated users), 152, 154–156, *155*, 160nn31–35; and pilot production, 182–184; and post-network reflexivity, 140, 143–156, 158nn21–24; and preproduction, 177, 181–184; and producers-as-audiences, 152–154, 158nn21–24; and *Star Trek*, 36–38; and *Star Wars*, 77; and T-Mobile ad, 73; and transmedia, 7, 135–136, 139n64, 244–245, 250, 252–265; and Transmedia Producer job title, 7, 135–136, 139n64, 261; and user-generated content, 73–74, 76–79, 81, 83–85, 88–89, 91, 93; and video games, 33–35, 38, 43–44, 46–49, 55–57, 59, 62–63, 67;

producers/production (*continued*)
and Videomaker, 81, 83–84; and "You Write It," 110. *See also* producers/produsage; showrunners; *names of producers*
producers/produsage, 34–35, 42–49, *45*; and *Battlestar Galactica*, 34–35, 42–49; and *Beyond the Red Line*, 44–46; and *Diaspora*, 35, 44–49, *45*; open-source, 35, 42, 44, 46
Project Greenlight, 155
promotions, 4, 9–10, 12–16; and *Battlestar Galactica*, 24, 45–46, 48–49; cross-promotion, 11, 33, 122, 145, 250; and Kellogg's, 90–91, 93, 97n81; and *Kick-Ass*, 92–93; and licensing, 41, 134; and "The Lost Experience," 25, 119–120, 122–125, 127–130, 134–135; and *The L Word*, 98, 100, 106, 109, 111–112, 115n5; and MyNetwork TV, 223–224, 228–232, 234, 236; and post-network reflexivity, 145; and *The Simpsons*, 24; and transmedia, 246–250, 252–254, 260; and *24*, 24; and "Ultimate Lost Fan Promo Contest," 92–93; and user-generated content, 80, 83–84, 89–92; and Videomaker, 80, 83–84, 89; and YouTube, 21
propaganda, 169, 201–202
proprietary ownership, 35, 42, 46, 48, 114
prosumers, 3, 12; and "going pro," 10–14
The Protestant Ethic and the Spirit of Capitalism (Weber), 190–191
Provencio, Marla, 127, 136n2
public good, 23, 25; public input on licensing decisions, 203; public service model, 186, 189
Puerto Ricans, 202–203, 209, 218n3, 231
Pugh, Sheenagh, 85–86
Pushing Daisies, 221n45
Pussycat Dolls, 72

Queer as Folk, 181
queer culture, 75–76, 97n67, 112; and FanLib, 112–113; and *The L Word*, 99, 101, 104, 107, 109, 113–115, 116n15. *See also* slash fiction

racial diversity. *See* ethno/racial diversity
racism, 64, 212
Radical Entertainment, 55, 67
Radiohead, 162
Radio Times, 85
RAI (Italy), 186
ratings, 13, 27n7; and MyNetwork TV, 223–225, 236–237; and neo-platoon shows, 198, 207, 215, 220n35; Nielsen ratings, 102, 115n10, 116n15, 193, 236; and transmedia, 247; and video games, 54
Raving Fans (Blanchard), 163–164
Razorfish, 5

Reagan, Ronald, 204–205, 215
reality shows, 3, 120, 137n17, 147–148, 155; and Canada, 263; and global market, 176–177, 181, 183; and MyNetwork TV, 225, 227, 230, 235–237; and neo-platoon shows, 206–207, 213, 220n35. *See also* ARGs; *titles of reality shows*
Reaper, 221n45
Red Hat, 172
Redstone, Sumner, 145
Reed, Kristan, 59–60, 62
reflexivity, 140–156, 157n6; and consumerism-as-production, 152, 154–156, *155*; corporate reflexivity, 143–150; and distributed cognition, 152–153, 158n16, 158nn18–19; and *Entourage*, 25, 140–143, *142*, 152, 156; and global market, 25, 177; "industrial self-reflexivity," 8–9; and "The Lost Experience," 4, 128, 130; and *The L Word*, 104, 109, 111; and *The Office*, 25; and OurChart.com, 109; and PGUs (producer-generated users), 152, 154–156, *155*, 160nn31–35; and post-network industry, 8–9, 140–156; and producers-as-audiences, 152–154, 158nn21–24; and regeneration/legitimation, 150–152; and unruly audiences, 146–149; and unruly technologies, 146–148; and unruly work worlds, 146–148; and video games, 67; worker reflexivity, 143–150; and "You Write It," 111
Reisman, David, 1, 190
Reiss, Jon, 9
residuals, 120–124, 136n11, 148, 182, 187
retransmission policy (1992), 23–24
Reveille, 177–178, 181–183, 194
reviews: and *Beyond the Red Line*, 45; evaluative criteria of, 55; and video games, 54–55, 57, 59–67
Revlon hair color, 235
Revolution, 13, 29n34; "Revolution Revealed Series," 29n34
Revolution Studios, 55
Rhimes, Shona, 122, 211, 214, 221n40
Rhoda, 202
The Riches, 213
Rio de Janeiro, 263–264; RioContent Market, 264
Roberts, Gareth, 85
Roberts, Kevin, 165–166, 170
Robinson Crusoe, 182
RockYou! 19
Roddenberry, Gene, 11, 36–37
Rogers, Thomas S., 130
Rookie Blue, 137n17
Roper, Chris, 60
Roper, Tim, 138n46

Rose, Frank, 248
Rosen, Hilary, 108
Rosenberg, Jordan, 119, 124, 126, 128, 130, 134, 137n31
Ross, Andrew, 5
Ross, Sharon Marie, 24
Roswell, 221n45
royalties, 3, 36, 147–148
Rubio, Kevin, 84–85
Ruggill, Judd, 55
Russo, Julie Levin, 25, 74–76, 80, **98–117**, 271
Ryan, Shawn, 122

Saatchi & Saatchi, 165
SAG (Screen Actors Guild), 4, 121, 148, 206
Salir de Noche, 231
Sanchez, Eduardo, 258
Sander, Ian, 12, 254–258, *255, 256*
San Diego Comic-Con, 258–259
Sands of Iwo Jima (film), 202
Sandvoss, Cornel, 164
Sanford and Son, 203
Santo, Avi, 36, 56, 249–250
Santo, Matthew, *132*
satellite carriers, 23, 180, 186, 205, 224
Saturday Night Live, 20
Saving Grace, 221n45
Sayre, Daniel, 60
SCE Studio Cambridge, 54
Schamus, James, 160n31
Schneider, Andrew, 76–77
Schrag, Ariel, 116n17
Schudson, Michael, 201
"scientific management," 150, 192
Sci-Fi Network, 42–43, 80–83, 89, 95n26. *See also* SyFy Channel
SCMS (Society for Cinema and Media Studies), 118
Scott, Suzanne, 77, 95n24
Scratchware Manifesto, 144, 157n3
ScreenPlay: Cinema/Videogames/Interfaces (Krzywinska), 58
screenwriters: and post-network reflexivity, 140, 144, 148, 153, 156; and "You Write It," 98, 109–111
Secrets, 230
Sega, 67
Seinfeld, 213
Selling Television (Steemers), 186–187
Seven (film), 10
sex, 16–17. *See also* gender
Sex and the City, 16
Shankar, Avi, 127–128
sharing, 24, 26, 31n69, 145, 148

Shark, 198
SharonMustLive, 89
Shearman, Rob, 85
Sheen, Martin, 218n3
Shelmedine, Guy, 131, 138n46
She-Ra, 251
Sherlock, 263
Sherry, John F. Jr., 171
The Shield, 122, 198
Shillace, Darren, 119
Sho.com, 114
Shoot-out, 155
Short, Thomas, 148
"Shout" (Lulu), 72
showrunners, 3, 12–15; and *Battlestar Galactica*, 33, 41, 47, 49; and *Ghost Whisperer*, 258; and *Grey's Anatomy*, 211; and licensing, 24, 33, 41; and *Lost*, 210, 215; and "The Lost Experience," 12, 118–120, 122, 124, 127, 129, 133–134, 136, 139n61; and *The L Word*, 100; and neo-platoon shows, 210–211, 214–215, 221n40; and *The Office*, 182; and post-network reflexivity, 149; and transmedia, 258. *See also names of showrunners*
Show Sold Separately (Gray), 24
Showtime, 7, 9, 16–18, 25, 221n40; and *The L Word*, 99–100, 104, 106–108, 110–111, *113*, 114, 116n15; "No Limit," 16
Sierra Entertainment, 44–45
Silicon Valley, 3–6, 10–11, 13, 18–20, 134
Silverman, Ben, 181–182, 187–188, 193–194, 199, 240
The Simple Life, 159n25
Simpson, David, 61, 64
Simpson, Nicole Brown, 211
Simpson, O. J., 211
The Simpsons: "The 1895 Experiment," 176–177, 194; and global market, 176–177, 183, 194; "Homer3," 61; and neo-platoon shows, 213; as parody, 176; "Treehouse of Horror VI," 61; and user-generated content, 73–74; and video games, 24, 53, 55, 59–64, 66–68. *See also* entries beginning with *The Simpsons*
The Simpsons Hit and Run Game, 55, 61–64, 66–68
The Simpsons: Road Rage, 67
The Simpsons Skateboarding, 67
The Simpsons Wrestling, 67
The Sims, 55, 138n34; *The Sims 2*, 66–67
Sins of the Heart, 240
Sin Tetas, 182
sitcoms, 55; and global market, 183, 227; and neo-platoon shows, 198–199, 204–206, 214–215, 218n3, 221n38, 221n40. *See also titles of sitcoms*

Six Degrees, 212–213, 215
slash fiction, 12, 74–75, 77–78, 87; and fan fiction (fanfic), 75, 87, 97n67; and FanLib, 78; and "Master and Apprentice," 75; and *Star Wars*, 75, 77
Slocum, Chuck, 19, 119, 122
Smallville, 12, 14, 40, 125, 221n45, 260; "Smallville Legends," 12
Smallwood, Brad, 5–6
smart phones, 13, 24
Smith, Anna Nicole, 237
Smith, Beau, 39
Smith, Quincy, 14, 19–20; as "Energizer Bunny," 20
Smits, Jimmy, 218n3
soap operas: and *Desperate Housewives*, 55, 61; and fan fiction (fanfic), 85; and *The L Word*, 104, 108, 115n5; and MyNetwork TV, 223–228, 233–234, 237–240; and neo-platoon shows, 25, 197–198, 204, 208, 220n20
Social Ethic, 190–191
Socialist Writers' Conference, First (1934), 201
social media, 2–3, 5, 7–10, 12–13, 17–19, 24, 27n7, 137n17; and consumer-as-fan, 166–170, 172; and digital natives, 26, 31n69; and *Heroes*, 12; and "The Lost Experience," 123, 134, 136; and "Lost Untangled," 5; and social media gurus, 167; and transmedia, 245, 259. *See also* names of social media companies
The Social Network (film), 10
social networking, 35; and *Hey! Nielsen*, 102, 116n15; and "Join the Fight," 43–44, 46, 89; and Kellogg's, 90; and "The Lost Experience," 123–124, 127; and *The L Word*, 25, 100, 103–109, *105*; and Mashable, 108; and MyNetwork TV, 236; and OurChart.com, 25, 100, 104–109, *105*, 111, 114–115, 116nn15–17, 116n21; and transmedia, 258–259; and *True Blood*, 258–259; and user-generated content, 89–90, 93, 107. *See also* names of social networks
Solis, Brian, 167
Sons of Anarchy, 16
Sony, 206
Sony Computer Entertainment, 54
The Sopranos, 16, 213; "Seven-Minute Sopranos," 14
Sorkin, Aaron, 141
Soul Food, 214, 221n40
Soviet Union, 201–202
So You Think You Can Dance, 137n17
Spielberg, Steven, 141, 251, 262
Spigel, Lynn, 24
sports: and blackouts, 23; boxing matches, 23; and consumer-as-fan, 163–164, 166, 168–169; football games, 23, 235; integration in, 220n35; sports fans, 163–164, 166, 169
Spotwelder, 138n46
"spreadable media," 123, 137n18
Spreadable Media (Jenkins, Ford, and Green), 24
Sprite, 119, 122, 126–127, 129–133; and "Sprite Lab," 131–132, *132*; and Sublymonal.com, 119, 126, 130–133, *131*, *132*, 138n46, 139n55
Staples, Inc., 181
Starlight Runner Entertainment, 7–9, *8*, 26, 135, 248, 261; and cable networks, 16–17; and licensing, 41
Star Trek, 11, 34, 36–38, 41, 44, 48; and consumer-as-fan, 170; and licensing, 34, 36–38, 41, 44, 48; and neo-platoon shows, 202, 220n22; and transmedia, 251. *See also* entries beginning with *Star Trek*
Star Trek: Deep Six Nine, 220n22
Star Trek Office, 37
Star Trek Online, 38
Star Trek: The Next Generation, 37, 220n22
Star Trek: Voyager, 220n22
Star Wars (film franchise), 38–39, 53; and transmedia, 251–252; and user-generated content, 74–75, 77–78, 84
Star Wars Episode I: Racer, 56
Steemers, Jeanette, 186–187
Stein, Louisa, 59
Steinberg, Marc, 252
St. Elsewhere, 25, 204–205, 208
Stephens, Geordie, 138n46
Stewart, Sean, 262
Stories of a New Life, 264
storylines: and *Battlestar Galactica*, 32; and *The Blair Witch Project*, 128–129; and global market, 189; and *Hill Street Blues*, 204; and *Lost*, 14; and "The Lost Experience," 118, 124, 126, 128–130, 135, 139n64; and MyNetwork TV, 224, 228–229, 232, 239; and neo-platoon shows, 204, 211; and *The Office*, 189; and "Shattered Armistice," 47; and transmedia, 261; and *Ugly Betty*, 239
storytelling, 3, 7–10, *8*, 12, 16, *18*, 24–27; and backstory, 37, 40, 76, 251, 253–254; and *Battlestar Galactica*, 37, 40, 47, 51n32; and consumer-as-fan, 170–171, 173; and "continuity mining," 40; and "Dawson's Desktop," 77; and *Diaspora*, 47; "dispersed storytelling," 258; and fan fiction (fanfic), 87; and fan filmmaking, 73–75; and flashbacks, 40, 51n32, 62, 251; and *Flashforward*, 92; and *Ghost Whisperer*, 255, 257; and *Heroes*, 40–41; and licensing, 40, 47; and "The Lost Experience," 25,

118–119, 124, 126, 128–130, 137n16; and *The L Word*, 104, 106, 108–109; and MyNetwork TV, 224, 226, 228, 234, 237; "rabbit holes," 13–14, 47–48; and *Star Trek*, 37; and story arcs, 224, 226, 228, 234, 237, 250, 253, 257; and transmedia, 26, 244–255, 257–262; and *True Blood*, 258–260; and user-generated content, 76, 80, 87; and video games, 47, 54, 58–59, 62–63, 65–66, 68; and *Zarek*, 40. *See also* fan fiction (fanfic)
Storyworld, 245
story worlds, 7, 13, 26; and "Dawson's Desktop," 77; and *Flashforward*, 92; and *The L Word*, 108; and Sublymonal.com, 119; and video games, 54, 56, 58–59, 61, 63, 65–68; and Videomaker, 80
Strange, Niki, 24
Strasberg, Rob, 138n46
streaming sites, 17, 19–22, 121–122, 135, 177. *See also* names of sites
strike, WGA (2007–2008), 4, 10–12, 15–16, 19; and global market, 182; and "The Lost Experience," 119–125, 133, 135; and *The L Word*, 98; and *The Office*, 182; picket line, 12, 122, 133; and post-network reflexivity, 147–148
Strong, Brenda, 61
Strong Medicine, 198
Studio 60, 141
studios, 5–9, 13, 17, 24, 122; and licensing, 133, 135; and post-network reflexivity, 150–152, 156, 157n14, 158nn23–24; and transmedia, 246–247, 253, 257, 262; and tree-lighting ceremonies, 151, 157n14; and video games, 62. *See also* Hollywood; *titles of films*
Subaru, 115n12
Suber, Howard, 202–203
subjectivity: as labor, 100, 103, 109; and *The L Word*, 99–100, 102–106, 108–109, 114
Sublymonal.com, 119, 126, 130–133, *131, 132*, 138n46, 139n55
Sullivan, Pat, 249
Summers, Cynthia, 101
Super Eight (film), 13
Superman, 250
Super Mario Brothers, 244
Supernatural, 221n45, 253
supernatural genre, 213, 221n45, 254–258, *255*
Survivor, 210, 213, 220n35
Swarth, Brian, 17–18
sweatshops, 5, 121, 184
Swedish Americans, 202
Sweeney, Anne, 19
SyFy Channel, 17, 25, 80–83, 95n26, 155, 257
S/Z (Barthes), 172

Tapscott, Don, 10, 34
Taxi, 198, 203, 208
Ted X Transmedia, 245
Teenage Mutant Ninja Turtles, 244
Telefonica Foundation's Transmedia Lab, 245
Telemundo, 224
telenovelas, 26, 224–226, 228–239, 241n19; and dubbing, 239; and familiar physical locales, 225, 238, 240; and translations, 228; and *Ugly Betty*, 233, 239–240, 243n53. *See also* titles of telenovelas
Telesilla (fan author), 79, 88, 93
Television after TV (Spigel and Olsson), 24
Television and New Media (Gillan), 24
Television as Digital Media (Bennett and Strange), 24
television industries, 1–3, 6–9, 16–17, 25, 27; and appointment-based model, 246; and blurred distinctions, 9, 25, 128–129, 140, 143, 145, 149, 152; and children's programming, 251–253; and engagement-based models, 246; and erosion of job classifications, 144–148; and global market, 25, 176–194; and imitation, 176; and neo-platoon shows, 197, 199–200, 202–203, 205–206, 208, 213–215; new business model for, 177–185, 188–189, 194, 227; and post-network reflexivity, 141–156, 157n15, 158n21; and transmedia, 246–247, 249, 251–254, 261–262, 265. *See also* networks; post-network industry; *titles of programs and series*
"Television's Interchangeable Ethnics" (Suber), 202–203
The Television Will Be Revolutionized (Lotz), 24
Televisionwithoutpity.com (TWOP), 149, 154
televisuality, 153; and licensing, 60, 62–63, 66, 68; and neo-platoon shows, 197–198, 204, 208
Terranova, Tiziana, 34, 42, 99
Textual Poachers (Jenkins), 162
Thank God You're Here, 227
TheForce.Net, 75, 77
A Theory of Adaptation (Hutcheon), 57
30 Rock, 3–4, 15; "The Donaghy Files" webisode, 4
thirtysomething, 204
Thomas, Samantha, 134, 139n61
Thompson, Kristin, 56
Thundercats, 251
Till Death Do Us Part, 227
Time Warner, 36, 206, 225
Tin Gods, 264
Tipton, Franklin, 138n46
TiVo, 130, 193
T-Mobile, 72–74, 93
TMZ.com, 149

TNT, 155
Tolkien, J.R.R., 248
Top Model, 148
Torchwood, 263
Touched by an Angel, 221n45
toy manufacturers, 7, 36, 251–252
trademarks, 36
Transformers, 251
Transformers (toys), 7
transgenders, 25, 75–76, 107, 116n21
Transmedia, 4
transmedia, 6–10, *8*, 12–13, 244–265, 268n50; and cable networks, 16–18; and Campfire, 258–261; and consumer-as-fan, 171; East Coast model, 248, 262; and fourth wall, 125; future of, 260–265; and *Ghost Whisperer*, 254–258, *255*, *256*, 260; and "high concept" style, 250–251, 253; and immersion, 260; and "The Lost Experience," 118–119, 124, 128–129, 135–136, 139n64; and *The L Word*, 100, 106; and marginalization, 257; and media mix model, 252–253; and mothership model, 26, 246–248, 255, 261–262, 265; and multimodality, 245; prehistory of, 248–254; and radical intertextuality, 245, 253; as total engagement experience, 254–258, *256*; "transmedia czar," 134, 136; and Transmedia Producer job title, 7, 135–136, 139n64, 261; "transmedia supersystem," 251–252; "transmedia team," 40–41; transmedia today, 245–248; and *True Blood*, 258–261; and video games, 54, 59, 67–68; West Coast model, 248, 262–263
Transmedia Europe, 263
Transmedia Hollywood, 245
Transmedia Storytelling (Giovagnoli), 260
Transmedia Television (Clarke), 6–7, 133
transsexuals, 212
Trendle, George, 249–250
Tribeca Film Festival, 261; "Future of Film," 261
Tron (film), 7, 9
Troops (film), 84
True Blood, 17, *18*, 221n45, 258–261
Truman, Harry, 202
Tumbleweeder, 84
TV Guide (periodical), 14
TVWF ("Television Without Frontiers"), 186; Chapter IV, 186
Twelve O'Clock High (film), 202
20th Century Fox, 7, 35
Twentieth Television, 229, 233, 241n18
24: and comics, 35, 39; and licensing, 24, 35, 39; and neo-platoon shows, 218n3; and video games, 24, 54, 60–63, 65–67. *See also* entries beginning with *24*
24: Conspiracy, 159n25
24: The Game, 54, 61–63, 65–67
Twitter, 2, 5, 10, 24; and "The Lost Experience," 123–124, 127, 135; and Sho.com, 114; and transmedia, 258; and user-generated content, 93
2K Games, 54
227, 205

Ubisoft, 54–55, 133
Ugly Betty, 124, 148; and MyNetwork TV, 233, 239–240, 242n32, 243n53; as neo-platoon show, 26, 198, 208, 212, 214–215, 221n52
uncertainty, 140, 146, 182–183, 258
Uncharted: Drake's Fortune, 66
Undercover, 13
Unification Church, 165
unions, 147–151; and global market, 189; and non-union workers, 150, 226, 241n18; and *The Office*, 189. *See also* labor guilds
The Unit, 122, 198, 221n52
United Kingdom, 93, 119, 159n25, 179–189; and *Absolutely Fabulous*, 224; Communications Act (2003), 184; Conservative Party, 93; Department of Culture, Media and Sport (DCMS), 185; elections in, 93; GDP of, 185–186; and global market, 179–189, 227; Labour party, 93; and neo-platoon shows, 210; Parliament, 184; Producers Alliance for Cinema and Television (PACT), 185; and *Till Death Do Us Part*, 227; and transmedia, 263
Univision, 224
uploads: and fan fiction (fanfic), 87; and FanLib, 77–78; and post-network reflexivity, 144–145, 147–149; and user-generated content, 42–43; and Videomaker, 83
UPN (United Paramount Network), 14; and MyNetwork TV, 224–226, 229; and neo-platoon shows, 205–208, 214–215
upsell, 79, 86
USA Networks, 18, 181, 258
USA Today (newspaper), 215
user-generated content, 4, 19–21, 73–94; and *Batman: The Brave and the Bold*, 92; and British elections, 93; and "Dawson's Desktop," 12, 14, 76–78; and distributed cognition, 152–153, 158n16, 158nn18–19; and fan fiction (fanfic), 33, 74, 76–89, 93–94; and fan filmmaking, 73–75, 77, 80–85, *81*; and FanLib, 74, 76–81, 84, 86, 88–89, 93–94, 95n16, 111; and Farmville, 91, 93; and *Flashforward*, 92; and gender, 73–90, 95nn24–25, 96n47; and

"going pro," 76, 79, 84–88, 96n47; and Hotel City, 91; and *How to Train Your Dragon*, 92; and Kellogg's, 90–91, 93, 97n81; and *Kick-Ass*, 92–93; and licensing, 34, 42–44; and Lost University, 91; and *The L Word*, 98, 105, 107–109, 111; and Mafia Wars, 91; as "networked externalities," 152–153, 158n16; and *The Office*, 89–91; and OurChart.com, 105, 107–109; and "Oxo Factor," 92; and PGUs (producer-generated users), 152, 154–156, *155*, 160nn31–35; and post-network reflexivity, 148–149, 152–156, *155*, 160n35; and *Star Wars*, 74–75, 77–78; and "Ultimate *Lost* Fan Promo Contest," 92–93; and vidding, 75; and Videomaker, 43, 74, 80–84, *81*, 89, 91–93, 95n24
Using the Force (Brooker), 77
Utopian Entrepreneur (Laurel), 246
utopianism, 42, 49, 58, 110; and neo-platoon shows, 206; and platoon films, 201, 203; and transmedia, 260

V, 135
Valarltd, 88–89
Vargas, Lucila, 238
Verizon, 23, 120, 127, 159n25
Verrone, Patric, 122
VH1, 149
Viacom, 37, 116n15, 145, 149, 205–206
Via Domus. See *Lost: Via Domus*
Victorian novels, 248
vidding, 75, 95n24
video games, 6, 8, 10, 53–68; and authenticity, 61, 63, 67; and *Battlestar Galactica*, 24, 33, 35, 39, *39*, 42–49, *45*; as blockbusters, 53; and canonicity, 47–48, 62; and cast members, 61–62, 66; design of, 46, 53, 55–59, 61–63, 66, 68; and double jump, 57–58; and experiential core, 58–59, 63, 66, 68; and fan creep, 162; financial success of, 53, 60; and immersion, 58–68; and licensing, 24, 33, 35, 38–39, 42–49, 53–68, 133; literature on, 55–59; and "The Lost Experience," 123; and *The Matrix*, 245; and post-network reflexivity, 152; and repetition, 64, 66; reviews of, 54–55, 57, 59–67; and scratchware game, 144, 157n3; and *The Simpsons*, 24, 53; and *Star Trek*, 38; and *Star Wars*, 38–39, 53; temporality/spatiality of, 55, 63–68; and tone/style/pace/timbre, 59, 63, 68; and transmedia, 251–252, 257, 262–264; and voiceover work, 61. *See also* titles of video games
Videomaker, 43, 74–76, 80–84, *81*, 89, 91–93, 95n24; and BSG Videomaker Toolkit, 43, 80, *81*; website of, 80, 82–83

viewers. *See* audiences
violence, 16–17, 65
viral marketing, 13, 73, 137n18; and *Entourage*, 141–143, *142*; and post-network reflexivity, 141–143, *142*, 145, 147, 156
Vivendi, 183; Vivendi Universal Games, 55
The Voice, 25
Volition, 44

Wachowski brothers, 59, 253
Wahlberg, Mark, 141
Waisbord, Silvio, 186, 188, 226–227, 238
Wake Island (film), 200
The Walking Dead, 16
Walliams, David, 61
Wall Street Journal (newspaper), 109–110, 228
Ward, Josh, 93
war films. *See* platoon films
Warner Bros., 68
Warner Communications, 253
Warshaw, Mark, 4, 12, 16, 40–41, 51n34, 125
Washington, Denzel, 204–205
Washington, Isaiah, 211–212
Watch Over Me, 236
The WB (WB Television Network), 11, 14, 76; and MyNetwork TV, 224–226, 229; and neo-platoon shows, 205–207, 214
The Weakest Link, 181
Web 2.0, 1–2, 16, 99, 127, 135, 153, 247
Weber, Max, 1–2, 178, 184–185, 190–192
webisodes, 4, 12, 29n34, 120–121; and *Battlestar Galactica*, 17, 245; and *Ghost Whisperer*, 255; and global market, 182; and *The Office*, 182; and *Smallville*, 40; and transmedia, 245, 255. *See also* titles of webisodes
weblets, 11, 26
Weeds, 198, 213
Weiler, Lance, 9, 260, 261
Weise, Matthew, 58
Weissenberger, Daniel, 66–67
Welcome Back, Kotter, 198, 203
West, Cornel, 160n31
West Indians, 210
The West Wing, 218n3
WGA (Writers Guild of America), 3, 15–16; and ABC-Apple deal, 19; and global market, 182; guidelines of, 15, 124, 137n22; and "The Lost Experience," 119–122, 133, 135–136; MBA (Minimum Basic Agreement), 3, 124; and MyNetwork TV, 241n18; Negotiating Committee, 121–122, 125; and *The Office*, 182; WGA West, 120, 123. *See also* strike, WGA (2007–2008)
What's Happening!! 203

Whedon, Joss, 257
Whoopi, 218n3
Who Wants to Be a Millionaire, 181, 227
"Why So Serious?," 90, 262
Whyte, William, 1, 190–191
Wicked Wicked Games, 236
wikinomics, 34–35, 48–49
Wikinomics (Tabscott and Williams), 10
wikis, 11, 19–20, 24; and "The Lost Experience," 123; and Sho.com, 114
The Wild Bunch (film), 202
William Morris talent agency, 181
Williams, Anthony D., 10, 34
Williams, Chris, 78–79, 86, 89, 95n16
Williamson, Diana, 90
The Wire, 16, 198
Wired magazine, 17, 19, 99
Without a Trace, 198
The Wizard of Oz (Baum), 248–249
Wolf, Matt, 68
Wolf, Michael J., 164–166
Woodside, D. B., 218n3
word-of-mouth campaigns, 12, 127, 141, 258
workers: and Autonomist Marxism, 102; and benefits, 5, 7, 189; contract workers, 6–7, 151; craft workers, 140, 143–144, 146–151, 157n1; and distributed cognition, 152–153, 158n16, 158nn18–19; and erosion of job classifications, 144–148; and global market, 184, 188–190; and hot spots, 145; and "Join the Fight," 44; and *The L Word*, 98, 102, 112; non-union, 150, 226; and *The Office*, 25, 189–190; and post-network reflexivity, 143–152, 157n14; as spoilers, 149–150; and strikes, 12, 15; and "studio family" identities, 151, 157n14; and tree-lighting ceremonies, 151, 157n14; and unemployment/underemployment, 146, 150; and unruly work worlds, 146–148; and "work-for-hire" contracts, 7, 44. *See also* labor; *types of workers*
workplace, 2, 20; and global market, 189–192; and *The Mary Tyler Moore Show*, 203; nonhierarchical workplace sharing, 145; and *The Office*, 189–192; and post-network reflexivity, 145
World War II platoon combat films, 198, 200–203; and Office of War directives, 201; resurgence of, 202
Wrather Corporation, 250
Wright, Bob, 183
writers, 3, 7, 10–16, 124; and *Alias*, 12; and *Battlestar Galactica*, 33–34, 39–40; and *Diaspora*, 47; and *Doctor Who*, 85–86, 96n55; and *Flashforward*, 92; and global market, 182–183; and *Heroes*, 12, 40; and licensing, 33–35, 37–40, 47; and "The Lost Experience," 4, 12, 25, 119–126, 129–131, 133–135, 139n61; and *The L Word*, 100, 109; and MyNetwork TV, 226, 241n18; and neo-platoon shows, 214; and *The Office*, 182–183; staff writers, 40; and *Star Trek*, 37–38; and WGA strike, 3, 15–16, 119–125; and "writerly texts," 172; writer-producers, 15–16, 122, 124, 134; writers' assistants, 15, 40, 124; writers' rooms, 12, 121, 133–134. *See also* fan fiction (fanfic); *names of writers*
Wyatt, Justin, 250–251

Xbox Live Arcade *Battlestar Galactica* game, 39, *39*
xenophobia, 188
X Factor, 92
The X-Files, 11–12, 210, 213, 221n45
X-Wing, 53

Yahoo! 5, 21
Yankee Group, 159n25
Yazıcıoğlu, Taçlı, 168
Yo Soy Betty, la Fea, 233, 239–240, 242n32
The Young Indiana Jones Chronicles, 251
YouTube, 2, 5, 20–22, 24, 110, 230; and copyrighted material, 20–21; creators of, 10, 28n27; and fan fiction (fanfic), 87; Google-YouTube, 19, 21; and illegal downloads, 20–21; and lawsuits, 145, 147; "Lazy Sunday" rap-parody on, 20; and "The Lost Experience," 127, 134–136; and neo-platoon shows, 213; "The Office Is Closed," 122; 100 channel partnerships, 21, 136; and post-network reflexivity, 145, 147, 149, 153, 156, 160n35; and transmedia, 259; and *True Blood*, 259
"You Write It," 25, 100, 109–115, *113*; as fanisode, 109–110, 112, *113*; and "Lez Girls," 110–111; and Molly (winner), 110–111; and PDF e-zine, 109, 112, *113*

Zarek, 40
Zeroes, 73
Zigler, Vivi, 12, 15, 19
Zorro, 250
Zucker, Jeff, 15, 20, 22, 182
Zynda, Thomas H., 201, 203